The Nuts and Bolts of Materials Science

Christoph Steinbrüchel

Copyright © 2013 by Christoph Steinbrüchel. All Rights reserved.

No part of this publication may be reproduced, stored in a retrieval system, or transmitted, in any form or by any means, electronic, mechanical, photocopying, recording, scanning, or otherwise, without the prior written permission of Christoph Steinbrüchel.

Published by:
Nuts and Bolts Publications
7148 Somerset Farms Drive
Nashville, TN 37221
U.S.A.

http:\\www.nutsandboltsmaterials.com

ISBN-13: 978-0-9910127-1-8
ISBN-10: 0991012712

Contents

Preface

This is a new introductory textbook on Materials Science for Engineers, but as you well know, it certainly is not the first of its kind. Thus, two questions will come to your mind immediately:

- Why Materials Science for Engineers?
- Why another book on Materials Science for Engineers?

The true scope of the answers to these questions will reveal itself, I hope, as you work through this book. However, here are answers in brief:

The answer to the first question seems evident to me but, in any event, my book will try to make the case that Materials Science is the enabling science for most of engineering and technology which relies on hardware in some way. By hardware I mean anything made out of "stuff", i.e. different materials.

I am always amazed at how materials-based technologies have evolved and changed since I entered college, and I am confident that your generation of students will come to the same conclusion in due time. This course of events is not a recent phenomenon, only perhaps the rate at which it has occurred recently. Keep in mind that advances in the production and use of materials have been at the root of human development, for better or worse, since ancient times, and there is no indication that this will stop anytime soon. Thus I have no doubt that if you are at the start of your career in Materials Engineering now, your work will always be exciting and full of interesting twists and turns.

The answer to the second question is multi-faceted. Again I will be succinct here but will elaborate further below: I wanted to write a book that was light in weight, but not light-weight.

Many excellent introductory Materials textbooks are available at present. I know most of them well and have used several extensively in my own teaching. Almost without exception, their approach is encyclopedic and comprehensive: They present everything one might possibly want to include in an introductory course. Consequently, they present far more than one could possibly cover in a typical one-semester course, and students are often at a loss to find a thread they can follow readily through such a book.

My book follows the opposite approach: It seeks to present the nuts and bolts of materials science for engineers. By this I mean materials science pared down to its essentials. My selection of topics with this approach, of course, will be personal, and I am sure one can have many arguments about what I have left out. It is clear to me that one or the other missing topic may be dear to a particular reader. However, I do think that once students have learned the material presented in this book, they will be well-prepared to learn whatever else they need to know about materials.

To the students

Here are a few specific words of advice, as you embark on learning materials science. I am not trying to preach to you, but you should be aware of a couple of things that my experience tells me will be useful to you:

1. Be prepared to learn new vocabulary, new concepts, new procedures and, yes, even to memorize some facts. As you are on the way to becoming an expert in something, acquiring expert knowledge, you cannot look up everything just in time all the time. At the very least, it takes a certain amount of organized mental background to know what to look up, and how to connect it.

2. An important question you should ask yourself regularly is: How do I know what I know? A good gauge of knowing something is that you are able to talk about what you think you know to a friend, and your friend nods his or her head, tells you back in different words what you just explained, and then you both nod your heads in agreement. Or, perhaps you do not nod your heads in agreement, and then you will have an argument before you come to an agreement.

3. In order to help you along, I spell out explicitly what I call learning outcomes at the beginning of each chapter. These are things you should know how to do once you have worked through the chapter. I have also tried to provide a narrative with a clear thread about where we are going with our arguments. In order to learn the material and test how you are getting on, it will be indispensable that you spend some serious time doing homework. The homework comes in two forms: ***Exercises*** and ***Problems***. Exercises are meant to be doable for you just with what has been covered in the text. However, I realize that what looks straightforward to me may not be so for you. Problems are more open-ended and challenging, and will require some research.

4. The prerequisite for getting the most out of this book is a basic course in chemistry. In particular, you should have:

    - Familiarity with some basic concepts of physical chemistry (atom electronic structure, types of chemical bonding in molecules, properties of elements and systematics of the periodic table),
    - Familiarity with some simple physics concepts from mechanics, thermodynamics, electrostatics and dynamics,
    - A certain fluency with numerical calculations, algebra, and a little bit of calculus,
    - Familiarity with physics and chemistry units (e.g. Joules (J), Newtons (N), moles (mol), electron volts (eV), calories (cal), temperature (°C and K), etc.),

- Facility converting from one set of units to another, e.g. cal to J, cal/mol to eV/atom,
- Knowledge of Greek letters,
- Some knowledge of names of chemical elements. The more you know, the better.

In Chapters 1 and 2, I will review briefly some ideas from chemistry that are basic for the study of materials, but primarily for the purpose of establishing a common vocabulary. If you have not seen at least some of this stuff before, here is not really the place to learn it for the first time. That said, with a bit of effort, you should be able to pick up what it takes along the way in order to follow our discussion of materials science, which will begin in earnest in Chapter 3.

In order to help you learn new concepts or refresh your memory about old ones that may not be current in your mind, I have set in bold type certain terms where they and the corresponding definitions or explanations appear for the first time.

To the teachers

Let me mention a few additional points here that I think distinguish my book from others.

1. I have tried to provide a strong, coherent, and brief introduction of materials science for engineering students, with content built on what I believe students cannot do without and what they can reasonably learn in one semester. Thus, I hope that you will not find a lot of things you can leave out when teaching this type of a course with my book.

2. At the same time, I have tried to structure my narrative so that it should be fairly easy for you to insert your own supplementary material if you find that the book has left out something important.

3. I have put emphasis on what students do, i.e. LEARN, rather than on what teachers do, i.e. TEACH. With this goal in mind, I have formulated learning outcomes rather than teaching objectives at the beginning of each chapter. Of course, I hope for our sake that there will be a strong direct correlation between the two activities. I am also a strong believer in space learning and repetition, i.e. learning in intervals, over time. Therefore, I often introduce a concept at an elementary level first, then pick it up again later in order to fill in details.

4. My focus is on strong fundamentals, a close tie-in with everyday experiences, and at the same time exposure to modern topics with exciting prospects for the future.

5. I hope that I have provided some new ideas, examples, explanations. These include especially: Original illustrations and homework, numerous pictures from the research literature applying modern imaging techniques to the study of materials, and extensive web-based references.

6. My book contains no design in the traditional sense. I am fully aware that design is what distinguishes engineering from other science-based activities. My reasons for leaving out design are, first, that I am not a design expert, even though I think I have a thorough appreciation of its importance to engineering. And second, in the interest of conciseness, I wanted the book to focus on the science essentials for materials engineering.

7. The **homework** at the end of each chapter comes in two categories:

   The **Exercises** are meant to be fairly straightforward. The students are asked to practice certain calculations or procedures, or answer word questions, on the basis of what has been described in enough detail in the text. At the same time, I have tried to minimize "get-a-number" type of questions in favor of "how-to" and "why" questions.

   There are also the so-called **Problems**. These are open-ended and require independent investigation. First an examination, and decision, needs to be made of what is really the question (or the possible questions), followed by some research, analysis and discussion going beyond just what is in the book. Typically there may be more than one "correct" answer. In other words, the solution of a Problem will often involve a design-like approach, if not real design.

The organizing principles of my book

A focus on learning outcomes is something I have come to appreciate from having been involved in two ABET program reviews. (For the students: ABET is the organization which accredits undergraduate engineering programs at universities in the U.S.A.). With regard to materials science in particular, one focus of ABET has been to describe the field in terms of the interplay of the four concepts of structure, processing, properties, and performance. These are represented as the corners of a tetrahedron; hence the acronym SP3.

My point of view is somewhat different. I do not think that the symmetry implied by the metaphor of the tetrahedron provides the proper description of the relationship between structure, processing, properties, and performance. My personal view, and the first organizing principle of my book, is that the ***structure*** of a material is the ***central concept***, as illustrated in Fig. 1.

That is: Composition and processing give rise to the structure of a material, and structure in turn determines material properties and performance. If I were to pick an acronym for this point of view, it would be: **cpSp2** .

The second organizing principle of my book is that I am interested in providing the simplest idea, rationale, or model, to explain a certain phenomenon,

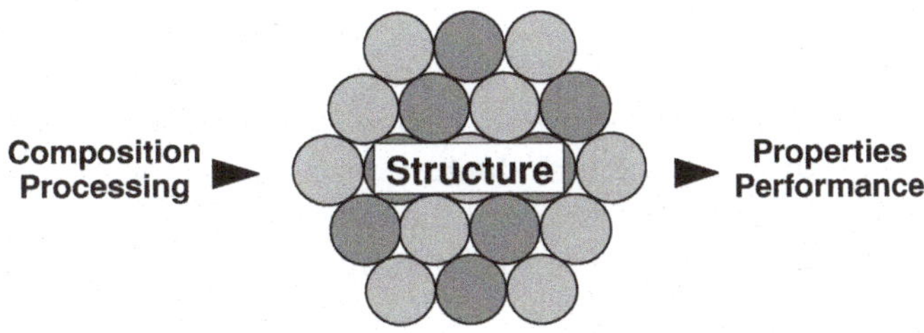

Figure 1: The central position of material structure: **cpSp2**

effect, or rule. This means that I will try to hone in on the essence of the issue at hand, rather than describe it with all its possible variations and ramifications. Thus, I will settle generally for a zeroth-order explanation, which in some instances may even be a little over-simplified, but this will be in the interest of helping the student ***develop a useful intuitive understanding*** of the hierarchy of importance, the relevant orders of magnitude, what I like to call ***engineering common sense***. For cases and topics which I consider of special interest, I will suggest resources for further study at the end of a chapter.

My third organizing principle in dealing with material properties is that I will use what one might call a kind of ***successive approximation***, or a ***widening circle***. I will first introduce and describe certain observed phenomena at an elementary level, or with a reduced set of variables. At a later time I will pick up the same theme again and add more details. This is in support of what I called space learning above. Ultimately, my goal is to use, as much as possible, ideas and models based on the modern quantum picture of the behavior of electrons in atoms, and bonding between atoms to form molecules and solids, in order to explain experimental observations and engineering material properties.

The outline of my book

I will begin with an introduction to the main classes of materials and some of their important properties. After briefly revisiting concepts of atomic and molecular electronic properties and bonding, I will present the structure of solids in detail (both crystalline and amorphous), including motivating examples of how structure determines material properties. This will be followed by a description of defects, mainly in crystalline materials. Structural concepts will then be applied to the discussion of mechanical properties of materials. The role of temperature will be outlined in the following chapters on diffusion, equilibrium phase diagrams, and phase transformation kinetics. The last part of the book entails more detailed discussions of properties of the different classes of materials, electronic properties, and a final chapter with a personal selection of new and exciting developments in materials science and engineering.

Acknowledgments

I am grateful for many conversations I have had with my colleagues while teaching and coordinating the introductory materials course in the Materials Science and Engineering Department at the Rensselaer Polytechnic Institute in Troy, NY. My writing of this book has been informed strongly by this experience.

I have learned a great deal from the students in my classes, trying to help them learn materials science and appreciate the scope and excitement of the field. There is evidence that I have had at least some partial success in my endeavor, from time to time. On the other hand, I heard from one student that I have a funny way of pronouncing the word "iron". How could that be, with a real materials engineer? I have harbored doubts about my calling ever since.

Last but really first, I thank my wife Pat for coming up with the title of the book. It captures perfectly what I tried to do in it. She has also let me have the time and space to pursue this project, even though we could have cruised the world instead.

Christoph Steinbrüchel
Nashville, TN

2nd Printing (2016)

Following suggestions by several reviewers, the main changes in the 2nd printing are:

1) additional material in Chapters 5 and 7 on viscoelastic deformation,
2) an expanded index of key words referring to page numbers rather than chapter numbers.

In addition, I have made minor adjustments in a number of figures and in the text in various places. Also, crowd-sourcing identified a number misprints in the first printing, which I hope to have eliminated this time around.

Chapter 1
Introduction

Overview

The periodic table lists all elements in a systematic fashion, so that elements with similar characteristics appear underneath each other. Different types or classes of materials can also be correlated with the positions of the corresponding atoms, or combinations of atoms, in the periodic table. We will enumerate and discuss briefly some of the main properties of the major classes of materials: metals, ceramics, inorganic glasses, polymers, and semiconductors.

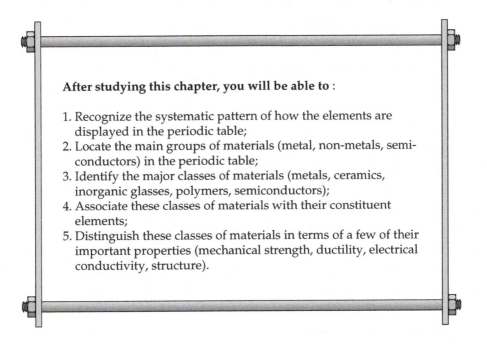

After studying this chapter, you will be able to :

1. Recognize the systematic pattern of how the elements are displayed in the periodic table;
2. Locate the main groups of materials (metal, non-metals, semiconductors) in the periodic table;
3. Identify the major classes of materials (metals, ceramics, inorganic glasses, polymers, semiconductors);
4. Associate these classes of materials with their constituent elements;
5. Distinguish these classes of materials in terms of a few of their important properties (mechanical strength, ductility, electrical conductivity, structure).

The five major classes of materials

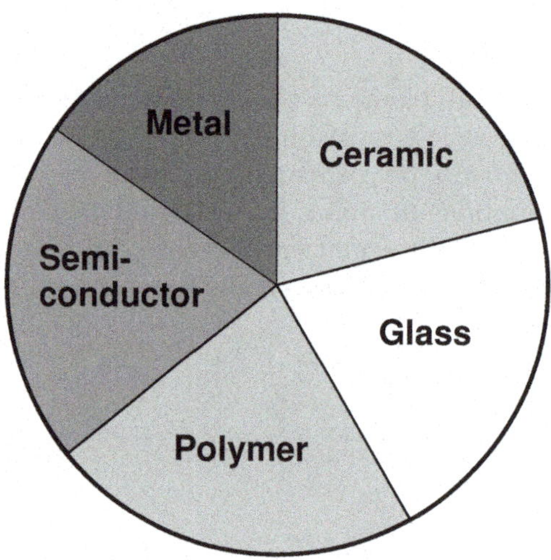

The graph identifies the five major classes of materials, which will be the subject of this book.

1 Introduction

We begin our discussion of materials with a look at the periodic table. Its great achievement is that it displays the chemical elements, and thus elemental materials, in a systematic way so that elements appearing in the same column have strikingly similar chemical properties.

IA	IIA											IIIA	IVA	VA	VIA	VIIA	0
H 1																	He 2
Li 3	Be 4											B 5	C 6	N 7	O 8	F 9	Ne 10
Na 11	Mg 12											Al 13	Si 14	P 15	S 16	Cl 17	Ar 18
K 19	Ca 20	Sc 21	Ti 22	V 23	Cr 24	Mn 25	Fe 26	Co 27	Ni 28	Cu 29	Zn 30	Ga 31	Ge 32	As 33	Se 34	Br 35	Kr 36
Rb 37	Sr 38	Y 39	Zr 40	Nb 41	Mo 42	Tc 43	Ru 44	Rh 45	Pd 46	Ag 47	Cd 48	In 49	Sn 50	Sb 51	Te 52	I 53	Xe 54
Cs 55	Ba 56	La 57	Hf 72	Ta 73	W 74	Re 75	Os 76	Ir 77	Pt 78	Au 79	Hg 80	Tl 81	Pb 82	Bi 83	Po 83	At 85	Rn 86

Metals | Semicon-ductors | Non-Metals

Figure 1.1: Simplified periodic table

The elements in the table are listed with their chemical symbols and **atomic number**, that is, the number of protons contained in their nuclei. As you know, atomic nuclei also contain neutrons, and the nuclei are surrounded by electrons. Furthermore, protons carry a unit of positive charge, neutrons are uncharged, and electrons carry a unit of negative charge.

Another quantity listed in the periodic table is the **atomic weight** of an element. This is for practical purposes equal to the weight of the protons + neutrons, and is usually given in terms of atomic mass units, or **amu**, i.e. in terms of units of the real proton mass. The atomic weight is of interest because nuclei with a given number of protons can have different numbers of neutrons. Such atoms would be called **isotopes**. For example, helium (He) can have either 1 or 2 neutrons. These two isotopes are referred to as ^{3}He and ^{4}He and have atomic weights of 3 and 4 amu. Note that ^{3}He is much rarer in nature than ^{4}He. Also, for the lighter elements, the number of neutrons is typically about the same or a little higher than the number of protons.

Atomic weights and amu are not just atomic quantities, but are in fact very useful for dealing with macroscopic quantities of elements, molecules, and compounds in general. Protons have, of course, a real mass we will call m_p, which can be measured in units of physics, here grams:

$$m_p = 1.66 \times 10^{-24} g \tag{1.1}$$

This means that the inverse of this equation,

$$1/m_p = 6.02 \times 10^{23} g^{-1} \tag{1.2}$$

tells us that 1 g of protons contains 6.02×10^{23} protons. You will recognize the number 6.02×10^{23} as **Avogadro's Number** N_{Av}.

It follows that if you pick a certain element, say C with an atomic weight of 12 amu (equivalent to 12 protons, but representing 6 protons and 6 neutrons), then 12 g of C will contain Avogadro's Number of C atoms. This number of atoms (or molecules) is also called **one mole** of said atoms (or molecules). In some texts Avogadro's Number of atoms is called a gram atom, but we prefer to use the term of one mole uniformly for atoms, molecules, and compounds, so that e.g. one mole of NaCl refers to sodium chloride containing 6.02×10^{23} Na and Cl atoms.

Note also the roman numerals on top of the periodic table. They are labels for the columns, often called groups of elements. Some of these groups have special names, such as alkali metals for Group IA (except H, hydrogen) and halogens for Group VII. (The groups listed in the figure are the ones that will be most important for us. The elements in the lower middle have also been assigned to groups by column, e.g. Group IB, IIB, etc., but the order is not simply ascending from left to right).

We will describe in more detail in the next chapter the other striking feature of the periodic table, namely that similar chemical properties of elements are a consequence of similar configurations of the electrons. At this time I will simply point out that the metals occupy the left side and the middle of the periodic table, and the non-metals occupy the right side, especially the upper right side. The two elements silicon (Si) and germanium (Ge) are the only elemental semiconductors. The column at the outermost right contains the noble gases.

Be aware that the periodic table as shown is incomplete. Note the gap in the atomic numbers between 57 (La, lanthanum) and 72 (Hf, hafnium). However, the elements shown are, for the most part, the ones you are likely to come across when studying materials.

Furthermore, even though we draw boundaries between metals, non-metals and semiconductors, it should be clear that properties of elements will vary in a more or less continuous fashion across the periodic table. Therefore, these boundaries may be not be so well-defined in all places. For example, crystalline carbon (C), known as diamond, which we identified as a non-metal, may also be thought of as a special semiconductor.

With help of the periodic table, we are now in a position to identify the major classes of materials:

1.1 Metals

Let us note first that the metals can be divided into two sub-groups: On the left of the periodic table are the so-called **simple metals** (Groups IA and IIA plus B (boron) and Al (aluminum)), and in the lower middle the so-called **transition metals**. Again the differences in their properties can be related to their electron configurations. By and large, transition metals are the most important engineering metals, and it is with those in mind that we will list important properties of metals below. The simple metals are often used as additives or components in multi-element materials, although of course Al is an important metal in its own right.

Be aware that metallic materials need not be elements, but may be formed by combination of two or more elemental metals, in which case one refers to the material as an **alloy**. Below are listed a few major properties of engineering metals:

Elemental metals:

- medium mechanical strength,
- high mechanical ductility,
- high electrical conductivity (electrical **conductor**),
- regular arrangement of atoms, crystalline material

Metallic alloys:

- high mechanical strength,
- lower mechanical ductility,
- lower electrical conductivity,
- regular arrangement of atoms, crystalline material

Without going into details, I will just mention here that **strength** designates the ability to withstand a mechanical load, and **ductility** the ability to be deformed without breaking. In addition, high, low, higher, and lower refer to a comparison of an elemental metal and an alloy with the same base metal, for example pure copper (Cu) and brass (Cu with some zinc (Zn) in it).

1.2 Ceramics

Ceramics are compounds, specifically compounds of a metal and a non-metal. Examples are NaCl (sodium chloride; sodium Na + chlorine Cl), MgF_2 (magnesium fluoride; magnesium Mg + fluorine F), Al_2O_3 (aluminum oxide; Al + oxygen O). The main properties of ceramics can be summarized as follows:

- high mechanical strength,
- no mechanical ductility,
- zero electrical conductivity (electrical **insulator**),
- regular arrangement of atoms, crystalline material

1.3 Inorganic glasses

Inorganic glasses resemble ceramics in most respects. They are also compounds of a metal, or perhaps a semiconductor, with a non-metal, and they may even have the same composition as a ceramic. For example, SiO_2 (Si + oxygen) may crystallize to form quartz, or it may form a silicate glass, silicon dioxide. The main difference is that in the **glass**, atoms are arranged in an irregular, more or less random fashion, so as to produce an **amorphous** material:

- high mechanical strength,
- no mechanical ductility,
- zero electrical conductivity (electrical insulator),
- irregular arrangement of atoms, amorphous material

1.4 Polymers

Polymers are a special class of materials, in that they consist mostly of carbon (C). Although they are of course made up ultimately of atoms, it is most convenient to consider polymer molecules as building blocks for polymer solids. These molecules typically have a backbone of -C-C- bonds forming a long chain to which other atoms such as H, Cl, etc. are attached. A simple example is polyethylene, with the chemical formula C_nH_{2n+2}, where n is of the order of 10^3 or more. The properties of polymers can be summarized as follows:

- low mechanical strength,
- high mechanical ductility,
- zero electrical conductivity (electrical insulator),
- amorphous or partially ordered arrangement of molecules

1.5 Semiconductors

The elemental semiconductors of practical importance are Si and Ge (germanium). Certain compounds can also have semiconductor properties, e.g. GaAs (gallium + arsenic, called gallium arsenide). The properties of semiconductors can be summarized as follows:

- high mechanical strength,
- zero mechanical ductility,
- low, and highly variable, electrical conductivity,
- mostly regular arrangement of atoms, crystalline material; amorphous possible.

Even though we list mechanical properties above, semiconductors are usually of interest because of their unique electronic properties, especially the electrical conductivity which can be adjusted over many orders of magnitude.

The purpose of the remainder of this book is to guide you towards an understanding of the origin of these differences in materials and their properties, and towards an appreciation of how materials engineers have become more and more skilled at tailoring such properties to their needs.

References

Throughout this book, I will refer to the following books for comparison, using the authors' names:

Callister, Jr., W. D. and D. G. Rethwisch, *Fundamentals of Materials Science and Engineering*, 3rd edition, John Wiley & Sons, New York, 2008.

Shackelford, J. F., *Introduction to Materials Science for Engineers*, 7th edition, Pearson Prentice Hall, Upper Saddle River, NJ, 2009.

Smith, W.M., *Foundations of Materials Science and Engineering*, 3rd Edition, McGraw-Hill Higher Education, New York, NY, 2004.

Van Vlack, L.H., *Elements of Materials Science and Engineering*, 4th Edition, Addison-Wesley, Reading, MA, 1980.

Guy, A. G., *Essentials of Materials Science*, McGraw-Hill, New York, 1976.

Callister and Rethwisch, Shackelford, and Smith are current examples of the comprehensive type of textbook.

Van Vlack and Guy are older books, quite detailed and with some wonderful explanations and examples, but at a somewhat more advanced level than the current introductory books. Guy's text in particular will remind you of whom the recent authors are indebted to.

Exercises

1. What is your best guess for the atomic weight of chlorine (Cl) atoms ?
2. Al and Si have atomic weights of 27 and 28, respectively. What does that tell you about the number of neutrons in their nuclei ?
3. What type of a material is vanadium (V) ?
4. What type of a material is cesium chloride (CsCl) ?
5. You have determined that a certain unknown material is an electrical insulator. What type of a material could it be ?
6. What type of a material is barium titanate ($BaTiO_3$, barium + titanium + oxygen) ?

Problems

1. Can the electrons be neglected in determining atomic weights ?
2. You will find in a more extensive periodic table than the one in Fig. 1.1 that the atomic weight of naturally occurring Cl is about 35.5 amu. What are the natural isotopes of Cl ? What about the isotopic composition, i.e. how much of each isotope is there naturally in Cl ?
3. What type of a material is silicon carbide (SiC) ?
4. What type of a material is gallium nitride (GaN, gallium + nitrogen) ?
5. Which metals are referred to as noble metals ?
6. Which group of metals is called alkali earth metals ?
7. With respect to the insulator in Exercise 5 above, what other test, or tests, could you do to narrow down your choices ?
8. What other property (i.e. other than the ones listed in the text) distinguishes most metals from polymers ?
9. Why did I say "most metals" in the previous problem ?
10. Identify five parts of your iPhone or other smartphone which are made out of materials from the five materials classes.

Chapter 2
Atomic Bonding in Solids

Overview

Quantum theory explains how electrons in atoms only exist in discrete energy states. Combined with the Pauli Principle, this implies that the lowest energy configuration of an atom is achieved by electrons filling energy states one-by-one, from the lowest state on up. The properties of the outermost electron states suggest a natural explanation of the systematics of the periodic table. Electrons also exist in discrete states in molecules. These molecular electron states, in turn, determine the nature of the bonding within and between molecules. Bonding between atoms in solids can be understood by an extension of the principles of atomic and molecular electron states, together with the notion of the extent of electron sharing among many atoms. Thus a natural path emerges to a description of bonding in the different classes of solids. A universal atom-atom bonding curve provides a unified description of such phenomena as the bond strength of solids, bond stiffness, thermal expansion, and melting point.

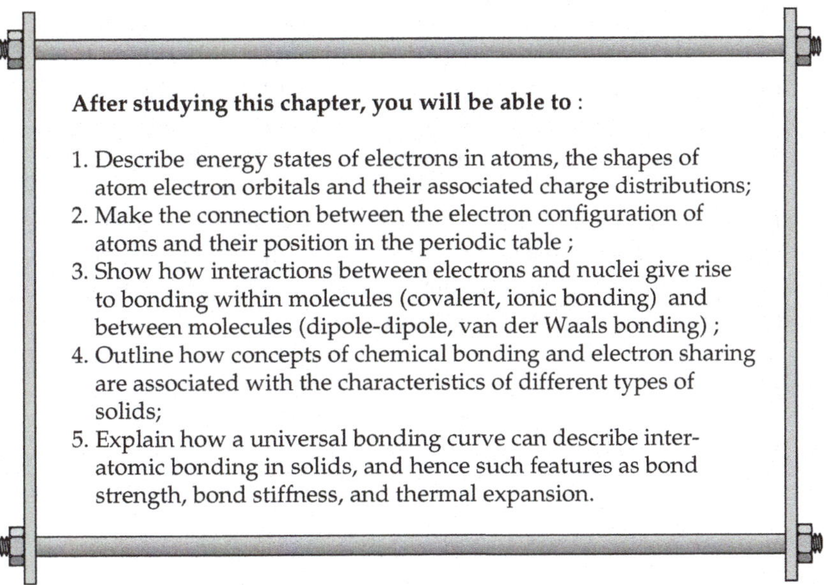

After studying this chapter, you will be able to :

1. Describe energy states of electrons in atoms, the shapes of atom electron orbitals and their associated charge distributions;
2. Make the connection between the electron configuration of atoms and their position in the periodic table ;
3. Show how interactions between electrons and nuclei give rise to bonding within molecules (covalent, ionic bonding) and between molecules (dipole-dipole, van der Waals bonding) ;
4. Outline how concepts of chemical bonding and electron sharing are associated with the characteristics of different types of solids;
5. Explain how a universal bonding curve can describe inter-atomic bonding in solids, and hence such features as bond strength, bond stiffness, and thermal expansion.

Atomic bonding between carbon atoms in diamond

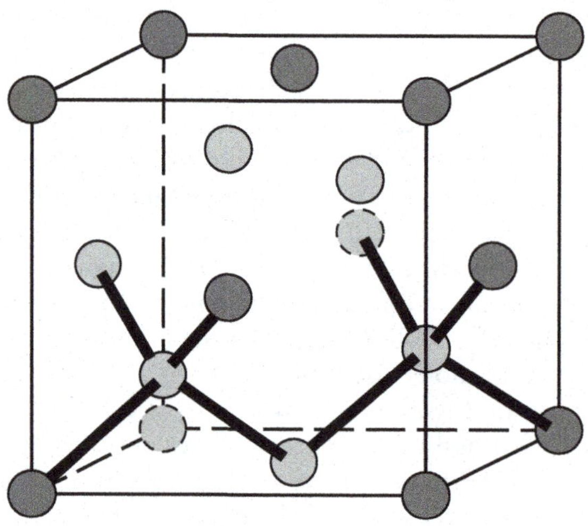

The picture illustrates the bonding between carbon atoms in diamond. Each C atom is connected by covalent bonds to four other C atoms in a tetrahedral geometry. (Only two tetrahedra are shown). Note that all C atoms are equivalent. It so happens that if you focus on the positions of the C atoms, they are in a cubic arrangement. As a visual aid, the C atoms of the cube facing the viewer are drawn as dark gray and the C atoms inside the cube or in back as light gray. In a diamond crystal, the tetrahedral, or cubic, arrangement of C atoms extends by many atomic distances in all three directions.

2 Atomic Bonding in Solids

In this chapter, we review briefly some pertinent facts and properties of the allowed quantum states for electrons in atoms and molecules. These considerations lead to a natural understanding of the systematics behind the periodic table and the chemical bonding in molecules. With some minor modifications and extensions, these arguments also serve to outline a unified framework for the different types of interatomic bonding which are present in the different classes of solids.

2.1 Electron States in Atoms

The next two sub-sections will explore some of the consequence of the quantum picture of electrons in atoms for the properties of elements.

2.1.1 Quantum States and the Periodic Table

Around the beginning of the 20th century, it had become evident from a large amount of experimental data, mainly optical spectra, that individual atoms absorbed or emitted light only at discrete wavelengths. Since these processes were due to electrons in atoms changing their energies, this meant that the electrons could only exist in discrete energy states. It was the monumental achievement of modern quantum theory to explain why this is so in general, and what the nature of these discrete energy states is in the different kinds of atoms across the periodic table.

We will not delve into the fundamentals of quantum theory here. We will simply take it as a fact that electrons in atoms are only allowed to be in states with discrete energies, and we will take note of some features of these states as they are important for our discussion of atom-atom bonding in solids and, eventually, other aspects of the structure and properties of solids.

A simplified diagram of the allowed energies for electrons in a generic atom is shown in Fig. 2.1. You should note first that individual electron states are labeled as $1s, 2s, 2p$, etc. The numerals $1, 2, 3$, etc. are called **principal quantum numbers** and delineate groups of energy levels. Each group is also referred as an electron **shell**. The letters s, p, d, etc. identify levels within a group and are referred to as **subshells**. They represent the quantum-mechanical angular momentum of the electrons. (Be aware that for the sake of being precise one should always make the distinction between a state and its associated energy, or energy level. We may not always be so rigorous and use the two terms interchangeably, but we will assume that when it really matters, we will know the difference).

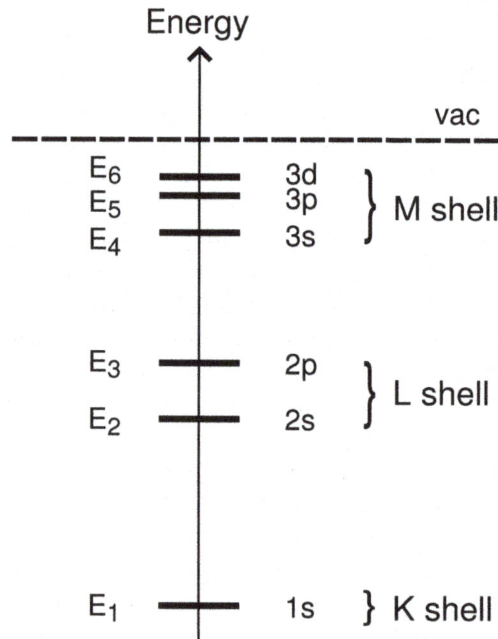

Figure 2.1: Generic diagram of electron energy levels in an atom. The zero of the energy scale is taken as the energy of an electron in vacuum, meaning far removed from the atomic nucleus. The energies of individual levels are not drawn accurately to scale but represent qualitatively typical energy differences. Note that in this scheme, electron energies are more or less negative.

In real atoms there are many more energy levels than those shown in Fig. 2.1. The series continues with 4s, 4p, 4d, 4f, .., and the levels come closer and closer to each other and approach the vacuum level yet always remain below it.

The principal quantum numbers are most important for the electron states in that they distinguish these states broadly by their energies. The interesting feature of subshells is that they are closely related to the spatial configuration of electrons (see Subsection 2.1.2 below).

The diagram of Fig. 2.1 is simplified in yet another respect. Strictly speaking, all subshells having a label other than s are subdivided further, into a number of **substates** if you will. For example, each p subshell is composed of three separate substates, and each d subshell is composed of five substates. (An s subshell consists of a single substate). In the context of this book, in general it will not be necessary to make such a fine distinction. We need it here mainly for keeping track of the total number of electrons in the different substates. However, we will remember that when we talk simply about an electron state, we refer ultimately to a particular substate in a certain subshell.

The final piece of the periodic table puzzle is the **Pauli Principle**. It postulates that each substate can hold at most 2 electrons, and if it does, the two electrons must have opposite **spins**. You can think of spin as another quantum degree of freedom, which represents the intrinsic magnetic moment of an electron either pointing up or down.

If one adopts this most detailed point of view, the number of distinct elec-

Subshell	Number of states	Max. no. of electrons in subshell	Max. no. of electrons in entire shell
1s	1	2	2
2s	1	2	
2p	3	6	8
3s	1	2	
3p	3	6	
3d	5	10	18

Figure 2.2: Number of states and maximum number of electrons in subshells

tron states in atoms, and hence the maximum number of electrons that can be accommodated in those states according to the Pauli Principle, work out as shown in Fig. 2.2.

The Pauli Principle also provides the prescription of how to realize the **ground state** in a certain type of atom, i.e. the electron configuration with lowest total energy. One needs to put two electrons into the various substates, beginning with the lowest energy state, and proceeding step by step to higher energy states, until all electrons have been accounted for.

I want to point out that you should make a clear distinction between states and electrons in states. Consider the following analogy: An atom is like a tall chest with different types of drawers stacked on top of each other. The s-type drawers have one box in them, the p-type drawers three boxes, d-type drawers five boxes, and so forth. The lowest drawer is s-type, and so is the second lowest. The third drawer from the bottom is p-type, the fourth drawer is again s-type, etc. You have a bunch of electrons to store in boxes, you are only allowed to put one or two electrons in each box, and you have to fill the boxes in the drawers from the bottom up. When you are done, a certain number of the lowest boxes will be filled, mostly, and the upper boxes will be empty.

Now we are in a position to make sense out of the periodic table Fig. 2.3. Note that in the first two rows, the ground state electron configurations are written out completely. From the third row on, they are given in an abbreviated form, so that e.g. for K (potassium) the complete notation would be $1s^2 2s^2 2p^6 3s^2 3p^6 4s^1$. Also keep in mind that after the element Ca (calcium) in the 4th row, in the middle of the periodic table the transition metals appear. These have more complex electron configurations involving d subshells.

H $1s^1$				He $1s^2$
Li $1s^2\ 2s^1$	Be $1s^2\ 2s^2$	O $1s^2\ 2s^2\ 2p^4$	F $1s^2\ 2s^2\ 2p^5$	Ne $1s^2\ 2s^2\ 2p^6$
Na .. $2p^6\ 3s^1$	Mg .. $2p^6\ 3s^2$	S .. $3s^2\ 3p^4$	Cl .. $3s^2\ 3p^5$	Ar .. $3s^2\ 3p^6$
K .. $3p^6\ 4s^1$	Ca .. $3p^6\ 4s^2$	Se .. $4s^2\ 4p^4$	Br .. $4s^2\ 4p^5$	Kr .. $4s^2\ 4p^6$

Figure 2.3: Periodic table showing elements with electron configurations

Fig. 2.3 illuminates several of the regularities in the periodic table. First, if you go across in a row, you see how one electron is added at a time from one element to the next. If you go down a column, you see that all the elements have the same type of highest subshell. For example, in column 1 with the alkali metals, the highest subshells are all of the s^1 type. In column 2 with the alkali earths, they are of the s^2 type, and in Group VIIA with the halogens, they are of the p^5 type. We will explore these regularities a little further shortly. Here we note that it is the common type of subshell that gives elements in a certain column their similar chemical reactivity.

2.1.2 Electron Orbitals and Charge Distributions

Above I emphasized that modern quantum theory is able to describe accurately the discrete energies of electrons in atoms. A further conclusion of the theory is that electrons in well-defined energy states in atoms do not have a well-defined position. Rather, one should think of them as being distributed with a certain effective probability density, or simply density, around the nucleus. It turns out that this distribution is different for each substate.

Furthermore, when one multiplies this probability density with the electron charge e (which equals 1.6×10^{-19} Coulomb, or 1.6×10^{-19} C), one can obtain an effective (negative) charge density around the positively charged nucleus. This charge density has a shape that is characteristic of the substate which it derives from. The charge density shapes most relevant for us are those of the s type and p type substates, especially 1s, 2s, 3s, and 2p.

This is a good point to introduce three additional terms. First, chemists often use the term **orbital** interchangeably with what we called substates above.

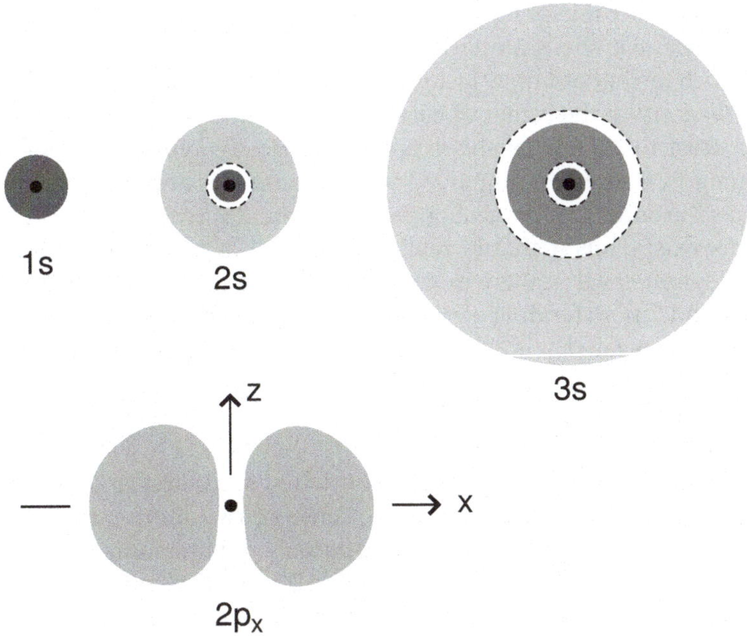

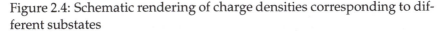

Figure 2.4: Schematic rendering of charge densities corresponding to different substates

We will use this term as well, perhaps somewhat loosely, but primarily for a substate occupied by electrons when we are interested in the shape of its charge distribution. Second, the electrons in the outermost shell may be referred to as **valence electrons**. These are typically the ones interacting most strongly with electrons on other nearby atoms, thus leading to the formation of chemical bonds. And third, the remaining inner electrons, those in shells closer to the nucleus than the valence electrons, may be called **core electrons**, so that an atom can be thought of as an inner core surrounded by the valence electrons.

The shapes of the various orbitals are illustrated schematically in Figure 2.4. The relative sizes of the distributions in Fig. 2.4. are qualitatively accurate, and the brightness is a measure of the average charge density in each region. However, you should be aware that the boundaries between regions are not sharp as drawn. Instead you should think of the regions as charge clouds with diffuse boundaries. The black dots in the middle of each diagram represent the atom nucleus.

Several features of these shapes are of interest to us. First, all the s orbitals are spherically symmetric. The 1s distribution consists of a central spherical core. The 2s distribution has a spherical core with what appears like a ring around it, but the ring is really a spherical shell. (This is, of course, not the same as the shell we mentioned earlier, in connection with sets of quantum states identified by their principal quantum number). The 3s distribution has a core with two concentric shells around it. Also note how the diameter increases: 2s extends much farther out than 1s, and 3s much farther than 2s. It is understood that a nucleus sits at the center of each distribution.

By contrast to 2s, the 2p shape is *not* spherically symmetric. Rather, you should imagine it as extending along a coordinate axis and being cylindrically symmetrical around that axis. For example, the 2p orbital in Fig. 2.4 points along the x-axis. Thus it should really be labeled $2p_x$, and then it follows for reasons of symmetry that there must be two more equivalent distributions just like it, $2p_y$ and $2p_z$, extending along the y- or z-axis, respectively. These are the three shapes associated with what we called earlier the three substates in the 2p subshell. As an aside, the 2p orbitals are sometimes referred to as having a dumbbell shape. Also note from Fig. 2.4 that a $2p_x$ orbital extends farther out in the x-direction than a 2s orbital.

With this knowledge about shells, subshells, substates, and shapes of orbitals, we are now in a position to make sense of two more important trends for materials science in the periodic table, namely the first ionization potentials and the electronegativities. In the table in Fig. 2.5, the first ionization potentials in eV are listed on the left and the electronegativities on the right, underneath each element.

H 13.6 eV 2.2					He 24.6 eV --
Li 5.4 eV 0.98	Be 9.3 eV 1.6		O 13.6 eV 3.4	F 17.4 eV 4.0	Ne 21.6 eV --
Na 5.1 eV 0.93	Mg 7.6 eV 1.3		S 10.4 eV 2.6	Cl 13.0 eV 3.2	Ar 15.8 eV --
K 4.4 eV 0.82	Ca 6.1 eV 1.0		Se 9.8 eV 2.6	Br 11.8 eV 3.0	Kr 14.0 eV --
Rb 4.2 eV 0.82	Sr 5.7 eV 0.95				

Figure 2.5: Excerpt of Periodic Table with Ionization Potentials (in eV) and Electronegativities

The **first ionization potential** of an element refers to the energy it takes to remove the first electron from an element. For example, the first ionization potential of Mg (magnesium) is 7.6 eV, as it removes the first of the two 3s electrons from the Mg atom (cf. Fig. 2.5). This also means that the 3s energy level of Mg is 7.6 eV below the vacuum level (cf. Fig. 2.1). Note that removal of the second 3s electron from the Mg atom would take much more energy, i.e. the second ionization potential would be much larger than the first ionization potential. The reason is that the second electron is removed not from a neutral atom, but from an atom that has lost an electron already, i.e. from a singly positive ion. Due to Coulomb's Law, the positive ion exerts a much stronger attraction on the second electron than does the neutral atom on the first electron.

The **electronegativity** is an empirical number derived by Pauling that describes the relative tendency of an element in its neutral state to take on an extra electron, thus forming a negative ion.

There are two clear trends in the periodic table (Fig. 2.5) with respect to these two parameters. When you look at the first ionization potentials, you can see that they decrease going down in each column, and they increase going across a row from left to right. The trend in going down can be understood by taking the first column of the alkali metals as an example. You note that the outermost s-orbital, from which the electron is removed, becomes larger and larger. By Coulomb's Law this means that this electron is held less and less tightly. The trend going across a row can be understood on the same basis. As you go across, the diameter of the inner charge cloud, the one inside the outermost orbital, becomes smaller mostly due to the increasing positive charge of the nucleus. This in turn causes the electron in the outermost orbital to be held more tightly.

When you look at the electronegativities, a similar trend is evident. The electronegativities decrease going down in a column, and they increase going across in a row. This trend can be rationalized similarly, by the size of the relevant orbitals and electrostatic interactions according to Coulomb's Law. Also note that the highest electronegativities are with the elements of the halogens in Group VIIA; the highest among those is fluorine. This is sometimes explained by way of the special stability of the electron configuration of the noble gases. Adding an electron to a halogen means that it "completes a noble gas shell".

The two overall trends in the periodic table can be summarized as follows: The first ionization potentials decrease going down in a column and going across in a row, and the electronegativities increase diagonally from the lower left to the upper right. We will see shortly that these trends are among the principles at the root of understanding the formation, structure and properties of solid metals and ceramics especially, and other materials as well.

We conclude this section by noting that metal atoms are sometimes referred to as being **electropositive** because of their tendency to give off one or more electrons and become positive ions. By the same token, non-metals on the right side of the periodic table are referred to as being **electronegative** since they tend to form negative ions.

2.2 Electron States and Bonding in Molecules

In this section, we will examine briefly how ideas about atomic states and orbitals can be applied to a description of electron states in molecules and bonding within a molecule (i.e. between its atoms) and between molecules.

2.2.1 Atom-Atom Bonding in Molecules

In many ways, electron energy states in molecules are analogous to those in atoms. Only states with certain discrete energies are allowed. The Pauli Principles applies. States are filled with electrons from the bottom up. And so forth.

The major complication with molecules arises from the fact that one has to deal with two or more nuclei, i.e. two or more centers which attract the electrons. This makes the determination of the allowed energies and associated states/orbitals much more difficult.

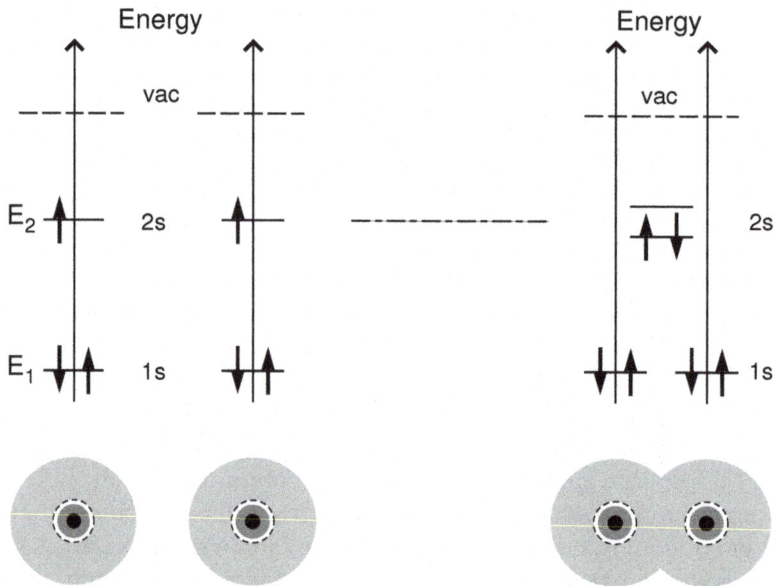

Figure 2.6: Energy states and orbitals of two Li atoms approaching each other

We will try to develop at least a qualitative sense of some of the important parameters and effects by looking at the example of the Li_2 (lithium) molecule. The ground state configuration of a Li atom is $1s^2 2s^1$ (cf. Fig. 2.5). Now observe what happens with the atomic electron energy states as two Li atoms approach each other (Fig. 2.6). The top of the figure shows the energy levels and the bottom the atomic orbitals. The little arrows indicate the electrons with their spin orientation.

As long as the two Li atoms are far apart, all the atomic orbitals are clearly separated. When the Li atoms are close enough, the 2s orbitals begin to overlap, and thus interact, while the 1s are still separate. This causes the two atomic levels at E_2 to shift up and down slightly. Now, in order to assume the lowest possible energy configuration, the two 2s electrons will go into the lower of the two 2s states, and you have made yourself a Li_2 molecule.

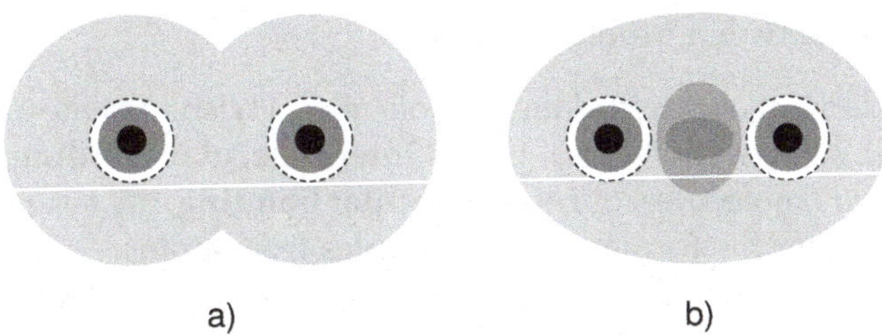

Figure 2.7: Charge distributions for a Li_2 molecule. a) Two overlapping 2s atomic orbitals; b) The corresponding molecular orbital. Note the increased charge density in-between the atoms.

As regards the charge distribution in the molecule, Fig. 2.7a with two overlapping atomic orbitals gives a rather inaccurate picture. As noted above, the atom-atom interaction gives rise to the formation of a molecular orbital occupied by two electrons with opposite spins, and with increased charge density between the two atomic centers (Fig. 2.7b). It is in effect this locally enhanced charge which brings about the chemical bonding in the Li_2 molecule. We can say that a **covalent bond** has been formed by two electrons shared by the two Li atoms. Again, be aware that this extra charge, and all charge in general, is distributed continuously, without sharp boundaries.

We will not need to delve into the intricacies of molecular orbitals. The most important orbitals for us will be of the bonding type, i.e. those that allow a pair of electrons to be shared between two atoms while at the same lowering their combined energy. There are also molecular orbitals of the so-called antibonding type which, if they were occupied by shared electrons, would raise rather than lower the energy of the electrons. An example is the upper of the two 2s-derived orbitals for Li_2 in Fig. 2.6. Furthermore, there may be molecular orbitals which are occupied by one or two electrons coming from the same atom. In the latter case one would speak of a lone pair of electrons.

When it comes to predicting the geometries of more complex molecules, containing three or more atoms, a useful principle to keep in mind is that electric charges generally try to arrange themselves so as to minimize the total energy (i.e. the most stable state is the one with lowest energy. In particular, electrons (or clouds of negative charge) will try to stay away from each other and, at the same, as close as possible to nuclei (positive charges). The same idea will be applicable in discussions of the bonding and structure of solids.

An example for this principle in action would be the geometry of a CH_4 molecule (methane). It is well known that the four covalent C-H bonds are completely equivalent, and the molecule has a tetrahedral geometry: the four H atoms are at the corners of a tetrahedron, with the C atom in the center. The reason for this is that the electrons in the orbitals forming each of the C-H bonds want to be as far from each other as possible.

2.2.2 Bonding between Molecules

In order to round out our discussion of molecules, I wish to make a few remarks about bonding *between* molecules, that is **intermolecular bonding**, as opposed to bonding *within* molecules, or **intramolecular bonding**, the latter having been the theme above for the formation of molecules from atoms.

Although all interactions between electrons and nuclei, between atoms, and between molecules are ultimately electrostatic in nature, as per Coulomb's Law, you should not forget that even entities with no overall net charge can have attractive or repulsive interactions, depending on how close they are. After all, a large number of substances that are gaseous under normal conditions can be liquefied, even solidified, at sufficiently low temperatures, and when they are in the liquid state, they are very hard to compress. We will have much more to say about atom-atom bonding in solids in the next section.

As an aside let us agree that when we talk about **normal** or **standard conditions**, we use the terminology of chemistry, meaning **room temperature**, i.e. 20°C or 293 K, and 1 atmosphere of pressure.

2.3 Atom-Atom Bonding in Solids

Here we will discuss in greater detail the various types of bonding between atoms that are relevant for the formation of solids. For this purpose, our discussions above will have direct applications, but we will also have to deal with a few new types of bonding. These arise due to the very large number of atoms interacting when they come together to form a solid material. We begin with an examination of some generic properties of atom-atom bonding in solids. Then we will focus on the different classes of solids and their associated atomic bonding.

2.3.1 Generic Features of Atom-Atom Bonding in Solids

Perhaps the most universal property of solids, as compared to liquids and gases, is that they maintain their shape or, speaking a bit more like an engineer, they resist deformation. In addition, regardless of how exactly they are deformed, if the deformation is small enough, solids will resume their original dimensions once the deforming force is removed.

At the atomic scale this implies that the atoms in a solid must have fairly fixed positions relative to each other and that it is not easy to change these positions.

Now, let us be clear that when we speak about atom positions in solids, our default assumption will be that these are **equilibrium** positions. By equilibrium we mean the configuration corresponding to the lowest possible total energy of the system. This is not to say that it is always easy for atoms in a solid to reach equilibrium. In general, this will take some doing, for example keeping the sample at an elevated temperature for a period of time. Moreover, as you prepare a certain material, you may have good reasons to keep it from reaching its true equilibrium configuration.

In addition, there are several ways in which a solid can be deformed, all involving a very large number of atoms. In order to study equilibrium and deviations from it due to a deformation, we will focus on the special case where the entire solid is either compressed or expanded uniformly. At the atomic scale this means that all atom-atom distances are changed in the same manner, i.e. either compressed or stretched. Under this condition, it seems reasonable to assume that we can study the overall deformation by looking just at an individual atom and one of its neighbors and the effective potential energy between them.

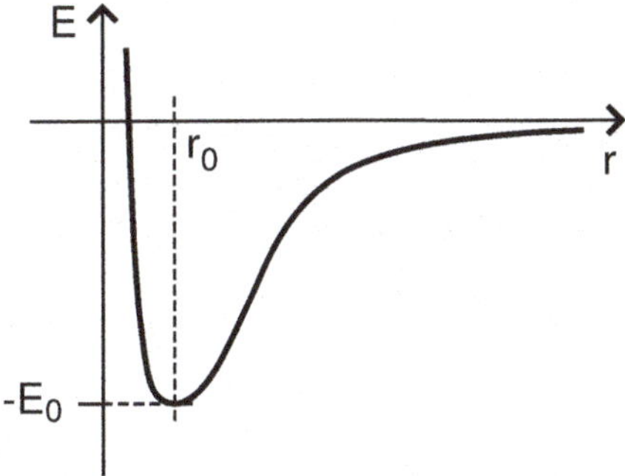

Figure 2.8: Potential energy $E(r)$ vs. interatomic distance r

Thus, from the resistance of solids against deformation, i.e. against moving atoms from their equilibrium positions, we conclude that the effective potential energy $E(r)$ between the two atoms as a function of the distance r between them must have the overall shape shown in Fig. 2.8. At the equilibrium separation r_0 the energy $E(r)$ has a minimum, whose value is $-E_0$. Any deviation from r_0 brings with it an increase in energy.

Equivalently, we can describe equilibrium in terms of forces between the atoms, due to the fact that the force $F(r)$ is equal to the first derivative of the potential energy $E(r)$,

$$F(r) = \frac{dE(r)}{dr} \tag{2.1}$$

or in graphical representation:

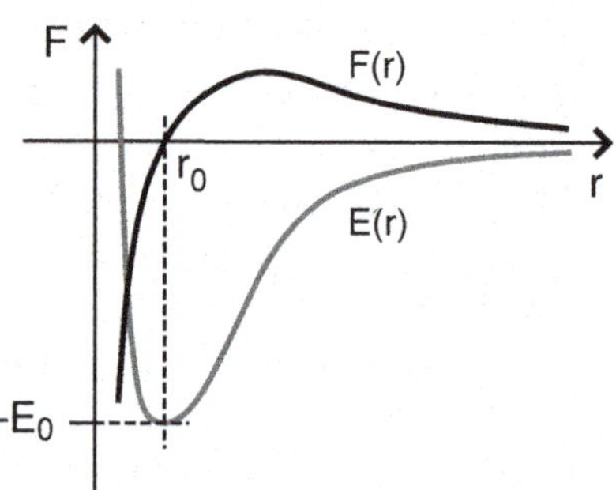

Figure 2.9: Interatomic force $F(r)$ vs. interatomic distance r. The gray curve is the potential $E(r)$ of Fig. 2.8.

We see from Fig. 2.9 that at shorter distances, $r < r_0$, the curve $F(r)$ is negative and falls steeply, indicating a strong repulsive force, and at longer distances, $r > r_0$, $F(r)$ is positive indicating an attractive force. Of course, at the equilibrium separation $r = r_0$ the force is zero.

Note also the interpretation of the parameter E_0 in Figs. 2.8 and 2.9, i.e. $E(r = r_0)$. Clearly, E_0 represents the atom-atom **bond energy** in the solid, that is the energy required to move two atoms from their equilibrium position of $r = r_0$ to infinite separation, $r = \infty$. (We take the value E_0 and thus the bond energy to be a positive quantity, whereas $E(r = r_0) = -E_0$ is a negative quantity).

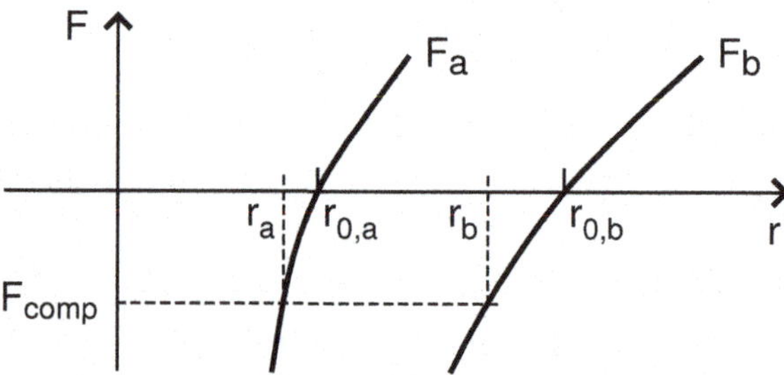

Figure 2.10: Expanded view of $F(r)$ curves near the equilibrium positions for two materials a and b. Note effect of compression by the force F_{comp}.

Given what we have outlined about solids subject to deformation, I think we can be quite confident of the general shape of the $E(r)$ and $F(r)$ curves, but there are two additional details we can extract from them. First let us envision two $F(r)$ curves for materials a and b (Fig. 2.10), and let us subject these two materials to the same compressive force F_{comp}.

As a result, their interatomic distances will be reduced to r_a and r_b, respectively. You can see that $r_{0,a} - r_a$ is smaller than $r_{0,b} - r_b$. We can say that material a was compressed less because of its stiffer atom-atom bonds. From the figure it is also apparent that this so-called stiffness is proportional to the slope of $F(r)$ at r_0 or, what amounts to the same thing, the stiffness is proportional to the second derivative of $E(r)$ at r_0, i.e. the curvature of $E(r)$ at r_0. (In physics this would be called the force constant of the atom-atom bond).

Now, looking again at the shape of the $E(r)$ curve (Fig. 2.8), we may even venture to guess that the deeper the $E(r)$ curve (i.e. the larger E_0), the sharper its minimum, and thus the stiffer the material. We shall see that this correlation is borne out well by experimental data when we discuss mechanical properties of materials in detail later in this book.

The other aspect of the shape of $E(r)$ curves we wish to discuss has to do with the symmetry of these curves relative to the equilibrium positions. We will call a curve symmetric about $r = r_0$, if its shape near the minimum is basically that of a parabola (Fig. 2.11b).

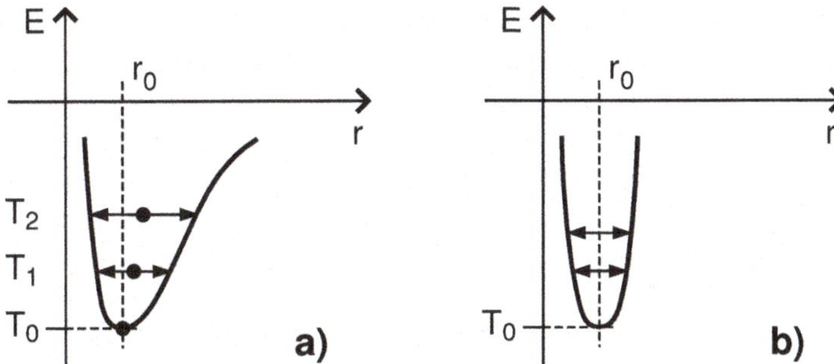

Figure 2.11: a) Asymmetric, realistic $E(r)$ curve; b) symmetric, hypothetical $E(r)$ curve. Note the black dots, indicating increased atom separation at elevated temperatures for the asymmetric curve. (Adapted from R.M. Rose, L.A., Shepard, and J. Wulff, *The Structure and Properties of Materials, Vol. 4, Electronic Properties*. Wiley, New York, 1966)

You should keep in mind that although we spoke of *the* equilibrium position of atoms in connection with the $E(r)$ curves, in reality the atoms do not stand still but vibrate slightly about their *average* equilibrium positions. It is only at very low temperature that the atoms can be envisioned to be essentially at $r = r_0$. For an asymmetric curve, as the temperature increases from T_0 to T_1 and T_2, so does the energy of an atom as the atomic vibrations become more intense. Consequently, the average equilibrium distance increases, and the solid expands (Fig. 2.11a). On the other hand, for a symmetric curve, there would be no dependence of the average equilibrium distance on temperature, thus no thermal expansion (Fig. 2.11b).

This effect of thermal expansion is exactly what is observed for practically all materials. For example, for metals the so-called **linear thermal expansion coefficient** (i.e. the ratio of change in length to original length due to an increase in temperature) is of the order of a few $\times 10^{-6}$/K for alkali metals to a few $\times 10^{-5}$/K for transition metals (i.e. a few to a few tens of parts per million and degree K).

2.3.2 Atom-Atom Bonding in Different Types of Solids

We are now in a position to examine atom-atom bonding in the different classes of solids. We will make a distinction between two broad groups of bonding: **primary** and **secondary** bonding. This distinction is based mostly on the strength of the bonding. We will also explain how each type of solid is associated with a particular type of atom-atom bonding.

Primary bonding: Covalent bonding

One of the primary types of bonding involves covalent bonds between neighboring atoms. We introduced this above for our discussion of the Li_2 molecule. In elemental solids it is observed in crystalline C (diamond), Si, and Ge, giving rise to a tetrahedral structure as illustrated in the figure at the very beginning of this chapter. Covalent bonding is also essential for the understanding of such compounds as GaAs or InSb (indium + antimony, indium antimonide).

Covalent bonds in solids are similar to those in molecules. They also come about via overlap of atomic orbitals, through shared pairs of electrons. With the elements from Group IVA, the interesting twist is that it turns out to be most appropriate to describe these covalent bonds not in terms of the individual s- and p-type orbitals in the outermost shell, but rather in terms of so-called hybrid s-p orbitals. These hybrid orbitals are combinations of the s- and the three p-orbitals atomic, and as a result they point from a certain fixed atom to its four nearest neighbors in the tetrahedral positions.

Please note that pure covalent bonding, i.e. covalent bonds that are completely symmetric with respect to the charge distribution between neighbor atoms, can only occur in elemental covalent solids. In compounds, a covalent bond will be more or less asymmetric, depending on the different affinities of the atoms for electrons. Therefore, covalent bonds are sometimes characterized as partially ionic, or having some ionic character. You can find an example in the exercises and more about it in the chapter on the structure of solids.

It should be apparent that in covalent solids the bonds are directional, meaning they are directed towards neighbor atom, so that they form in effect a strong and rigid network.

Even though we have focused on elemental solids and simple compounds here, you should be aware that covalent bonds are also an important component in polymers. As mentioned earlier, the building blocks of polymers are long chains of carbon atoms, and within each chain the carbon atoms are tied together by covalent bonds. However, the interaction *between* polymer chains involves another type of bond, which we will discuss in further detail below.

Primary bonding: Ionic bonding

The next type of bonding we will discuss, and in a way the opposite of covalent bonding, is called ionic bonding. In this case, no sharing of electrons takes place at all. Rather, an electropositive atom will give off one or more of its valence electrons to an electronegative atom. Hence, a positive and a negative ion are formed. Ions of opposite charge will attract each other strongly and will form an ionic solid.

This kind of complete charge transfer usually involves metal atoms on the one hand and non-metal atoms on the other hand. As mentioned before, the resulting compounds are called ceramics. Examples are NaCl, MgO, or Al_2O_3. With respect to how ions come together to form a solid, you should realize that even though positive and negative ions attract each other as per Coulomb's law, there will also be Coulomb repulsion between ions of like charge. The stable configuration of all these ions in a solid is due to a subtle interplay between all these forces.

Furthermore, even ions with the same charge eventually repel each other when they are close enough, due to the electrostatic repulsion of their positive cores. We may conclude that for ionic solids, just as for covalent solids, the change in potential energy as a function of the interatomic distance is well-described by the $E(r)$ curve displayed in Fig. 2.8.

Primary bonding: Metallic bonding

The third type of primary bonding is the one that prevails in metals. I would say that the nature of metallic bonding is more elusive than that of covalent and ionic bonding. So we will be content here with a somewhat vague description of its characteristics. The major property of metals that distinguishes them from covalent and ionic solids is that metals are excellent electrical conductors. This suggests that, somehow, at least some of the electrons inside a metal must be highly mobile, much in contrast to what we know about electrons in covalent and ionic solids.

For the time being, I will simply sketch how you should imagine metals and the mobile electrons inside (these are electrons that typically were valence electrons in the isolated atoms). In the chapter on electronic properties later on, we will develop a somewhat more detailed picture for electrons in metals. At this point we will envision a metal as a large number of positive ion cores embedded in a sea of negative electrons. Moreover, these electrons are shared universally, i.e. each of the mobile electrons is in effect spread out over the entire piece of metal. It is as if these electrons shield the ion cores from each other, thereby providing some sort of glue that binds them all together. This picture also implies that metallic bonding is non-directional. Still, solid metals tend to have highly ordered, crystalline structures in which atoms have very regular positions, in spite of non-directional bonding.

Secondary, or van der Waals bonding

Think of van der Waals bonding as the type of very weak bonding between overall neutral particles, atoms or molecules, that causes gases to liquefy and even solidify at sufficiently low temperature.

It may appear surprising at first that there should be an attractive force between electrically neutral entities. However, remember that an overall neutral particle can still have an underlying asymmetric charge distribution. The easiest case to understand is the attraction between two molecules with permanent **dipole moments**. Such molecules are also called **polar**, they have polar covalent bonds, and two or more electrical poles impart an electric dipole moment to the molecules. Examples are CO or H_2O.

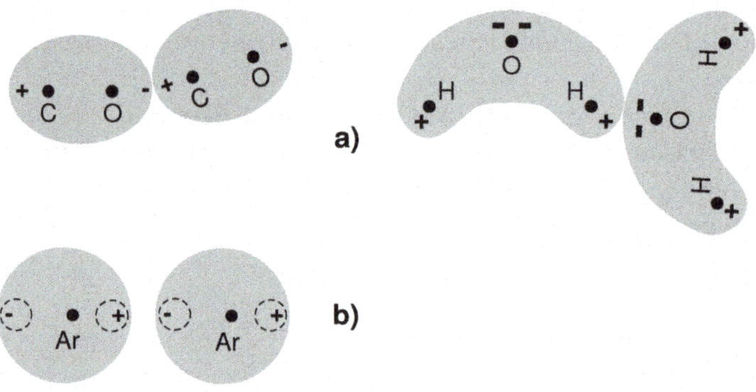

Figure 2.12: Schematic representation of dipole-dipole interactions.
a) Attraction between permanent dipoles; b) Attraction between induced dipoles; dashed little circles indicate temporary charge asymmetries

Even atoms such as the noble gases have a weak attractive interaction, in spite of being spherically symmetric and having no permanent dipole moment. However, short-lived fluctuations in the charge distribution on one atom can induce a temporary dipole moment on a neighboring atom. As noted, this type of interaction is very weak, but ultimately strong enough to lead to noble gases becoming liquid, and solid, at very low temperatures.

Van der Waals forces also play a role in the interaction between the chain molecules in a polymer. We will have more to say about that when we discuss polymers in detail.

In the table below we conclude this chapter with a brief summary of the classes of materials exhibiting primary bonding, the role of the electrons in their atom-atom bonding, and a few select materials properties.

Table 2.1: Major classes of materials and their properties

Material	**Bond**	**Electrons**	E_0 (eV/atom)	**Strength**
diamond	covalent	el. pairs	7.4	very high
Si	covalent	el. pairs	4.7	high
NaCl	ionic	transfer	1.7	high
MgO	ionic	transfer	2.6	high
Al	metallic	uni-share	3.4	high
W	metallic	uni-share	8.8	very high

The terms in the Electrons column are shortened but their meaning should be fairly self-evident. Uni-share means universal sharing of electrons. Note the unique correspondence between the type of bond and what happens to the valence electrons as the atoms form a solid. The quantity E_0 stands for bond strength as in the $E(r)$ curve in Fig. 2.8. It is understood that when we label something as high, this is a rather fuzzy description, which should be seen as a qualitative comparison among the items in the same category.

References

Both Shackelford, and Callister and Rethwisch list physical data of a number of elements. Callister provides additional data about atom electron configurations.

You will have no trouble finding various versions of the periodic table, as well as sites with information on properties of the elements, on the Internet.

For another example of pictures of atomic orbitals, see:
`http://users.aber.ac.uk/ruw/teach/237/shape.php`

For more on atomic energy levels in Li to F, and molecular orbitals and energy levels in Li_2 to N_2, see e.g.
`http://www.science.uwaterloo.ca/\~{}cchieh/cact/c120/molitof.html`

The bond length of Li_2 is 0.267 nm and the bond energy is 110 kJ/mol.

Exercises

1. What is the electron configuration of the highest subshell of a Cl (chlorine) atom?
2. What are the electron configurations of Ca^{2+} and S^{2-} ? Do they have anything in common?
3. What do the electron configurations of the noble gases have in common?
4. When we mentioned that the interaction of charges is important for the geometry of molecules, we left out something. What is it, and why can it be neglected?
5. What is the geometry of a CF_4 molecule? Why is it not flat?
6. Examine the $E(r)$ curves below: a) Which two, if any, have the same r_0;

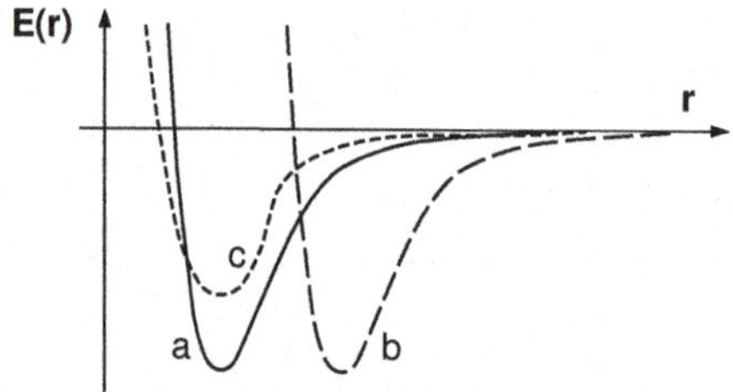

   b) which has the highest stiffness; which the lowest thermal expansion?
7. What is the material Invar? Where does its name come from? What is its unique property?
8. Why is the composition of aluminum oxide Al_2O_3 ?
9. An approximate measure of the ionic character of a covalent bond can be defined as follows:

$$f_i = 1 - \exp(-0.25(EN_A - EN_B)^2) \tag{2.2}$$

   where f_i is the fractional **ionicity** and EN_A and EN_B are the electronegativities of the atoms in the bond such that $EN_A > EN_B$. Note that $0 \leqslant f_i \leqslant 1$, so that $f_i = 0$ means a completely covalent bond, and $f_i = 1$ a completely ionic bond.

   Calculate f_i for the bond in CsF and in GaAs.

Problems

1. Look up the second ionization potentials of Na and Mg, $E_{i2,Na}$ and $E_{i2,Mg}$, and compare them to their first ionization potentials, $E_{i1,Na}$ and $E_{i1,Mg}$. Discuss the differences $E_{i2,Mg} - E_{i1,Mg}$ and $E_{i2,Na} - E_{i1,Na}$. Are they about the same, or if not, why are they different?

2. Extend the arguments we made in the text regarding the formation of the Li_2 molecule to the case of two He atoms approaching each other. Explain why a He_2 molecule does not form.

3. Find an example of a diatomic molecule formed by the combination, or overlap, of two atomic 2p orbitals and sketch the resulting molecular orbital. Note that when the two atoms approach each other, the shifts in energy levels may lead to interactions between 2s– and 2p– derived orbitals.

4. Look up the thermal expansion coefficients of Na and W. Why are they so different?

5. Why can ionic bonding be described as non-directional?

6. What type of bonding is present in SiC? How strong do you expect this material to be?

7. Are there any weak metals, i.e. metals with low E_0? If so, where in the periodic table are they, and why are they where they are?

Chapter 3
Structure of Solids

Overview

Solids are characterized first of all by the fact that atoms have fixed positions with respect to their neighbors. In addition, many solids are found most commonly in crystalline form, which means that atoms are arranged in a regular pattern that repeats itself over many thousands, even millions, of interatomic distances. The simplest and most prevalent types of solids involve ordered arrays of atoms with a cubic or hexagonal symmetry. The order of a solid structure can be determined by such methods as X-ray crystallography. We will examine some important examples of crystalline metals, ceramics, and covalent solids, and compare them to polymers and inorganic glasses with more or less random structure. The profound influence of structure on the properties of solids will become apparent.

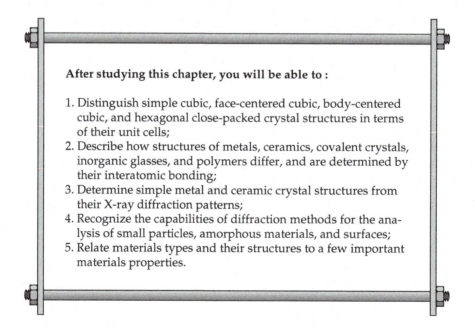

After studying this chapter, you will be able to :

1. Distinguish simple cubic, face-centered cubic, body-centered cubic, and hexagonal close-packed crystal structures in terms of their unit cells;
2. Describe how structures of metals, ceramics, covalent crystals, inorganic glasses, and polymers differ, and are determined by their interatomic bonding;
3. Determine simple metal and ceramic crystal structures from their X-ray diffraction patterns;
4. Recognize the capabilities of diffraction methods for the analysis of small particles, amorphous materials, and surfaces;
5. Relate materials types and their structures to a few important materials properties.

Field Ion Microscope image of a sharp tungsten tip

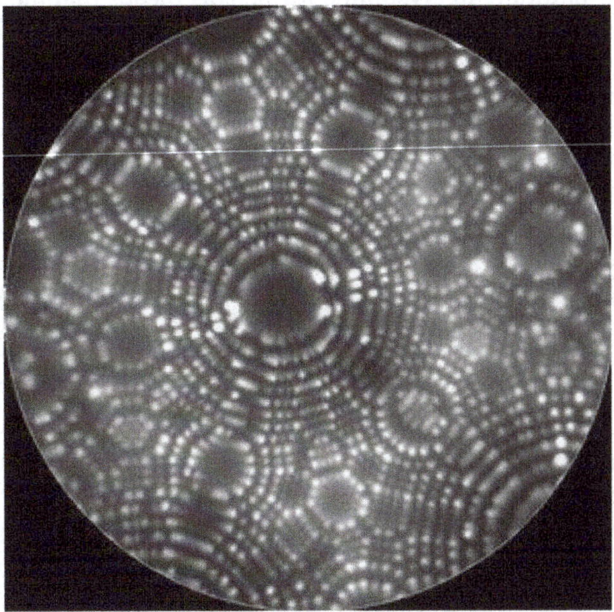

The picture above shows a Field Ion Microscope image of an atomically sharp tungsten tip. This tip is held at a very low temperature (around 10K). He atoms adsorb temporarily on the tip surface but get ionized by a very strong electric field. This field then accelerates the He^+ ions away from the tip and projects them onto a phosphor screen. The ion trajectories mirror the electric field very near the atoms on the surface of the tip, such that every bright spot on the screen corresponds to a surface atom. Note individual atoms, as well as ordered arrays of atoms forming small crystal planes. (Figure courtesy of Prof. K. Hono, National Institute for Materials Science, Japan).

3 Structure of Solids

The key structural feature of a large number of materials of most types is that they occur naturally in crystalline form. This means that the atoms are arranged in an orderly fashion, such that the local arrangement of neighbor atoms around a central atoms repeats itself over many atomic distances. We will study the types of crystal structures that are most important for metals, ceramics, and semiconductors. We will also compare these highly ordered structures to the amorphous, or random, structures occurring in inorganic glasses and in polymers. (Some polymers may be partially crystalline, meaning that they consist of ordered and disordered regions). We will see that the structures observed with different classes of materials are tied closely to the type of bonding present between the atoms involved. The fundamental principle guiding the formation of crystal structures is that atoms try to be packed as tightly as possible.

3.1 Structure and Properties: A First Glimpse

In this chapter we will dwell considerably on the structure of materials, in order to lay the foundation for our later discussion of their various properties. As I indicated earlier, the major theme of this book is how structure determines properties and performance of materials. Engineers go to great lengths to figure out, on the one hand, the relationships between structure and properties and, on the other hand, how to achieve a certain desired structure, leading to desired properties.

We surveyed the structure-properties connections briefly in Chapter 1 where we introduced the different materials classes. Here we will give two more examples, in oder to provide motivation for our examination of the structure of solids. These examples concern ductility and structural defects.

- *Structure vs. ductility*

  Remember our statement in Chapter 1 that metals are ductile. This is a simplification: Some metals have low ductility, and there is a direct correlation between crystal structure and ductility for metals.

  Also, in comparison to metals, polymers have equally high ductility, but ceramics, covalent crystals, and inorganic glasses have essentially zero ductility. Thus we call the latter materials **brittle** as opposed to ductile. At the end of this chapter, you will find it fairly straightforward to interpret this difference on the basis of structural differences between the classes of materials.

- *Structural defects vs. ductility*

  One finds that within the class of metals, in pure metals deviations from structural perfection have a great deal of influence on ductility, and on mechanical properties in general.

  In addition, alloys behave rather differently in comparison to their corresponding pure metals. By and large, alloys are stronger but less ductile than the pure metals. Think of an alloy as a base metal with other metallic impurities, i.e. a base metal with compositional defects.

We will encounter these and other structure-property connections over and over again, and in due time a fairly complete picture of them will emerge.

3.2 Crystals and Unit Cells

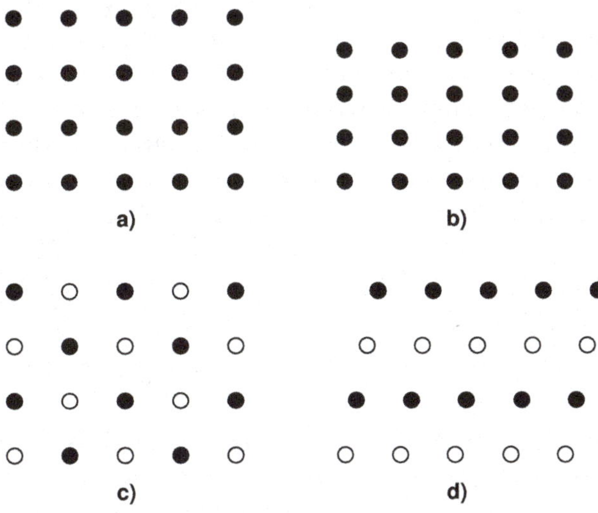

Figure 3.1: Examples of possible two-dimensional (2D) crystals

As we have indicated, many materials found in nature, or engineered by people, are in crystalline form. By **crystalline**, or being **crystals**, we mean that the atom positions in these materials form a regular array, so that you can move from a particular atom in various directions to another identical atom with the exact same surroundings. In other words: The local structure around an atom repeats itself over many thousands of atomic distance. Such a repeating arrangement of atoms in a crystal is also said to exhibit **long-range order**.

Fig. 3.1 gives a few examples in two dimensions, but of course in a real material you should visualize the regularity, or repeating property, as occurring in the third dimension as well. You can see readily 2D crystals in this figure, i.e. structures that repeat themselves. Clearly, the regularity could extend over a large distance, including many atoms. You could describe the regularity of the structure of Fig. 3.1a as square, Fig. 3.1b as rectangular but not square, Fig. 3.1c as square but with two kinds of atoms, and Fig. 3.1d as not rectangular with two kinds of atoms.

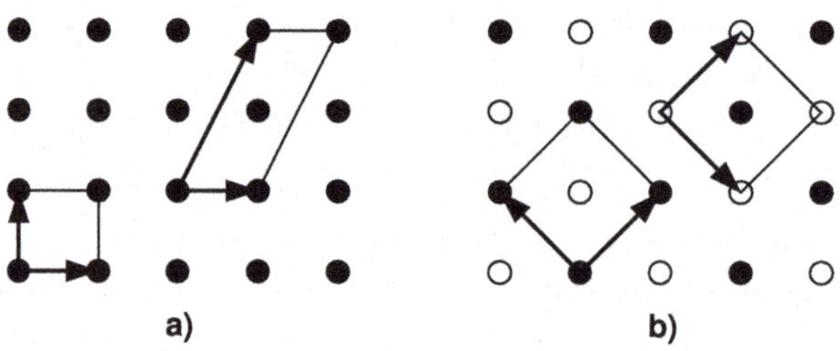

Figure 3.2: Examples of possible unit cells in 2D crystals

Now we need to become a little more systematic about dealing with the regularity of crystals. To this end, we introduce the notion of a **unit cell** (Fig. 3.2). This figure shows two of the structures from Fig. 3.1 before, each with two of many possible unit cells. The only requirement on a unit cell is that it allows us to generate the entire crystal. The unit cells shown obviously do.

Each of the unit cells also has two arrows, indicating what is called the **unit vectors** of the unit cell. When we want to describe the entire crystal, the rule is that when we move a unit cell, with its atoms, by a multiple of either of the unit vectors, we get to a position which is identical to the original one. Note that this means that the small square in Fig. 3.1a is a unit cell, but the same-sized small square in Fig. 3.1b is not. When you move the unit cell according to the rule above, black atoms have to end up on black atoms, and white atoms on white atoms.

Sometimes it is also of interest to speak of the **lattice** underlying a crystal structure. Think of the lattice as a geometrical abstraction, the grid on which the atoms are laid out. For example, we would say that the structures in Fig. 3.1a and Fig. 3.1b have the same type of square lattice, but the two lattices are populated by atoms in a different way. You will have ample opportunity to practice using these concepts when we talk about three-dimensional structures.

3.3 Close-Packing of Atoms

We noted in Chapter 2 that metallic bonding can be envisioned as the result of metal atom cores being immersed in a sea of highly mobile electrons. Furthermore, a key feature of this type of bonding is that it is non-directional. That is, the interaction between two atoms is independent of the direction in which they approach each other. This simple principle, together with some straightforward refinements, will allow us to account for essentially all crystal structures observed in metals.

Imagine that we build a metal crystal out of spheres stuck together. The spheres have an attractive interaction when they are close enough to each other, but they also have a given diameter that defines how close two spheres can come. Think of the spheres as billiard balls with mutual attraction.

The rule is that we ***build the configuration with the lowest possible energy***. This means that we will try for each sphere to touch as many other spheres as possible.

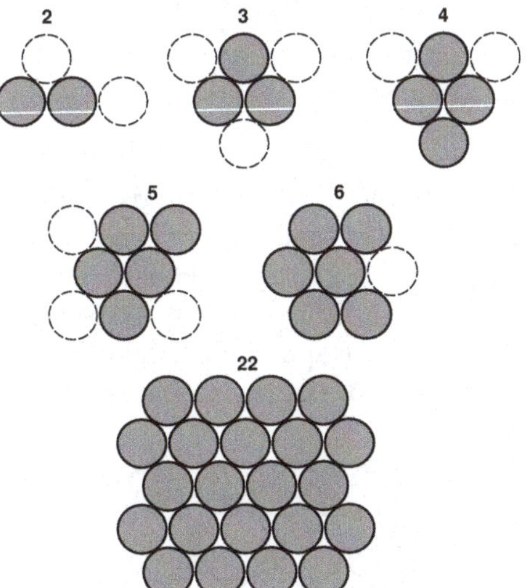

Figure 3.3: Atom-by-atom formation of a metal crystal. The numbers label clusters by their original number of atoms. With clusters 3, 4, and 5, the dashed circles indicate some of the equivalent positions for adding the next atom. For cluster 2, the two positions are not equivalent, and for cluster 6, the position shown is the best one.

Now let us put together a crystal, adding one atom at a time, using our engineering common sense. The first few steps are illustrated in Fig. 3.3. You have noted, of course, that the growth of this crystal was forced to occur in two dimensions. Beginning with the cluster labeled 3, there are even better positions if we allow for the third dimension to be used:

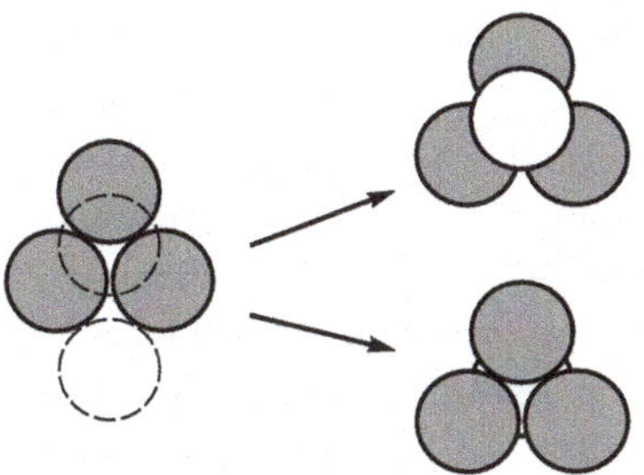

Figure 3.4: The two best positions for atom 4, on top or below the triangle of the first three atoms. In each case, the 4 atoms form a tetrahedron, and each atom touches the 3 other atoms.

It should be evident how we proceed to build a large crystals: Take layers of hexagonally arranged atoms, such as cluster 22 in Fig. 3.3, and put them on top of each other. The next figure shows two such layers, and the light gray one goes either on top or at the bottom of the dark gray one.

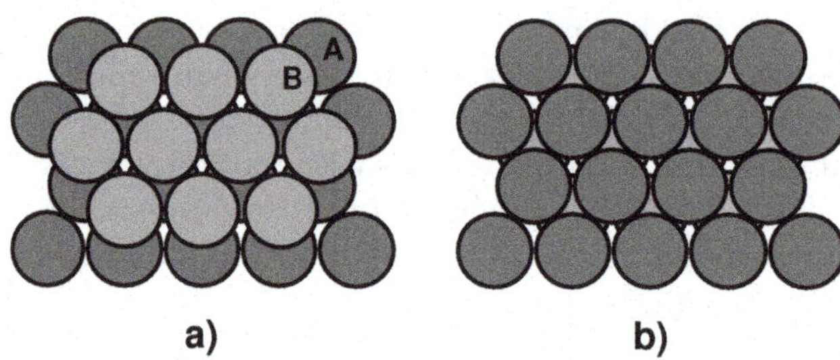

Figure 3.5: Two hexagonal atom layers put on top of each other. a) Second layer on top of first; b) Second layer at bottom of first layer.

The really interesting action happens with the third layer. Look carefully at the two graphics in Fig. 3.5, and you will realize that *once we have two layers, there are two different ways for putting down the third layer*! This is easiest to see when you focus on the light gray layer B in Fig. 3.5a: There are two types of sites for the third layer, depending on whether the spaces in-between the light gray atoms are open underneath or occupied by dark gray atoms.

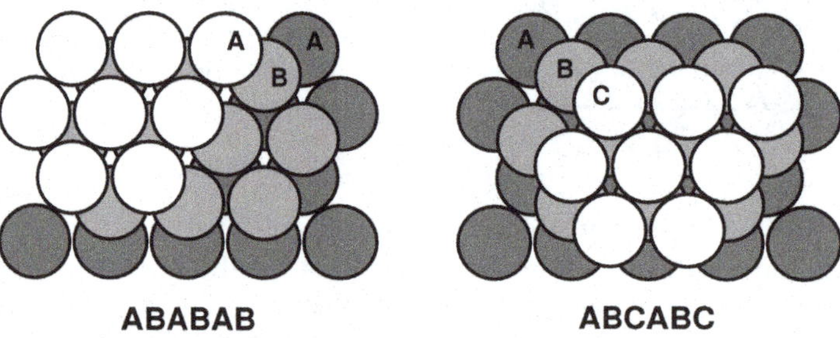

Figure 3.6: The two possible sequences for stacking three hexagonal layers. The layers marked A are exactly on top of each other, i.e. exactly in the same horizontal position.

Note that with the ABABAB stacking, vertical "channels" are left open between the atoms, but with the ABCABC stacking there are no open "channels".

Both these stacking sequences achieve what we will refer to as **close-packing**, and we will call the structures **close-packed**. The term should be clear: Hexagonal layers is how to pack atoms as closely as possible in a plane, and there are two ways to stack hexagonal planes. The resulting two close-packed structures are so important in materials science that they have special names:

- ***ABABAB sequence*** ⇔ ***hexagonal close-packed structure***
- ***ABCABC sequence*** ⇔ ***cubic close-packed structure***

The first name is evident. We will see below where the second one comes from.

Another way to describe close-packing is to realize that in both structures each atom has the highest possible number of **nearest neighbors,** i.e. each atom touches as many other atoms as possible. In a hexagonal layer each inner atom touches 6 other atoms within that layer (see Fig. 3.3, cluster 22), and then another 3 atoms on top and 3 atoms at the bottom (Fig. 3.5), for a grand *total of 12 nearest neighbors.*

A concise summary of how to pack metal atoms as efficiently as possible would be:

close-packing $\Leftrightarrow$ ***maximize number of nearest neighbors***

The structures thus realized turn out to be two of the three most important ones for metal crystals.

3.4 Structure of Metals

3.4.1 Face-centered Cubic Structure

We begin with the structure displayed in the figure below. This structure is called **face-centered cubic** or **FCC** for short. Its feature is a cubic unit cell with additional atoms in the centers of the cube faces.

A few examples of metals having the **FCC** crystal structure are **Al, Cu, Ni, Pd, Ag, Pt** and **Au.**

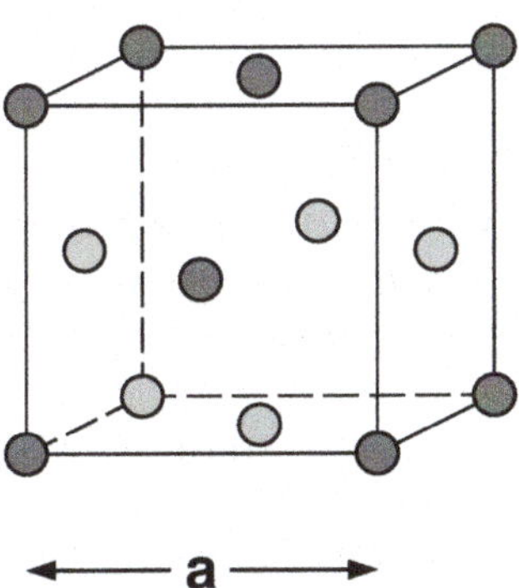

Figure 3.7:
Face-centered cubic (FCC) unit cell

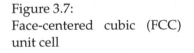

Note that all atoms in the cube are of course the same; the ones drawn in dark gray are those you would see if the cube were non-transparent, and the light gray ones are those that would be obscured.

But even more importantly, the atoms in the corners are not in any way different from those in the centers of the cube faces! You can convince yourself easily by looking at four unit cubes combined (Fig. 3.8). The cube with the dark lines is really identical in every way to the four cubes with lighter lines.

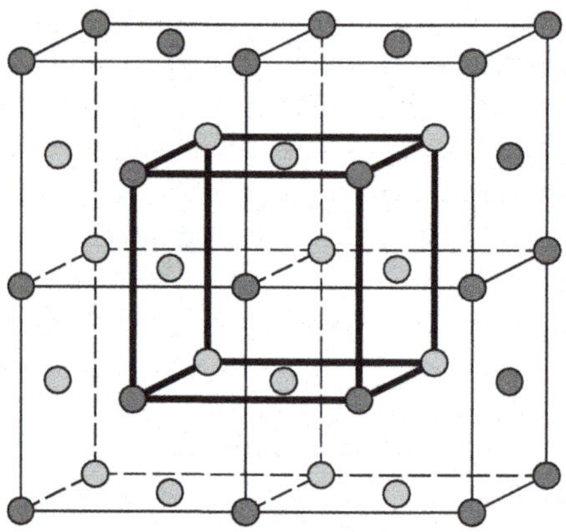

Figure 3.8: Four face-centered cubes stuck together

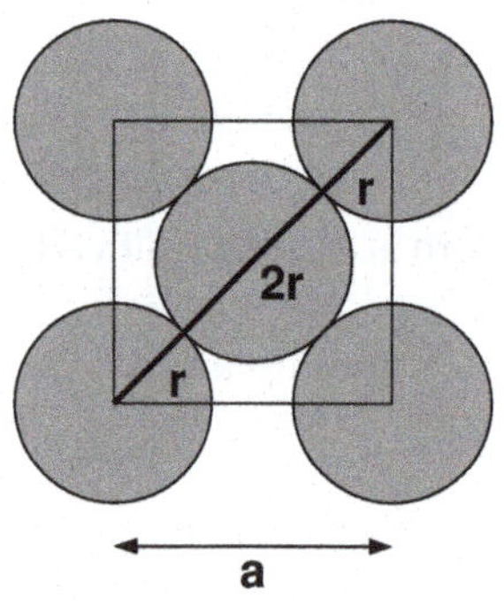

Figure 3.9:
A face of a face-centered cube, with atoms touching each other along the diagonal of the face. r is the atom radius, and a is the lattice constant.

If we represent the atoms as touching spheres, as we did above when discussing packing, we arrive at Fig. 3.9, which shows one face of one unit cube: The atoms touch along the diagonal of the cube face. The size, or length, of the unit cube is called the **lattice constant** a.

With a good sense of the geometry of this structure, we are now in a position to determine a few other quantities of interest. First, we would like to know the number of atoms that should be attributed to an individual unit cell. We note that there are 8 atoms on the corners, and 6 atoms in the centers of the faces. But 8 cubes come together at each corner, and 2 cubes at each face. Therefore, the count of atoms/unit cell is:

$$\text{Number of atoms/cube} = 8 \times \frac{1}{8} + 6 \times \frac{1}{2} = 4 \tag{3.1}$$

There are 4 atoms per FCC unit cell. We are also interested in what is referred to as the **coordination number**, i.e. the number of nearest neighbors or, equivalently, the number of atoms touching anyone particular atom. If you look at Fig. 3.8, you should be able to figure out that the number of nearest neighbors is 12.

Finally, we also need to know what is called the **atomic packing factor**, or APF. That means the fraction of the volume of a unit cube occupied by the atoms and parts of atoms inside the cube:

$$\mathrm{APF} = \frac{\text{vol. of atoms}}{\text{vol. of cube}} \tag{3.2}$$

More explicitly:

$$\mathrm{APF} = \frac{\frac{4}{3}\pi r^3 \times 4}{a^3} \tag{3.3}$$

From Fig. 3.9 it follows that

$$a = \frac{4}{\sqrt{2}} \times r \tag{3.4}$$

which when plugged into the equation for the APF yields

$$\mathrm{APF(FCC)} = \frac{\pi}{3\sqrt{2}} = 0.74 \tag{3.5}$$

At this point you will surely ask: This is all fine, but does it have anything to do with close packing of atoms? The answer is that it does indeed.

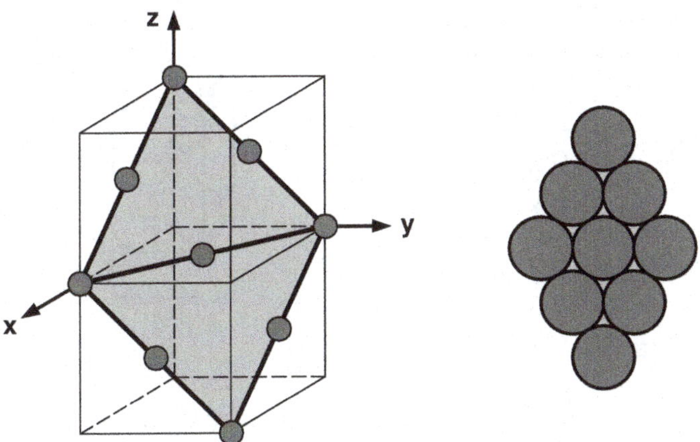

Figure 3.10: Two FCC cubes cut along a plane from the top back corner to the bottom front corner. Only atoms in the plane are shown. Note the hexagonal close-packing in the plane.

To uncover the connection, look at Fig. 3.10. Two vertically stacked cubes are sliced along a plane that extends from the top back corner of the upper cube to the bottom front corner of the lower cube. Only atoms in that plane are drawn on the left, and on the right you can see the plane by itself with the atoms touching each other. The hexagonal pattern in this plane is obvious.

Of course, the real crystal extends further, above and below this plane. In fact, the plane displayed on the right of Fig. 3.10 is what you would see looking at the FCC crystal in the direction of the body diagonal of a unit cell. Try it out with a crystal model. You would then see that the other planes above and below also have hexagonal symmetry and are stacked in the sequence ABCABC.

We may conclude that

cubic close-packed = face-centered cubic

This is really why the arrangement of hexagonally close-packed planes as in Fig. 3.10, stacked on top of each other in the sequence ABCABC, is called *cubic* close-packed.

3.4.2 Hexagonal Close-packed Structure

The next important structure for metals is the **hexagonal close-packed** structure, or **HCP** for short. We know it from above as the one with hexagonal planes stacked ABABAB.

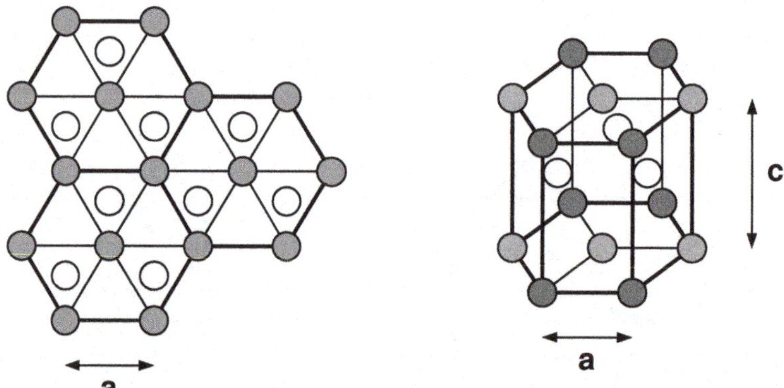

Figure 3.11: Unit cells of the HCP structure. The figure on the left shows a top-down view: Grey atoms are in the A planes and white atoms in the B planes. The figure on the right shows a perspective view of one large unit cell.

You will note that this is not a cubic, but a hexagonal unit cell. In fact, the figure provides an example of two equivalent unit cells. The large unit cell is the one with a hexagon as the base; the small unit cell is one third the size and has just a rhombus as the base. The plane with the hexagons is sometimes called **basal plane**.

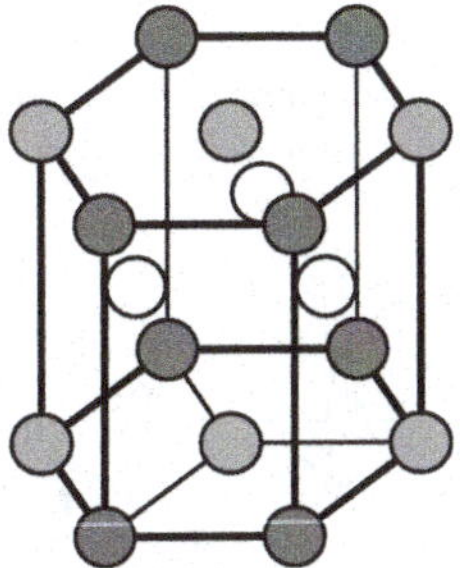

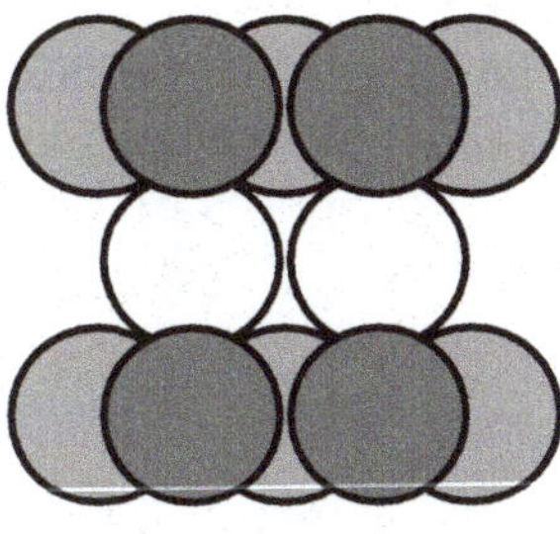

Figure 3.12: Unit cells of the HCP structure. The figure on the left shows a perspective view of one large unit cell. The figure on the right shows a side view, with nearest neighbor atoms touching each other.

Due to the non-cubic structure, we need two dimensions to characterize a unit cell: a for the side of a hexagon, and c for the height. These apply to both types of unit cells, large and small.

A few examples of metals having an **HCP** crystal structure are **Be, Mg, Ti, Co, Zn**, and **Zr**.

We will not analyze the HCP structure in any more detail. Due to its non-cubic nature, it is a little trickier than FCC. Of course, the coordination number is 12 and the APF is 0.74. These are the same values as for FCC.

3.4.3 Body-centered Cubic Structure

The third important structure for metals is the so-called **body-centered cubic** structure, or **BCC** for short.

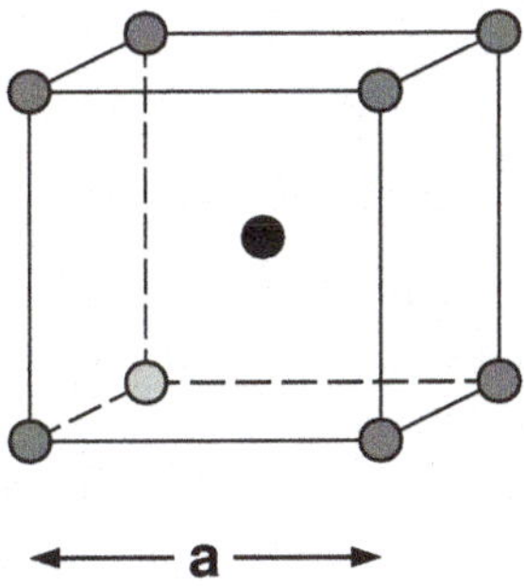

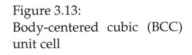

Figure 3.13: Body-centered cubic (BCC) unit cell

The characteristic feature of this unit cell is that it is cubic with an extra atom in the center of the cube. But realize, once again, that in a real BCC crystal all atom positions are entirely equivalent: Corners could be centers, and vice versa.

A few examples of metals having the **BCC** crystal structure are **Li, Na, K, Fe, Cr, Mo** and **W**.

In order to find the APF of a BCC crystal, we proceed as above for FCC, while observing differences in geometry between FCC and BCC:

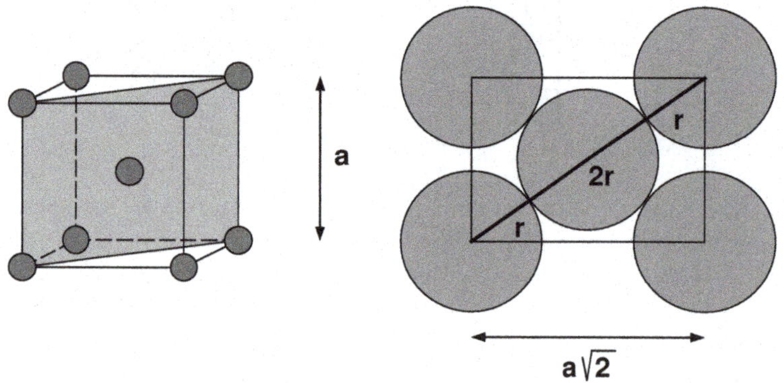

Figure 3.14: Geometry of the BCC unit cell.

For the number of atoms per cube:

$$\text{Number of atoms/cube} = 8 \times \frac{1}{8} + 1 \times 1 = 2 \tag{3.6}$$

In a BCC unit cube, the atoms touch each other along a body diagonal of the cube. Hence

$$a = \frac{4}{\sqrt{3}} \times r \tag{3.7}$$

This yields an APF of

$$\text{APF(BCC)} = \frac{\pi\sqrt{3}}{8} = 0.68 \tag{3.8}$$

We conclude that the APF of a BCC crystal is slightly lower than the APF of the close-packed FCC and HCP crystals. At the same time, the coordination number is lower as well: 8 in BCC vs. 12 in FCC and HCP.

The main geometrical properties of the three major crystal structures for metals can be summarized as follows:

Table 3.1: Properties of FCC, HCP, and BCC crystals

Crystal	Coord. Number	APF
FCC	12	0.74
HCP	12	0.74
BCC	8	0.68

At this point, bothersome questions may have popped up in your mind: When atoms want to be close-packed, how do they know whether to adopt an FCC or an HCP structure? And, why are there any BCC metals at all? Why aren't all structures close-packed, if that is the best one can do?

The answer is subtle: What these results reveal to us is that atoms are not just attracting hard spheres which know only about nearest neighbors. Apparently, interactions involving more distant neighbors are also important, and it is those interactions which determine ultimately whether a crystal is FCC, HCP, or BCC.

Let us conclude this chapter by noting briefly that metals, especially when used in engineering applications, are usually alloys. The presence of two or more metal elements will clearly complicate matters of structure, and we will address those in more detail as the need arises.

3.5 Structure of Ceramics

3.5.1 Rules for Building Ceramic Crystals

Ceramics, as we know, are compounds of a metal and a non-metal, in the simplest case. From our discussion in Chapter 2 it should be clear that such compounds will have ionic bonding, i.e. bonding from interactions by positive and negative ions according to Coulomb's Law. This type of bonding is non-directional, as with metallic bonding, but now there is an additional complication.

Due to the presence of the two types of ions, we cannot take it for granted that at close enough range there will always be an attractive interaction between two particles. In fact, there will always be repulsion, at any distance, between two positive or two negative ions. We will only see attraction between ions of opposite charge.

This has profound consequences for the formation of ceramic crystals. Once again we will try to pack ions as closely as possible, but there is now an important restriction: We must make sure that ions of the same charge will not touch each other.

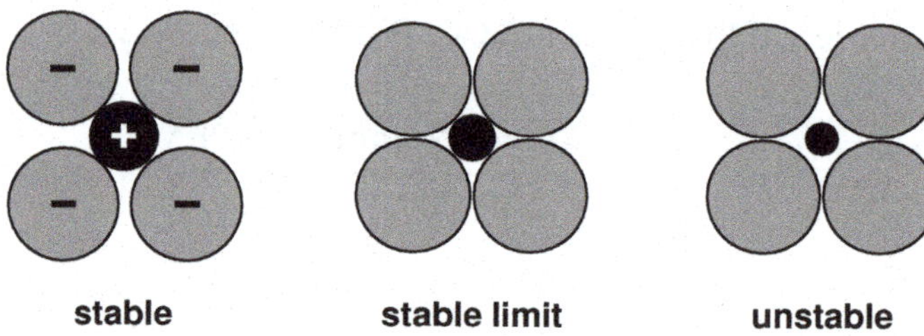

Figure 3.15: 2D illustration of the stability of ionic crystals, depending on the ionic radii.

The study of a vast number of ionic crystals has shown that most ionic structures can be understood and categorized on the basis of the rule formulated above, in combination with assigning each ion a unique radius, called its **ionic radius**. These two principles lead to a natural interpretation of what kind of an ionic structure is stable for a given pair of ions.

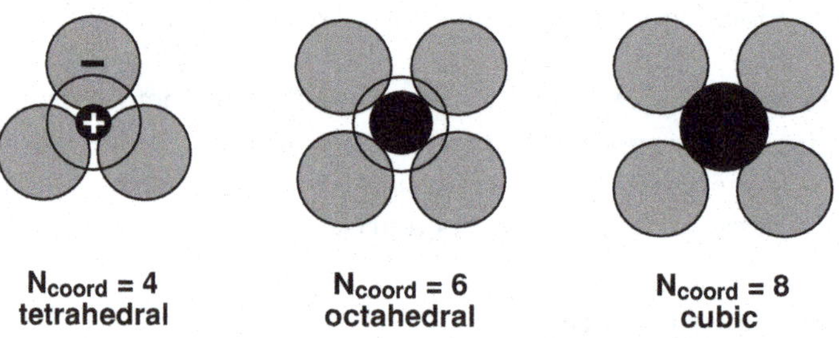

Figure 3.16: Important geometries, with ion coordination numbers, in ionic crystals.

We expect that the stability of an arrangement of ions, thus the allowed coordination number N_{coord}, depends on the ionic radii. In fact, the deciding factor is not so much the individual ionic radii but the *ratio of the ionic radii*, r^+/r^-. (This ratio is sometimes also called r_C/r_A where C stands for **cation** (the positive ion) and A for **anion** (the negative ion)).

As always, for real ceramics, the rules for forming crystals are played out in three dimensions. The most important cases are illustrated in Fig. 3.16 for **binary** compounds (i.e. compounds consisting of two ion species, one positive and one negative) with N_{coord} and the local geometry. Imagine that you look straight down onto the geometrical body, say the tetrahedron. The gray circles are the anions at the bottom of the tetrahedron. The open circle, is the anion on top of the cations, if the cation position is distinguishable from the bottom ions. For the case of N_{coord} = 8, the four remaining anions are exactly on top of the bottom anions shown.

In the following Periodic Table we list some ions with their ionic radii. The most common ion charge is given. Especially with transition metal ions, ions with different positive charge are possible. Note the trends in the ionic radii, and also that for most but not all combinations of ions $r^+ < r^-$.

H					He
Li^+ 0.068 nm	Be^{2+} 0.035 nm	N	O^{2-} 0.140 nm	F^- 0.133 nm	Ne
Na^+ 0.102 nm	Mg^{2+} 0.072 nm	P	S^{2-} 0.174 nm	Cl^- 0.181 nm	Ar
K^+ 0.138 nm	Ca^{2+} 0.100 nm	As	Se^{2-} 0.191 nm	Br^- 0.196 nm	Kr
Rb^+ 0.148 nm	Sr^{2+} 0.127 nm	Sb	Te^{2-} 0.221 nm	I^- 0.220 nm	Xe
Cs^+ 0.170 nm	Ba^{2+} 0.136 nm				

Figure 3.17: Common ions and their ionic radii.

The regions of stability of ionic structures with the various coordination numbers are given in Table 3.2. In each case, the lower of the two numbers for the ratio of ionic radii is the one where the anions begin to touch each other.

Table 3.2: Geometries of ionic crystals

N_{coord}	r^+/r^- range	geometry
4	0.225 - 0.414	tetrahedral
6	0.414 - 0.732	octahedral
8	0.732 - 1.0	cubic

3.5.2 Examples of Ceramic Crystals

We will now introduce specific examples of ceramic crystals with coordination numbers 8, 6, and 4 in turn.

a) CsCl structure: Coordination number = 8

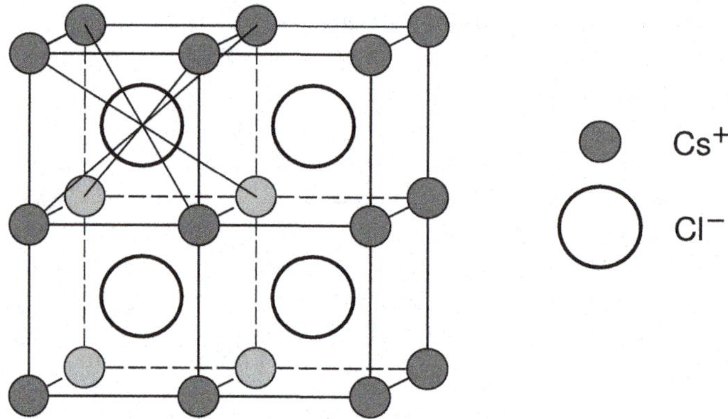

Figure 3.18: Crystal structure of CsCl. The Cl^- ions are in the centers of Cs^+ cubes. The darker Cl^- ions are in front, the lighter ones in back.

The figure shows four unit cells of the so-called cesium chloride structure, i.e. the ionic crystal structure with $N_{coord} = 8$. For clarity, the ions are not shown as touching along the cube body diagonal, as they really would.

Other examples of compounds with the CsCl structure are CsI, CsBr, and RbCl.

Please be aware that all Cs^+ ions are equivalent, of course, and so are all Cl^- ions. Also, I have drawn four unit cells in the hope that you can see that it is somewhat arbitrary where the cube corners are chosen. The four Cl^- ions could as well be the front square of a unit cube which would have a Cs^+ ion in its center. Just imagine four more cubes behind the ones shown in the figure.

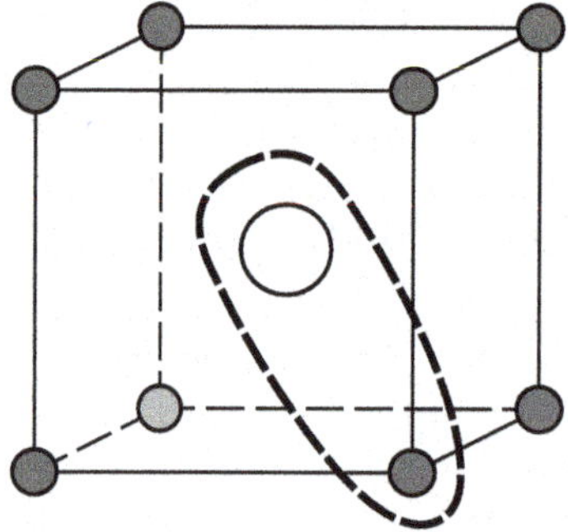

Figure 3.19:
Unit cell of the CsCl structure. Note the **basis** of this unit cell: a positive and a neighboring negative ion belonging to the same lattice point, as indicated by the dashed contour.

What about the unit cell of this structure? You may be tempted to say that it is BCC, but *it is not*! Remember that when you move a unit cell to an equivalent position, you must obtain the exact same configuration! Here, if you take the unit cell in Fig. 3.19 and move the positive ion in the lower right corner to the center, it will replace a negative ion. In other words, with that translation all ions change sign, and you do not reproduce the same configuration.

Instead, the way to think of the CsCl structure is to say that the unit cell is **simple cubic**, or **SC**, and two ions belong to each point of this lattice, as indicated by the dashed outline in Fig. 3.19. These two atoms are called the **basis** of the (simple cubic) unit cell. Now, when you translate the SC unit cell from one cube corner to another, the two ions of the basis move together, such that you will replicate the original configuration exactly.

b) NaCl structure: Coordination number = 6

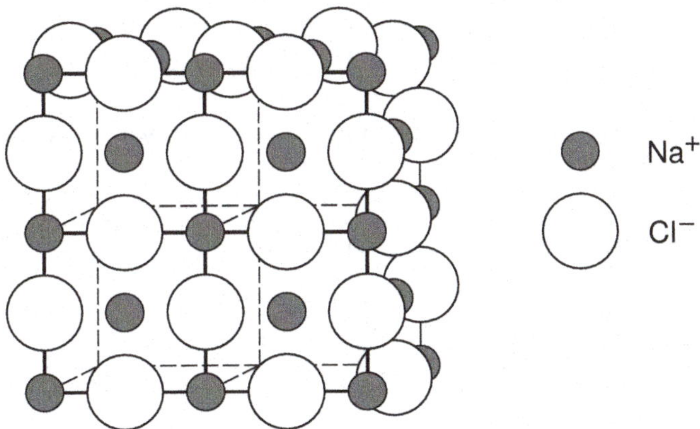

Figure 3.20: Crystal structure of NaCl. The Na^+ ions are on an FCC lattice, and the Cl^- ions are in the centers of the sides of the cube. Only ions on visible faces are shown.

The figure shows four unit cells of what is known as the NaCl structure, i.e. the ionic crystal structure with $N_{coord} = 6$. For clarity, the ions are not shown in proper size, i.e. they are shown as not touching within the faces of the cubes, as they really would.

Fig. 3.20 may be a little confusing because of the many ions, but the point is to prove again that either type of ion could be viewed as being at the corner of a cube. It is just a matter of whether you focus on the positive or the negative ions.

The next figure, with two equivalent versions of the face of a single NaCl unit cell, should clarify that this structure is indeed FCC, with a two-ion basis. Think of the basis as a corner ion plus a neighbor ion along one of the edges of the cube. Note also that the positive and the negative ions each occupy an FCC sublattice.

The FCC unit cell with a two-atom basis for the NaCl structure works out

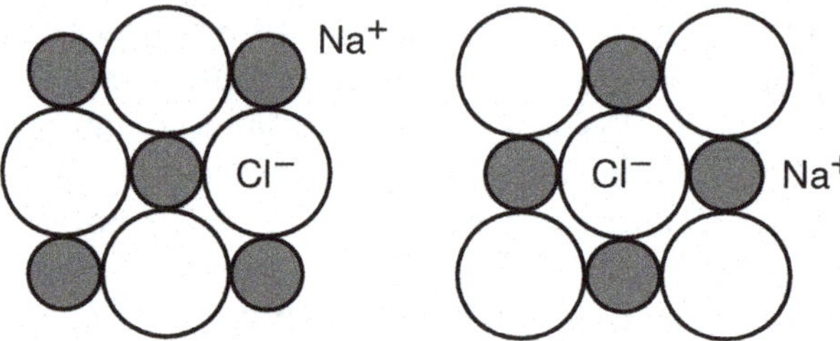

Figure 3.21: Two equivalent faces of a cube of the NaCl structure. This structure has a unit cell with a basis consisting of a positive and a neighboring negative ion.

the same way as above for the CsCl structure. If you translate the FCC unit cell of Fig. 3.21 with its basis from one FCC lattice point to another, the entire structure repeats itself.

Other examples of ceramic compounds with the NaCl structure are LiF, MgO, MgS, CaO, and NiO.

c) Zincblende structure: Coordination number = 4

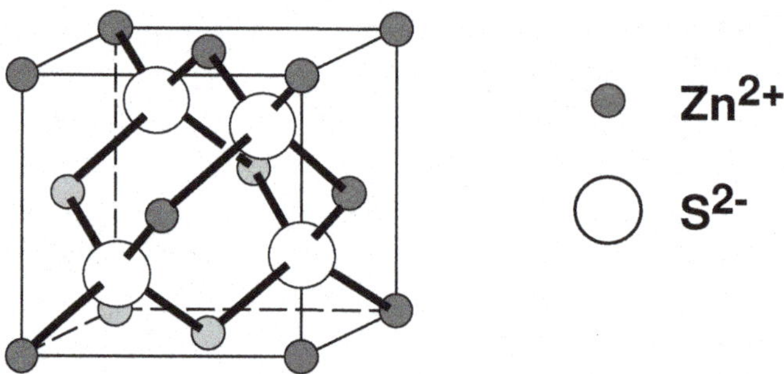

Figure 3.22: Crystal structure of ZnS. The Zn^{2+} ions and the S^{2-} ions occupy alternate sites on a diamond cubic structure.

The term Zincblende comes from the German word for the mineral ZnS (zinc sulfide), which exhibits this structure.

Fig. 3.22 displays the unit cell of a ceramic crystal with $N_{coord} = 4$. You can see the FCC arrangement of the Zn^{2+} ions. Furthermore, you can verify that all the ions of the same type are on an FCC lattice, and the ions of the second type are on planes exactly in-between the ions of the first type.

Keep in mind that the darker lines between opposite neighbor ions in Fig. 3.22 are not meant to suggest that there are true chemical bonds between those ions. The lines simply serve to clarify the four-fold coordination.

The Zincblende structure is quite prevalent among important electronic materials. Examples are so-called III-V compounds such as AlP, GaAs, and InSb, and so-called II-VI compounds such as ZnS itself, ZnSe, and CdS.

I should point out that with this discussion of ionic crystals, we have only scratched the surface. The wide variety of materials comes from the fact that we are dealing with compounds. The simple cases we have considered are of the form AB, i.e. binary compounds with an atom ratio of 1:1. Even among those there are more complex examples, e.g. BN and AlN. BN exists in several crystal structures, only one of which is cubic; AlN is hexagonal.

Many other binary compounds exist with different atom ratios, e.g. 1:2 as in SiO_2, or 2:3 as in Al_2O_3. Both of these are very important industrial materials. In addition, many **ternary** ceramics are known, ternary meaning composed of three kinds of atoms, e.g. $Al_xGa_{1-x}As$ or $SrTiO_3$. If you are interested, please check the references.

3.6 Structure of Covalent Crystals

3.6.1 Diamond and Silicon

The covalent crystal we know already is of course diamond, i.e. crystalline carbon.

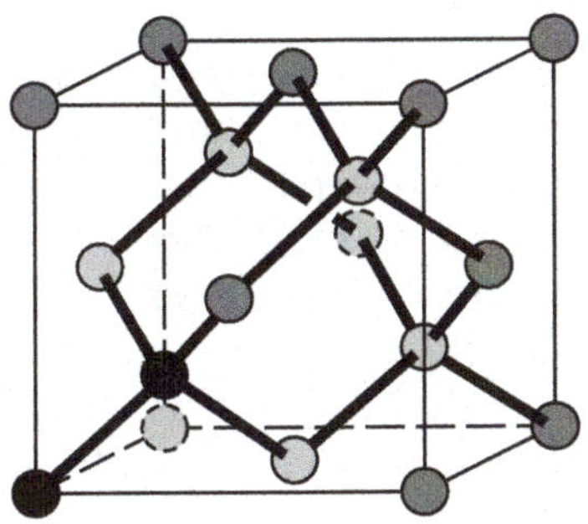

Figure 3.23:
Unit cell of diamond. The heavy lines indicate the C-C covalent bonds inside the cubic unit cell.

The heavy lines between two carbon atoms represent covalent bonds. This is an FCC structure with a two-atom basis, an example of which is indicated by the two black atoms in Fig. 3.23.

There are two additional elemental materials with the same structure and covalent bonding, namely silicon (Si) and germanium (Ge). Grey tin (Sn) below 13°C also has diamond-cubic structure but is a metal.

3.6.2 Other Forms of Carbon

Solid carbon has many **allotropes**, i.e. it comes in many structural forms. A few of those are shown in the next figure:

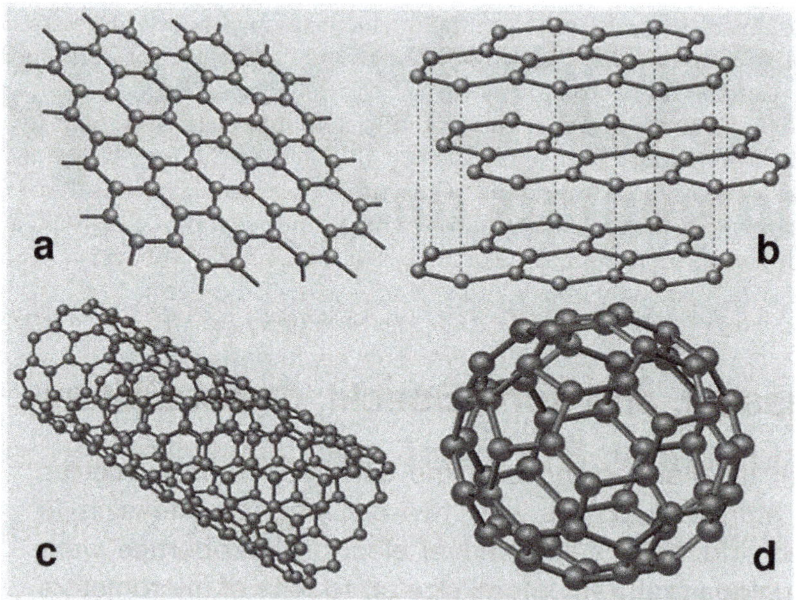

Figure 3.24: Allotropes of carbon. a) graphene; b) graphite; c) carbon nanotube; d) C_{60} molecule.

Fig. 3.24a shows a single flat sheet of carbon atoms in a hexagonal pattern, known as graphene. Fig. 3.24b shows graphite, in which hexagonal sheets are arranged in a sort of ABAB stacking, but this is not close-packed, and the vertical dimension of the unit cell is much larger than in HCP. The bonding in graphite is only covalent within the planes, but of the van der Waals type between planes. Fig. 3.24c shows a carbon nanotube, i.e. a single sheet as in a), rolled into a tube; Fig. 3.24d shows a C_{60} molecule, also know as a fullerene (named after R. Buckminster Fuller). These molecules can form solid films or even crystals.

You will note that in these materials, C atoms are not fourfold coordinated, or in a tetrahedral geometry. Rather, they are threefold coordinated, and in the carbon nanotubes and C_{60}, the three C-C bonds issuing from a central C atom are not in a plane.

3.6.3 Silicon Dioxide

SiO_2 is an interesting material for many reasons. One reason is that its structure is closely related to the structure of crystalline Si. Another reason is that SiO_2 is a so-called **polymorph**, i.e. it comes in several crystalline forms. Under the name of **quartz**, it is known as the most abundant mineral in the earth's crust. Also, it is often part of other minerals, which are then referred to as **silicates** (minerals containing SiO_2). A common short name for silicon dioxide is **silica**, and you should be sure to make the distinction between silica and silicon.

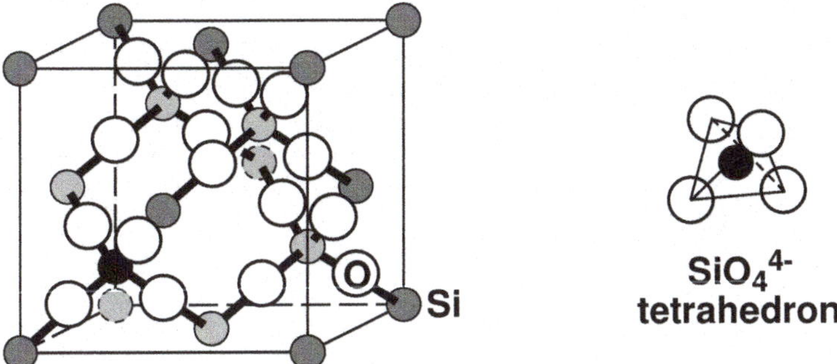

Figure 3.25: Unit cell of crystalline SiO_2 in the form of β-cristobalite. Atoms are not drawn to scale.

Let us now compare crystalline Si and a particular crystalline form of SiO_2 called β-cristobalite. The β-cristobalite form of crystalline SiO_2 is stable only at high temperature, above 1050°C.

The SiO_2 crystal structure can be viewed as similar to Si, but with an O atom between each covalently bonded pair of Si atoms. That implies, of course, that Si is covalently bonded to O, and O is bonded to the next Si. Consequently, in SiO_2 each Si atom has four O atoms as nearest neighbors, whereas each O atom has 2 Si atoms as nearest neighbors. This is exactly as it should be, given the **stoichiometry** (i.e. the composition) of the SiO_2 compound.

An alternate view is to say that the SiO_2 crystal structure consists of SiO_4^{4-} tetrahedra tied together at the O^{2-} corners. Fig. 3.25 shows these tetrahedra, but the tetrahedra joined together are more difficult to visualize because the figure displays only one unit cell.

A third reason why SiO_2 is interesting is that we are treating it here as a covalent solid but, being similar to a metal oxide, it could be regarded equally well as as a ceramic. Recall that in Chapter 2 we introduced the notion of ionicity, which is a number indicating *how ionic or covalent* a compound really is. The point is that pure ionic and pure covalent bonding represent the two extremes on a intrinsically continuous scale. Ionic bonding means that one or more electrons have been given over completely from one kind of atom to another. Covalent bonding means that pairs of electrons are shared equally between two atoms.

It follows that for equal sharing of electrons to occur, the participating atoms have to be the same. Otherwise, and by necessity, the sharing will be unequal, in which case you will have *polar covalent* bonding, or equivalently, *partially ionic* bonding. We illustrate this principle by calculating partial ionicities f_i according to Eq. 2.2 (See homework in Chapter 2. I have also used some electronegativities EN_A and EN_B which are not on Fig. 2.5):

Table 3.3: Bond ionicities f_i in sample compounds

Compound	EN_A	EN_B	f_i
NaF	0.93	4.0	0.91
NaCl	0.93	3.2	0.72
SiO_2	1.8	3.4	0.47
ZnS	1.6	2.6	0.18
SiC	1.8	2.5	0.12

Based upon the data in this table, we are led to conclude that SiO_2 is halfway between ionic and covalent. Note that ZnS is practically covalent yet we listed it among ionic compounds above for the purpose of illustrating fourfold coordination. On the other hand, SiC (silicon carbide) is covalent but is usually considered a ceramic material.

3.7 Structure of Inorganic Glasses

3.7.1 Overview of Glasses

If you review what we have discussed so far about the structure of solids, you will realize undoubtedly that it has all been about crystalline structures. This is not to say that most solids have a regular, repeating arrangement of atoms; far from it. However, with a good background in crystals, we are now in an excellent position to examine solids which are not crystals, i.e solids without long-range order. Such solids are called amorphous, as we noted in Chapter 1.

Of course, a question arises immediately: *Why glasses at all? Why don't all materials form crystals?* After all, we have emphasized repeatedly that on forming a solid, atoms try to achieve the configuration with the lowest possible total energy. This will involve some kind of close packing, and thus will result in long-range order.

But look at how solids are typically formed. You melt a certain material or mixture of materials and let it cool down so that it solidifies. This suggests that the argument of achieving a close-packed state is correct only if the atoms are allowed enough time to reach an orderly state, i.e. only if you cool down slowly enough. Otherwise, the disordered state of the liquid melt may get frozen in, and you end up with a glass.

With metals, cooling down slowly enough is usually no problem, since all atoms interact with each other equally. With metal-nonmetal compounds, things start to get more complicated. Not only do we deal with two or more kinds of atoms, but these atoms are charged. Hence, the formation of an ordered solid is constrained by the condition that like ions cannot be nearest neighbors. When it comes to forming something as complex as crystalline SiO_2, an additional complication arises because now we are dealing with the ordered arrangement of SiO_4 tetrahedra, rather than individual atoms or ions, and these tetrahedra tend to get in each others way.

Although we described an amorphous solid as lacking long-range order, it is important that you do not think of it as *completely* random. An amorphous solid usually has what is called **short-range order**. That is, if you focused on any given atom, then you could predict with high probability what its immediate neighbors are and at which distance they are likely to be.

In this section we will deal primarily with inorganic amorphous solids or, more concisely, inorganic glasses. I should point out that the generic term glass may refer to any type of amorphous material. However, if we do not specify what kind of a glass we are talking about, we will agree that we mean an inorganic glass. We may also use the word **glassy** on occasion, which will mean a material with a glass-like, i.e. amorphous, structure.

3.7.2 Silicate Glasses

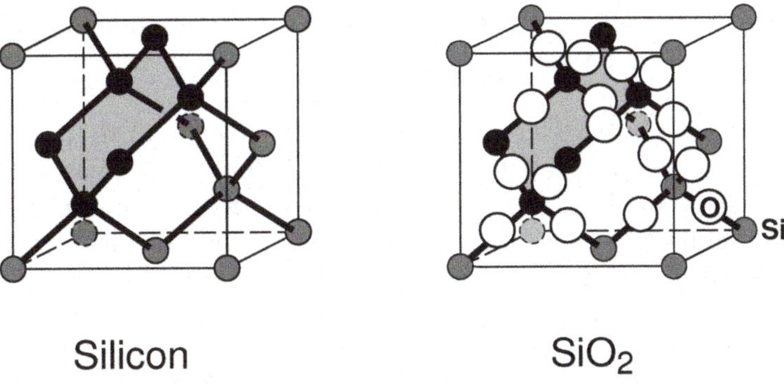

Figure 3.26: Comparison of unit cells of Si and β-crystobalite SiO_2. (We will use the short form β-quartz) for the latter.

Even among the inorganic glasses there are many different kinds. Here we will focus on one of these, namely silicate glasses, and more specifically on the simplest case, glasses based on amorphous SiO_2. For most people, this is what they mean when they mention glass (as in window glass).

First, let us take a step back and revisit the crystal structure of silicon, Fig. 3.26, which is the same as that of diamond, Fig.3.23. Now, let us adopt the point of view that the crystal structure of Si can be considered as a regular 3D network of interconnected hexagons. One of these is shown in light gray with black Si atoms in Fig. 3.26. Note especially that *these hexagons are not flat*! The angles in their corners are not 120° as in a planar hexagon, but 109.54°, which is the well-known angle between the center of a tetrahedron and two of its corners.

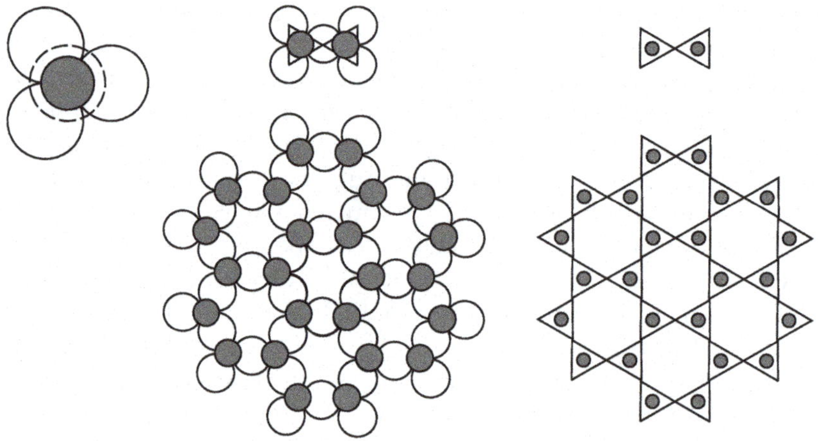

Figure 3.27: Schematic visualization of Si-O-Si hexagons in β-quartz. On top are, from left to right: a single SiO_4 tetrahedron with all atoms showing; two joined SiO_4 tetrahedra with atoms; two joined SiO_4 tetrahedra simplified as triangles. At the bottom are two representations of a set of seven Si-O-Si hexagons.

What is more, you can see the same kind of hexagon (also in gray) in the unit cell of β-quartz (Fig. 3.26), but in that case there are 6 Si atoms and 6 O atoms making up the hexagonal ring, with the Si atoms sitting in the corners. As regards the shape, think of the gray hexagons as similar to a normal flat one, but with one corner bent up and the opposite corner bent down.

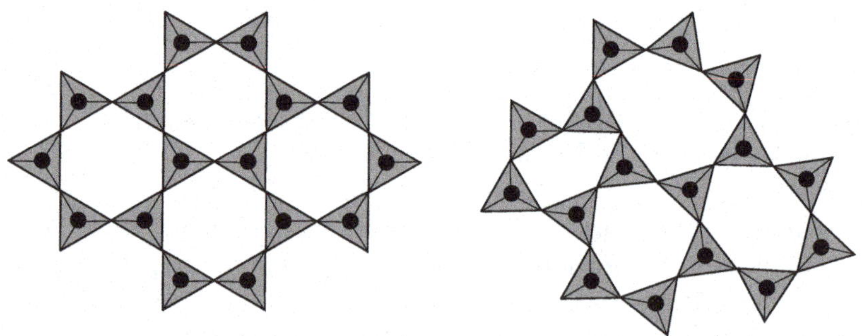

Figure 3.28: Ordered and disordered sheets of SiO_4 tetrahedra.

Please keep in mind that the graphs in Fig. 3.27 are schematic. Especially the seven hexagons are just meant to show the connectivity and the regularity, not the actual 3D structure. But now, using these SiO_4 triangles as building blocks, it is easy to see how disorder can be introduced into this structure. In the crystal, all rings are identical hexagons, whereas in the disordered structure we allow for rings with different sizes consisting of, say, distorted Si-O-Si pentagons and heptagons, as in Fig. 3.28.

Hence, a convenient way to keep track of the rings is to focus on the number of Si atoms in them. Also note that as the ring size changes, so do the bond angles. In the hexagonal structure, the Si-O-Si bond angle is 180° (see Fig. 3.26), and in the distorted structure this angle varies. It turns out that this bond angle variation does not cost very much in terms of increased energy of the structure (see the book by Putnis in the references).

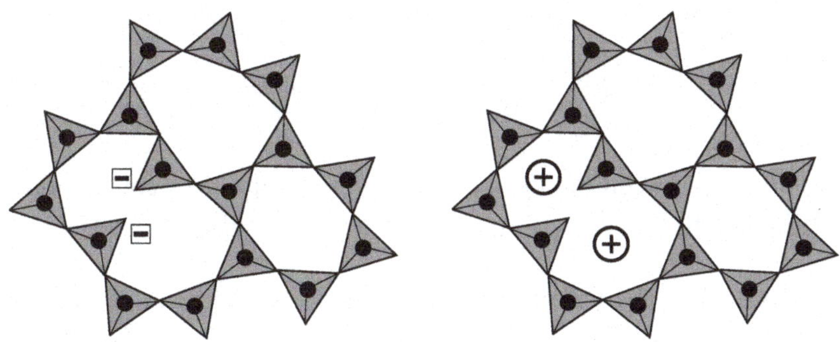

Figure 3.29: Left: Disordered SiO_4 tetrahedra with two unbalanced negative charges due to unconnected O^{2-} ions. Right: Charges balanced due to two extra positive ions from the network modifier.

It is also possible for rings to be opened up as illustrated in Fig. 3.29. However, in that case one would end up with two unbalanced negative charges, which would represent a rather high energy configuration. Normally, such negative charges are balanced by corresponding positive charges from ions such as Na^+ added to the glass.

As an aside, it is interesting to note a detail about how to keep track of charges in these SiO_2-type glasses. As you recall from above, we considered SiO_4 tetrahedra, or more explicitly SiO_4^{4-} tetrahedra, as the building blocks of these materials. The overall -4 charge of an individual SiO_4 tetrahedron comes from Si^{4+} and four O^{2-}. However, in a network two SiO_4 tetrahedra are joined at an O corner. The charge on the *shared* O atom must be counted as -1, not the -2 charge of an isolated O^{2-} ion. Hence the two minus signs for the two separated tetrahedra in the left graphic of Fig. 3.29.

The structure displayed in the graphic on the right in Fig. 3.29 is in fact not a flaw of the glass, but rather is put in on purpose by the addition of so-called

network modifiers such as Na_2O, K_2O, and CaO. These additives tend to make the glassy network less rigid and easier to work with.

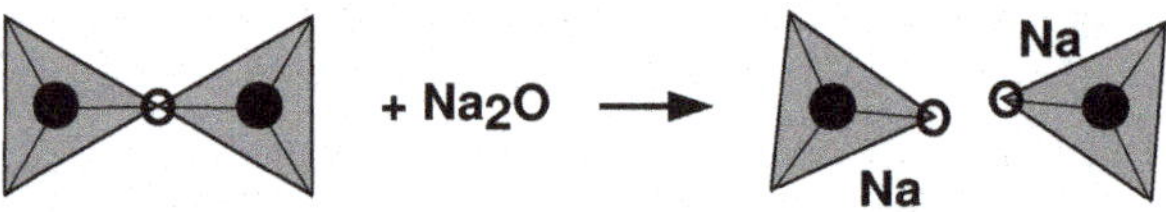

Figure 3.30: Effect of network modifier Na_2O on SiO_2 glass network. Triangles represent SiO_4 tetrahedra.

From a chemist's point of view, the effect of such an additive can be understood as a chemical reaction with the SiO_4 glass building blocks (see also Fig. 3.30 below). Two joined SiO_4 are written as Si_2O_7:

$$Si_2O_7 \quad + \quad Na_2O \quad \rightarrow \quad 2SiO_4Na \tag{3.9}$$

Some of the most commonly used silica glasses with their compositions are listed in Table 3.4:

Table 3.4: Typical silica glasses with compositions in weight percent

Glass type	SiO_2	Na_2O	CaO	Other	Use
Soda-lime	74	16	5	4MgO	Window glass
Silica glass	96	-	-	$4B_2O_3$	High T apps.
Borosilicate	81	4	-	$13B_2O_3$	Labware
Safety	55	16	2	$19Al_2O_3$, $4TiO_2$	tempered glass

Soda-lime glass is what we know as window glass. Borosilicate glass is widely used under the trade name Pyrex^{TM} for glassware in chemistry labs. In general, the purer the glass, the more stable it is at high temperature. We will discuss glass properties more extensively in a later chapter.

3.8 Structure of Polymers

This section will introduce polymers, a large class of organic materials. These turn out to have a mostly amorphous structure, although a few may have crystalline regions in them. Whether or not they do depends on the nature of the material and how it was prepared. An amorphous polymer would represent an organic glass, and a polymer with crystalline inclusions in a glassy matrix would be called **partially crystalline** or **semicrystalline**.

For the types of materials we have examined so far, we assumed that the natural building blocks would be atom, or perhaps small clusters of atoms as in the silica glasses. With polymers, this is a different story, as is most everything else. In many ways, polymers occupy a unique position among all materials.

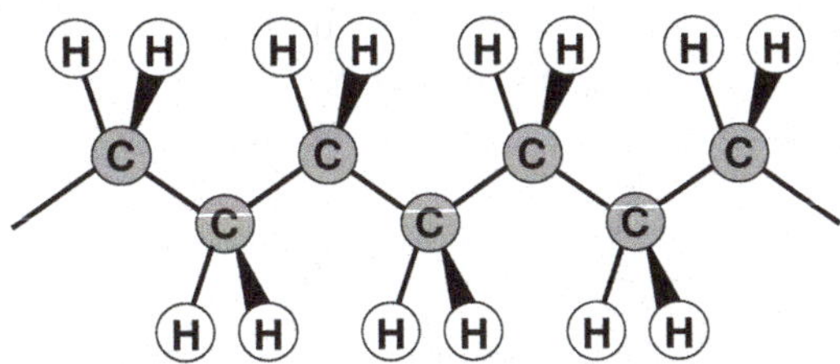

Figure 3.31: Small section of a polyethylene molecule.

The special feature of polymers is that they consists of long carbon-chain molecules. The C atoms within a molecule are held together by strong covalent bonds. At the same time, a major part of what holds together the molecules to form a solid are much weaker van der Waals forces. The simplest example of a polymer molecule is given in Fig. 3.31, where you should imagine the chain as continuing for several thousand carbon atoms. The molecule depicted is called **polyethylene**. The properties of a polymer material depend on a number of variables, but the length of the molecules is crucial. If the molecules were short chains with about 10 C atoms in them, the material would be a liquid.

When you look at the molecule in Fig. 3.31, you realize right away that the chains present a huge problem in forming a crystal. To do so, you would have to line up large numbers of molecules parallel to each other.

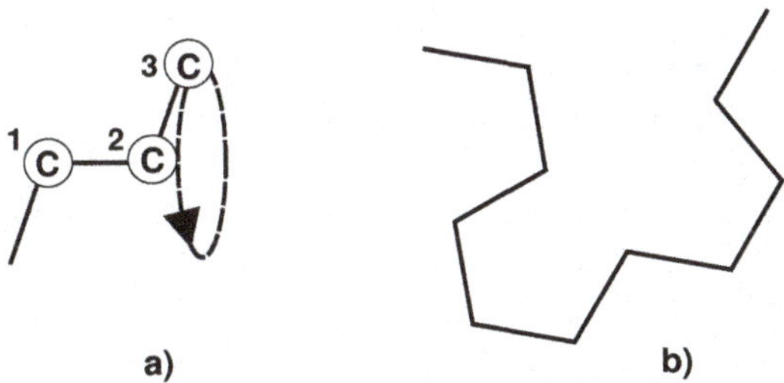

Figure 3.32: a) Rotation about a C-C bond; b) A more probable configuration of a polymer chain (only the C-C bonds are drawn).

Furthermore, the molecule is depicted in a most unlikely configuration: a straight chain. A polymer molecule behaves more like an ordinary chain with movable links. If you focus on a triangle of three carbon atoms and you keep the positions of atoms 1 and 2 and the bond angle at atom 2 fixed, you can rotate atom 3 freely about the 1-2 axis. This is only true, of course, if nothing else attached to the chain or in the chain's surroundings prevents such a rotation from happening. By contrast, a C=C double bond in the chain would not be rotatable, but would stiffen the chain.

In any event, one of many more probable configurations is shown schematically in Fig. 3.32b. All C-C-C bond angles are 110°. (I will use 110° as an abbreviation of the tetrahedral angle of 109.54°.) If you added the third dimension, the configuration would be even more complicated.

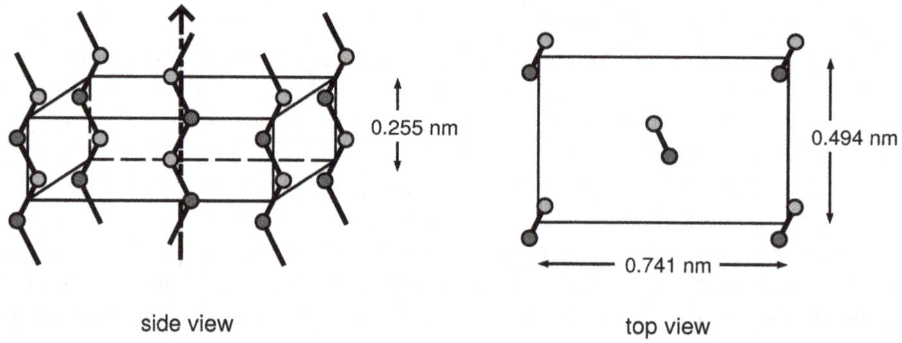

Figure 3.33: Unit cell of polyethylene. Note the five vertical polymer chains and their orientation. Only C atoms are shown. Dark gray atoms point towards the viewer, light gray atoms away from the viewer.

In spite of the problems mentioned, it is possible to make crystals with some polymers. A good example is polyethylene, whose unit cell is displayed here. This unit cell is of a new type: It is called **orthorhombic**. All its angles are 90,°, but the sides are different: Width a = 0.741 nm, depth b = 0.494 nm, height c = 0.255 nm. It is a rather complex unit cell, five polymer chains being part of it.

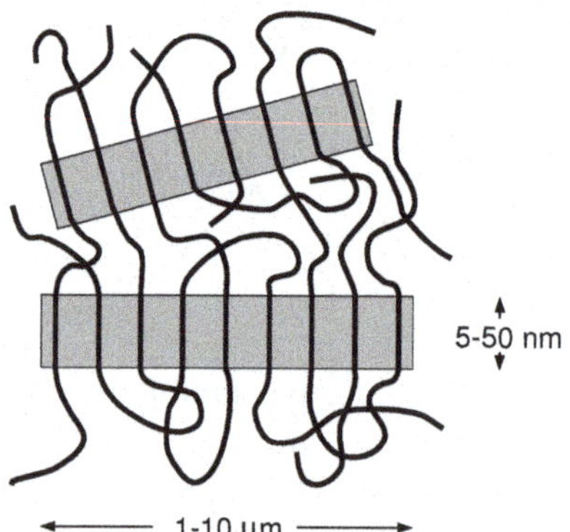

Figure 3.34:
Schematic picture of a partially crystalline polymer. Crystalline regions marked in gray for clarity. Note the amorphous regions surrounding the extremely thin crystal platelets.

It is difficult, however, to grow large polymer crystals. Even under the most favorable circumstances, a polymer can be made at best into a partially crystalline solid. For polyethylene, the crystalline regions may amount to as much as about 85%. A schematic picture of such a solid is shown in the Fig. 3.34. Crystalline regions tend to grow with aligned chains as very thin **platelets**, sometimes also called **lamellae**, which are surrounded by an amorphous matrix. Platelets are typically 10^3 times wider than they are thick.

It should be emphasized that polyethylene turns out to be the best polymer molecule for growing crystals. A structure more complex than a straight carbon chain, or atom groups sticking out to the side of the chain, may reduce crystallization or may even prevent it from occurring altogether.

We will revisit polymer structure and add more details when we discuss the materials properties of polymers in a later chapter.

3.9 Crystal Geometry and Solid Structure

Now that we have a good idea about crystals and the structure of solids at the atomic level, we need to explore further some of the geometrical properties of solids and their corresponding lattices. This will allow us to appreciate the usefulness of X-ray diffraction as a tool for determining crystal structures.

3.9.1 Atom Positions and Crystal Directions

Let us examine some basic geometrical properties of lattices, i.e. the grid of points which gets populated by atoms to form an actual crystal. The reason why geometry is important here is that X-ray diffraction turns out to determine *what kinds of atom planes* are present in a material. With that knowledge it is then possible to figure out the nature and size of the unit cell, as well as what kinds of atoms are in the unit cell.

Recall that the essential property of a crystal is that its atomic structure repeats itself over large distances (i.e. a large number of atomic distances). We also referred to this characteristic as long-range order.

The requirement of long-range order places severe restrictions on what kinds of unit cell are allowed. For example, it is not possible to have a truly repeating structure based on pentagons, in the same manner we saw a hexagonal pattern being able to fill a plane, and even three dimensional space.

It can be proved that there are exactly seven **crystal systems** , which we will define as seven shapes of unit cells, that can fill space in a repeating manner. We will not go into further details here, except to say that we know two of these systems, namely the cubic and the hexagonal system. In addition, there are exactly fourteen so-called **Bravais lattices** , which means, roughly, fourteen ways in which the seven types of possible unit cells can be populated with atoms. Again we will not go into further details, except to give an example we know already: The SC, FCC, and BCC lattices are the three possibilities that go with the cubic crystal system. The hexagonal system is also its own Bravais lattice.

We will focus our discussion on cubic crystals and how to describe systematically atom positions, atom directions, and atom planes in those types of crystals. Our starting point is to consider the cubic unit cell as defining a natural coordinate system for the description of a crystal. The *unit of distance* in this coordinate system will be the *length a of the unit cube.*

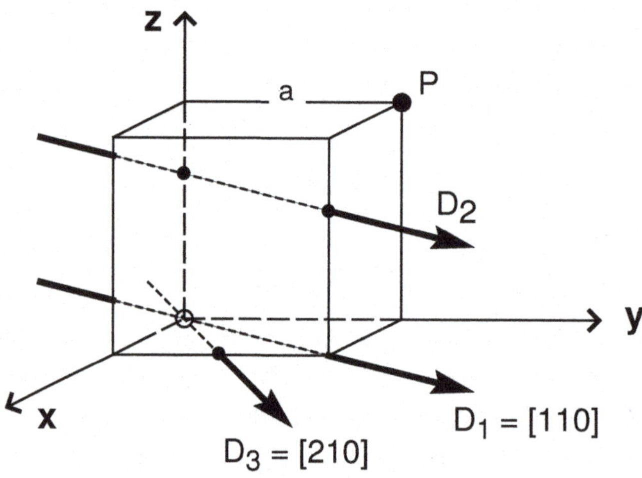

Figure 3.35: Coordinate system defined by the unit cube. One point P and three directions D_1, D_2, and D_3 with their indices are also drawn. The small black dots are meant to clarify the corresponding directions.

a) Atom positions

Atom positions are simply a set of three numbers giving the Cartesian coordinates of the position in units of a. So the point P in Fig. 3.35 has the position 0 1 1 . The center of the cube would be at $\frac{1}{2}\,\frac{1}{2}\,\frac{1}{2}$.

b) Crystal directions

A crystal direction, or crystallographic direction, is defined by the vector connecting two points. The indices of the said direction are given by the three Cartesian components of the vector. If one or more of these components are fractions, they are reduced to the smallest set of three integers. (Explicitly: You multiply the fractions with the smallest common integer that gives you three integers). The three indices are enclosed in square brackets.

For example, in Fig. 3.35 the direction D_1 would be written as [1 1 0]. Note that no commas are used. Just imagine the vector defining the direction as extending from the point 000 (the origin of the coordinate system) to the point 110. Now verify that the direction D_2 has the same indices [110]. In other words, parallel directions have the same indices.

For the direction D_3, if you take the vector from the origin to front edge of the cube, you will see that the vector has the components 1, $\frac{1}{2}$, and 0. Therefore, according to the rule above the indices of this direction are [210].

Directional indices can be negative, depending on the vector that defines them. A negative index is written as a number with a bar over it, e.g. $\bar{1}$. If you had chosen the vector for direction D_1 to go the other way, i.e. from the point 110 to the origin, you would have written the indices as $[\bar{1}\bar{1}0]$. You will note that in general a direction with all negative indices is opposite (antiparallel) to the original direction.

In the cubic system, there are sets of directions that are equivalent, for example the three coordinate axes. It is sometimes convenient to group equivalent directions into a family and designate them with a set of three indices enclosed in an angle bracket. For example, the family of $\langle 100 \rangle$ directions comprises $[100]$, $[\bar{1}00]$, $[010]$, $[0\bar{1}0]$, $[001]$ and $[00\bar{1}]$.

Even though the coordinate axes are equivalent, and thus you are free to choose these axes at your convenience, it is important that you make this choice explicitly upfront when analyzing directions, and also eventually planes, so that your designations are unambiguous.

3.9.2 Crystal Planes and Miller Indices

Just as directions are determined by two points, so are planes determined by three (non-collinear) points. The only questions is: Which three points? Look to the figure below for an answer.

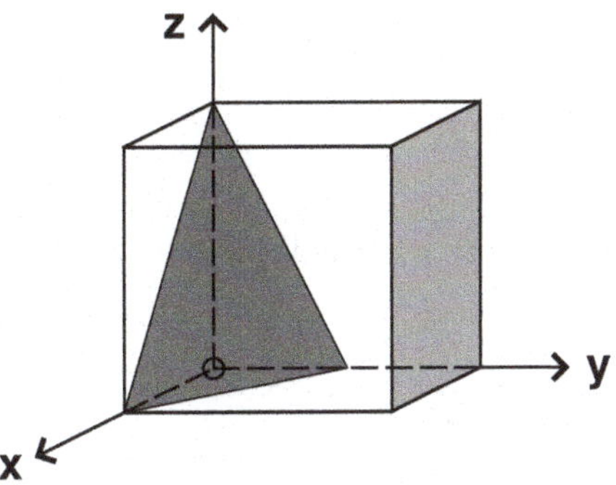

Figure 3.36: Coordinate system with unit cube and two sample planes.

In the most straightforward case of the plane on the left in dark gray, it is visually suggestive to use as the defining points the intersections, or intercepts, of the plane with each of the three axes. This idea runs into a bit of a snag, however, with the second plane in light gray. What should we use for its intercept with the x-axis, when it does not seem to have one? And the same problem shows up with respect to the z-axis intercept.

The solution is that in order to be consistent, we have to use intercepts of ∞ for axes that are parallel to a plane. (The closer to parallel to a coordinate axis a plane is, the farther out will be its intercept with that axis). But now we have to deal with infinite quantities! Fortunately, we can avoid that problem if we ***deal with the inverses of the axes intercepts***!

Here is the recipe for coming up with a consistent set of three numbers characteristic of a given plane, based on the axes intercepts of that plane. This is for cubic crystals with lattice constant a. It is understood that if a plane is parallel to a certain axis, the intercept with that axis is ∞.

1. Determine plane intercepts with the three axes in units of a.
2. Take reciprocals of the three numbers from 1).
3. Multiply the numbers of 2) with a common factor to obtain the smallest set of integers
4. Put the three numbers from 3) in parentheses.

The three numbers in parentheses are called **Miller indices** of the plane.

We will use the two planes in Fig. 3.36 to demonstrate the procedure. For the dark gray plane:

	x axis	y axis	z axis
Intercepts	1	$\frac{1}{2}$	1
Reciprocals	1	2	1
Reduce	1	2	1
Enclose in ( )		(121)	

Thus the dark plane has the Miller indices (121). For the light gray plane:

	x axis	y axis	z axis
Intercepts	∞	1	∞
Reciprocals	0	1	0
Reduce	0	1	0
Enclose in ( )		(010)	

The light gray plane has the Miller indices (010).

Miller indices can also be negative. This happens when the axes intercepts are negative. Negative numbers are written with bars over them, just as with the directional indices.

Occasionally, you may find that a plane contains, or goes through, a coordinate axis. In that case, you can either move the plane parallel to itself, preferably by a lattice constant, and then analyze it with respect to the original axes. Or you can leave the plane as is, move the set of coordinate axes, and then do your analysis of the original plane with respect to the moved axes.

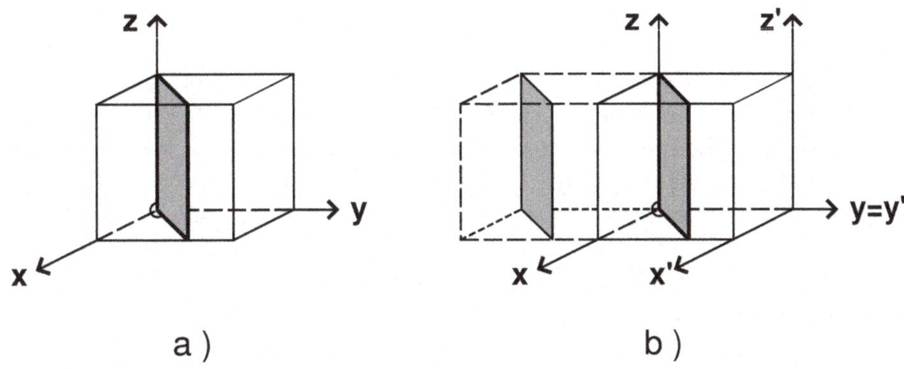

Figure 3.37: a) Original plane and coordinate system; b) Moved plane (use original axes x,y,z), or moved axes x'y'z' (use original plane).

We will work this out with help of Fig. 3.37. First, for the moved plane:

	x axis	y axis	z axis
Intercepts	$\frac{3}{2}$	-1	∞
Reciprocals	$\frac{2}{3}$	-1	0
Reduce	2	-3	0
Enclose in ( )		$(2\bar{3}0)$	

and then for the moved coordinate system:

	x axis	y axis	z axis
Intercepts	$\frac{3}{2}$	-1	∞
Reciprocals	$\frac{2}{3}$	-1	0
Reduce	2	-3	0
Enclose in ( )		$(2\bar{3}0)$	

The two procedures amount to the same thing, as they should. They both yield the same Miller indices $(2\bar{3}0)$. This is not surprising, since the original plane with the axes x'y'z' looks identical to the moved plane with the axes xyz (see Fig. 3.37). As a further check, you should go through the exact same procedure using the original plane but moving the coordinate system by one step (the lattice constant a) in the -y direction (see Homework).

Physically, this confirms what we know already: If we take a plane populated with atoms and move it by a lattice spacing a, or for that matter by a multiple of a, we get an identical plane. This is simply a consequence of the repeating property of the crystal.

3.9.3 Additional Properties of Crystal Planes

We conclude this section with a few miscellaneous remarks about properties of crystal planes:

1. If you change a certain set of Miller indices to their negative numbers, you will end up with an equivalent plane parallel to the original one, on the opposite side of the origin.

2. In the cubic system, certain planes with different Miller indices are equivalent in the crystal, e.g. $(1\ 1\ 1)$, $(1\ \bar{1}\ 1)$, $(\bar{1}\ \bar{1}\ 1)$, and $(\bar{1}\ 1\ 1)$ in an FCC crystal. These are all hexagonal close-packed planes (see Fig. 3.9). One may refer to the entire family of planes using the notation $\{1\ 1\ 1\}$.

3. Let us denote a generic set of Miller indices by (hkl) where h, k, and l are integers. Then we can say that the "higher" the Miller indices, the closer together are two adjacent parallel planes. By "higher" I mean that the expression $\sqrt{h^2+k^2+l^2}$ is larger. You can check this out with Fig. 3.38. The figure is only a 2D representation, but just imagine that you look straight at the xy-plane, i.e. in the $[0\ 0\ \bar{1}]$ direction (along the -z axis). The lines drawn are intersections of the noted planes with the xy-plane, so that it is easy to see the axes intercepts.

4. One can prove that in a cubic crystal with lattice constant a the distance d_{hkl} between two neighbor planes from the $\{h\,k\,l\}$ family is given by

$$d_{hkl} = \frac{a}{\sqrt{h^2 + k^2 + l^2}} \tag{3.10}$$

You can easily verify this relationship in Fig. 3.38 with a ruler for the two sets of planes shown there.

5. It can also be proved that the direction $[hkl]$ is perpendicular to the plane (hkl). For a simple verification, again use the sets of planes in Fig. 3.38 and draw the corresponding directions.

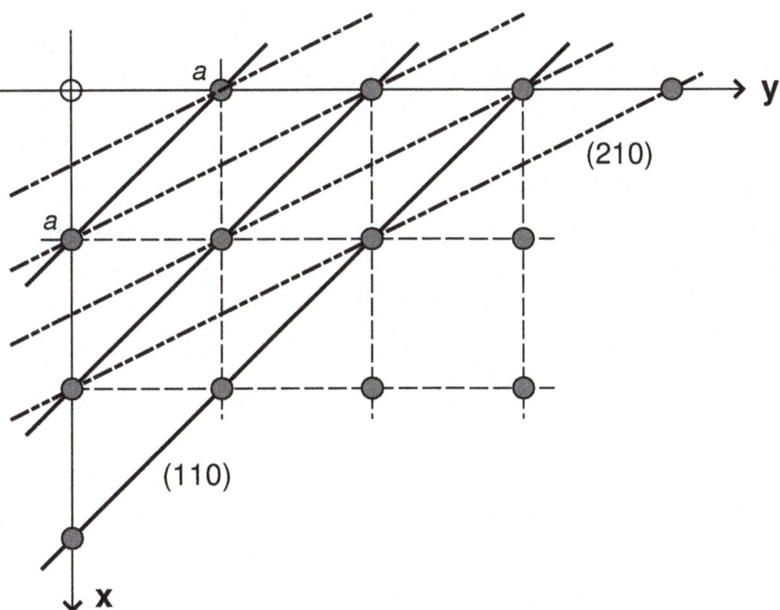

Figure 3.38: Examples of two sets of parallel planes. Note that within each set, all planes have the same Miller indices. The viewing direction is $[0\ 0\ \bar{1}]$, i.e. the -z axis.

6. It is sometimes useful to speak of the **planar density** of atoms. This simply means the number of atoms per unit area in a given plane. For example, if you had the (1 0 0) plane of a simple cubic crystal, its planar density would be $\frac{1}{a^2}$; for the (1 1 0) plane it would be $\frac{1}{\sqrt{2}a^2}$. You note that the "higher" the Miller indices of a plane, the lower is the planar density in that plane. You can verify that visually in Fig. 3.38. In the (2 1 0) planes the atoms are farther apart, but the planes are closer together, than in the (1 1 0) planes.

3.10 Structure Determination by X-ray Diffraction

Having discussed crystal structures extensively, we should now address the question how these structures are determined. This boils down to two issues:

- What type of unit cell is present in the material?
- What kinds of atoms are in that unit cell?

We will set ourselves the modest goal of answering these questions for cubic metal crystals. We would like to be able to tell whether a certain pure metal is simple cubic (SC), FCC, or BCC. For HCP metals, and for real materials such as alloys or ceramics, things are more complicated but the same principles apply. We will make some remarks about ceramics, however, towards the end of this section.

3.10.1 Principles of X-ray Diffraction

The technique we will discuss for our stated purpose is **X-ray diffraction**. It involves studying the reflections of X-rays off of a crystal. It is, as we shall see, a somewhat indirect but very powerful technique. For an example of an experimental technique which makes atoms visible directly, look at the figure leading into this chapter on p. 31.

Here is the principle of the technique of X-ray diffraction:

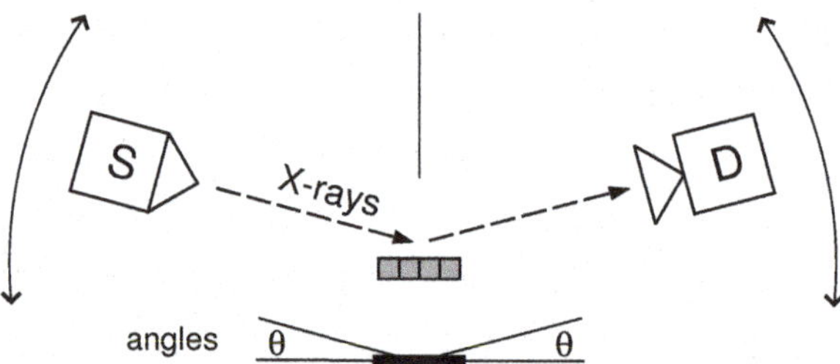

Figure 3.39: Schematic setup of an X-ray diffraction experiment

The crystal acts as a mirror for X-rays coming from the source S and being reflected into the detector D. Note that the angles on both sides are equal to θ so that the angle between the incoming and the outgoing directions is 2θ. Source and detector move together on a circle.

What you have to remember about X-rays is that they are a type of electromagnetic radiation, just like light, but with a much shorter wavelength λ. In fact, λ for these X-rays is of the order of an interatomic distance, typically about 0.1 nm. In addition, X-rays are highly penetrating for most materials. At the same time, whenever an atom is hit by an X-ray wave, it scatters away a tiny fraction of that wave. What one does in an X-ray diffraction experiment is to look at the total of all those tiny reflections from all atoms in a crystal.

It turns out that these reflections are very useful due to an effect called constructive interference. This is how it works:

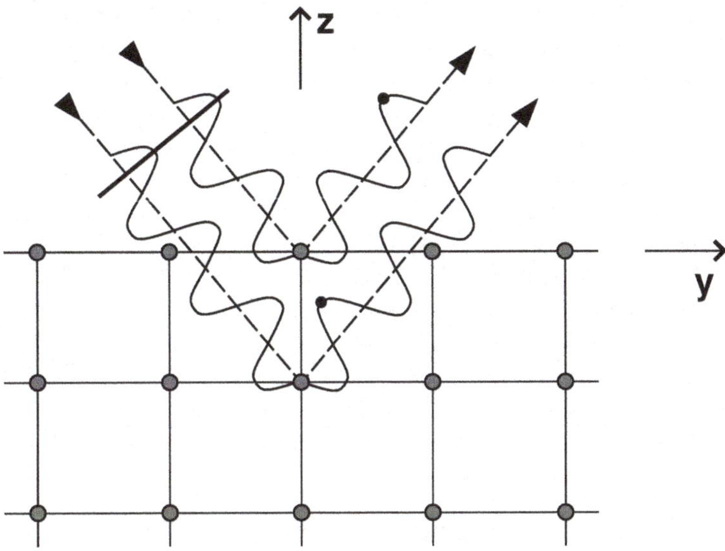

Figure 3.40: The principle of X-ray diffraction. The crystal is assumed to be simple cubic.

In Fig. 3.40 imagine you are looking in the $[\bar{1}\ 0\ 0]$ direction (the -x axis) at the crystal, and the X-rays impinge on the top (0 0 1) plane. A plane X-ray wave comes in and is scattered off of two atoms in the first and second layer. The angle θ and the wavelength λ are chosen such that after scattering the two waves are exactly "in synch" again. But if you focus on the two black dots on the waves, you will note that the lower wave is in effect behind the upper wave by 2λ. That is, the two scattered waves are in such a relationship that they reinforce each other. This condition is called **constructive**.

Of course, there are many more waves scattered off of atoms either in the top (0 0 1) plane or in parallel (0 0 1) planes farther down. The most important thing to realize is that *all these scattered waves are "in synch"* ! Waves scattered from atoms in the same plane have exactly the same path, and as you go down one plane by one, each scattered wave has its maxima aligned with the wave above it, and thus is "in synch" with all waves above it. Consequently, all scattered waves reinforce each other, and although the amount of scattering from an individual atom is minuscule, the sum total of all scattered waves gives rise to an observable X-ray signal in the detector.

By the same token, if the conditions on the incident angle θ and λ are off just a little, the various scattered waves will have an almost random relation to each other. This condition is referred to as **destructive interference**. In that case, the scattered waves will not reinforce each other, and thus we will see no signal in the detector.

The geometrical condition for constructive interference can be worked out quite easily. In the figure below we show only the paths of the waves, but not their oscillations.

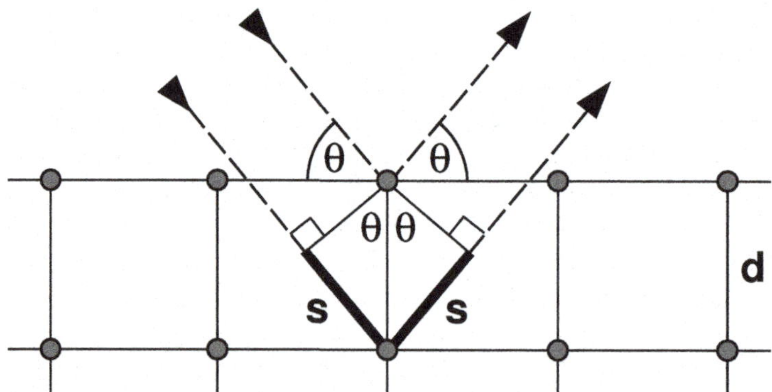

Figure 3.41: Select region of interaction between the X-ray wave and the crystal. The difference between the two path lengths is 2s

From the previous discussion we conclude that constructive interference will occur if the difference in path length between the first and second layer is an integer multiple of the X-ray wavelength λ. In the figure, the total path length difference is equal to 2s. Also note that we have labeled the distance between the two planes by d. It follows that

$$n\lambda = 2d \sin\theta, \quad \text{where} \quad n = 1, 2, 3, \ldots \tag{3.11}$$

This equation is known as **Bragg's Law**, after the British crystallographer W. L. Bragg. Eq. 3.11 requires that the parameters n, λ, and θ be in such a relationship that $\sin\theta < 1$. In particular, since n cannot be less than 1 and $\sin\theta$ has to be less than 1, one must have

$$\lambda < 2d \tag{3.12}$$

for diffraction to be possible.

We worked out our example using parallel (001) planes in an SC crystal. Of course, we could have used any other set of parallel planes, with the proper angle θ ! For example, pick (0 1 1) planes and figure out what the path length difference is as a function of θ in that case. Any set of parallel planes (h k l) has the property of giving constructive interference at the proper angle θ_{hkl}. The general Bragg condition can be formulated as follows:

$$n\lambda = 2d_{hkl}\sin\theta_{hkl}, \quad \text{where} \quad n = 1, 2, 3, \ldots \tag{3.13}$$

where d_{hkl} is given by Eq. 3.10. This is why the Miller indices are so important: Not only do they define a set of parallel planes in a unique fashion, but they also determine when the Bragg condition is fulfilled for those planes.

3.10.2 X-ray Diffraction of Simple Metals

The simplest real materials for X-ray diffraction are FCC and BCC metals. With these there is an additional complication that we must deal with. It turns out that we can use Eq. 3.10, but only with some restrictions. Eq. 3.10 gives a necessary but not sufficient condition for constructive interference in FCC and BCC crystals. The reason why this is so is easy to see from the following figure:

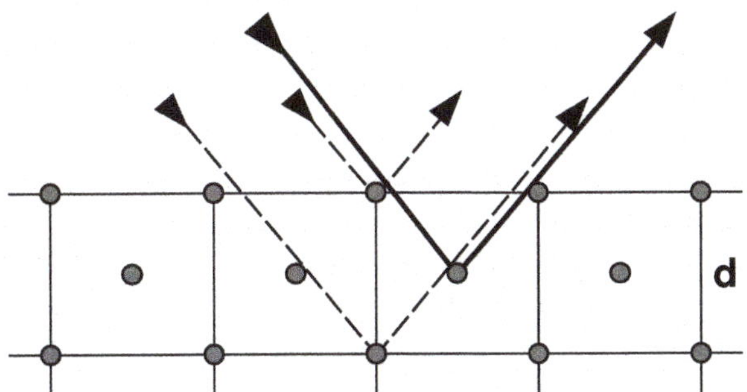

Figure 3.42: Schematic example of an extra scattering path introduced in an FCC crystal relative to an SC crystal.

Look at the new extra path in Fig. 3.42 that comes from reflections off of the atoms in the centers of the cube faces. If the two original paths from the SC crystal are "in synch", then the reflections of the top layer and the new layer in the middle are exactly "out of synch" because their path length difference is exactly one half of the original one! So, at this angle all reflections cancel each other, you have destructive interference, and there will be no measurable reflected X-ray signal.

The extra conditions for constructive interference from (hkl) planes in FCC and BCC crystals have been worked out and can be formulated in terms of the Miller indices as follows:

FCC:	h, k, l all even or all odd
BCC:	$h + k + l$ even

The first few allowed reflections are given in the table below in the order of increasing d_{hkl}:

Table 3.5: Allowed X-ray reflections for SC, FCC, and BCC crystals

SC	FCC	BCC
(1 0 0)	-	-
(1 1 0)	-	(1 1 0)
(1 1 1)	(1 1 1)	-
(2 0 0)	(2 0 0)	(2 0 0)
(2 1 0)	-	-
(2 1 1)	-	(2 1 1)
(2 2 0)	(2 2 0)	(2 2 0)

If we assume that all possible reflections are detected experimentally, a simple rule to distinguish FCC from BCC is to say that the second reflection will be (2 0 0), and the first reflection has to be consistent with FCC or BCC.

An example of an X-ray diffraction pattern is shown below:

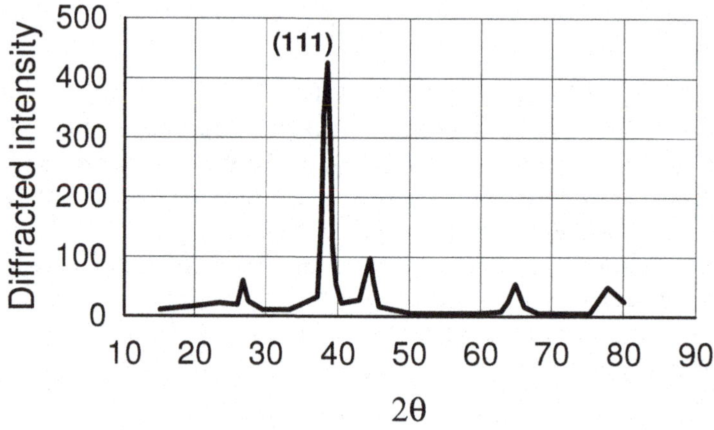

Figure 3.43: X-ray diffraction of a Au sample. Note that the second peak from the left, at about 38°, is the Au(111) peak.

The X-ray diffraction pattern in Fig. 3.43 is of a Au film on a quartz substrate. The pattern was taken with λ=0.154 nm. The second peak from the left, at about 38°, is the Au(111) peak, and the remaining peaks at greater angles 2θ are due to the other reflecting Au planes. You can verify this using the fact that for Au, the lattice constant a is 0.408 nm.

Also note here that X-ray diffraction patterns are frequently plotted as a function of the angle 2θ. This is due to the common experimental setup which measures 2θ directly (Fig. 3.39). On the other hand, analysis of the pattern according to Eqs. 3.10 and 3.13 requires that you use θ.

Another experimental detail worth mentioning is that most X-ray analyses are done with powder samples rather than with a single crystal. If you look at Figs. 3.39 and 3.40, you should be aware of the assumption that the (0 0 1) direction of the crystal and the vertical direction of the apparatus are aligned. This assures that the incoming and the outgoing angle for the X-ray beam are equal. (That is, the angles are θ on both sides. See also Fig. 3.41). Such alignment is rather difficult to achieve, especially if it is not obvious from your sample what the (0 0 1) direction is. If the alignment is not perfect, you will see no X-ray reflections at any angle! A crystal is indeed a very selective X-ray mirror.

However, if you make your sample into a *powder*, the alignment problem disappears. By powder we mean a macroscopic amount of *fine particles with arbitrary, random orientation* relative to the X-ray beam. This ensures that among that very large number of particles there will be enough with just the right orientation relative to the X-ray beam so that we can observe X-ray diffraction.

We finish up this section by showing a direct relationship between atomic quantities describing the unit cell of a material and the macroscopic quantity of the material's mass density.

Let us designate by ρ the macroscopic mass density, i.e. the mass per volume (in g/cm^3).Then we can express ρ as follows:

$$\rho(\text{mass/cm}^3) = (\text{mass/mol})\ (\text{Nr. of moles/atom})\ (\text{Nr. atoms/cm}^3) \qquad (3.14)$$

Now you can see that the first term on the right-hand side is the atomic weight A (in g/mol) and the second term is the inverse of Avogadro's number N_{Av}. The third term is the one that contains atomic information, in that we can express it in terms of properties of the unit cell as

$$\text{Nr. atoms/cm}^3 = \frac{n}{a^3} \qquad (3.15)$$

where n is the number of atoms per unit cell and a^3 is the volume of the cubic unit cell. This allows us to rewrite Eq. 3.14 as

$$\rho = \frac{An}{N_{Av}\, a^3} \tag{3.16}$$

which can be solved easily for the lattice constant a:

$$a = \sqrt[3]{\frac{An}{\rho\, N_{Av}}} \tag{3.17}$$

Alternatively, if one assumes that the atomic data are known, then Eq. 3.16 can be used to calculate the theoretical mass density.

We will work out an example according to Eq. 3.17 using aluminum as the material. Al is an FCC metal, thus has n = 4, A = 26.98 g/mol, and ρ = 2.71 g/cm^3. If we plug these values into Eq. 3.17 we get

$$a = \sqrt[3]{\frac{26.98 \times 4}{2.71 \times 6.02 \times 10^{23}}} = 4.05 \times 10^{-8}\text{cm} = 0.405\text{ nm} \tag{3.18}$$

This is in fact equal to the literature value of the lattice constant for Al.

3.10.3 X-ray Diffraction of Ceramics

X-ray diffraction of ceramic materials can be dealt with similarly to metals. Now the complication is that X-rays scatter off of atomic planes containing different kinds of atoms, and these atoms scatter X-rays differently. Therefore, obtaining constructive interference is more complicated, and Bragg's Law, Eq. 3.13, will surely look different.

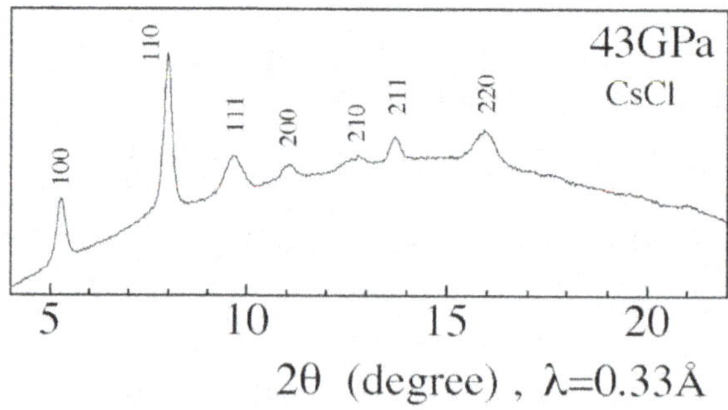

Figure 3.44: X-ray diffraction pattern of CsCl. Note that this is an SC pattern.

Instead of working this out in general, I will make a plausible qualitative argument using CsCl and KCl as examples. You recall that we described CsCl as simple cubic (SC) with a basis of two atoms, even though on casual inspection it looked like BCC (Fig. 3.19). So we would expect the diffraction pattern to be that of an SC structure. And indeed it is (Fig. 3.44).

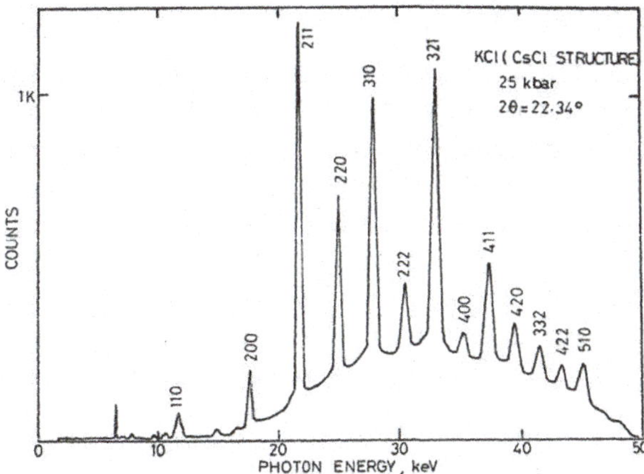

Figure 3.45: X-ray diffraction pattern of KCl. It looks like BCC: Only peaks with $h + k + l$ even are present.

But what about KCl? It also has the CsCl-type structure. In this case, it turns out that the K^+ and Cl^- ions are practically identical when it comes to scattering X-rays. This is, of course, because they have identical electron configurations. Therefore, the X-ray diffraction pattern of KCl does look like BCC, and indeed some of the SC peaks are missing!

Note that in Fig. 3.45 the diffracted beam intensity is plotted as a function of the X-ray photon energy, which means indirectly as a function of λ, at fixed θ. The X-ray source in this experiment was not a regular laboratory source, but a so-called synchrotron, which allows to vary λ continuously. If you want to check Bragg's Law, you can use the relationship

$$\lambda = \frac{hc}{E_{ph}} \tag{3.19}$$

where h is Planck's constant, c is the speed of light, and E_{ph} is the X-ray photon energy.

It is possible to work out X-ray diffraction from such structures as CsCl and KCl in general. The approach is first to consider the lattice, which determines the type of unit cell, and then to add the effect of having several, possibly different, atoms in a unit cell. The effect of these atoms is expressed in the form of what is called the structure factor. In this way, diffraction patterns of very complex structures have been analyzed and indexed.

3.10.4 Small Particles, Surfaces, and Amorphous Materials

We wrap up our discussion of materials structure by examining a few special topics we are prepared to appreciate now. All are areas of active research.

a) X-ray diffraction from small particles

When we discussed the principals of X-ray diffraction, one point we emphasized was that in order for interference of reflected beams to yield a sharp diffraction pattern, diffracted beams from a large number of planes (and atoms) had to interfere constructively. Conversely, this means that if the sample is small, the observed peaks may not be as sharp. Here is an example:

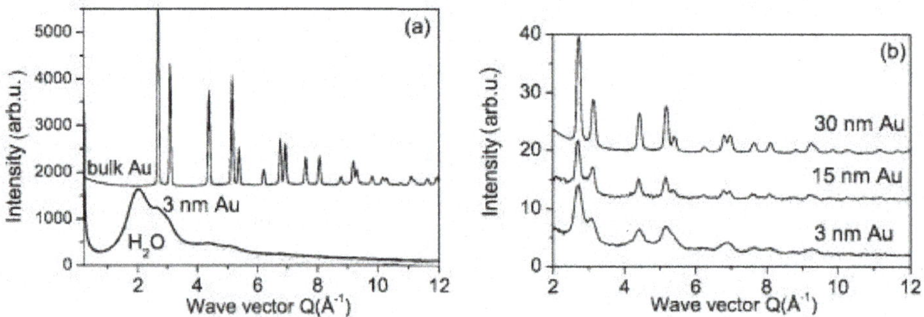

Figure 3.46: X-ray diffraction patterns of bulk Au, and small Au particles suspended in water.

Fig. 3.46a shows the diffraction patterns for bulk Au on top and for 3 nm Au particles suspended in water at the bottom. The diffraction peaks due to the Au particles are small modulations of the H_2O background curve. Fig. 3.46b shows diffraction patterns for particles with three different sizes, after the H_2O background has been subtracted. (The x-axis labeled Wave vector Q is essentially the same as sin θ). For particles, the small number of Au atoms scattering the X-rays not only causes a greatly reduced scattered X-ray signal, but also much wider diffraction peaks. (See also Fig. 3.43). With proper analysis, the width of the diffraction peaks can be used to determine the size of the particles.

Similar size effects can be observed in X-ray diffraction from very thin films of materials.

b) X-ray diffraction amorphous solids and liquids

After our focus on crystalline materials, you may be tempted to think that X-ray diffraction is useless for amorphous materials. That is not so: Amorphous materials do lack long-range order, but that does not mean that that there is no structure whatsoever over the short range, especially between a certain atom and its closest neighbors.

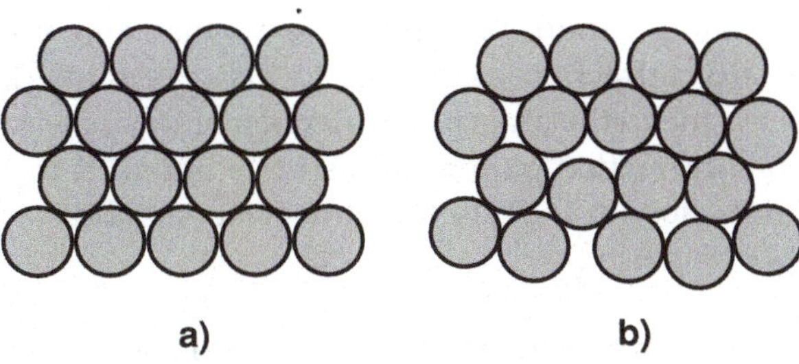

Figure 3.47: Schematic drawings of a hexagonal close-packed plane in a crystal and the slightly disordered structure of a liquid.

Fig. 3.47 illustrates the point: In the hexagonal close-packed plane, every atom has six nearest neighbors, and the distance between a central atom and all its neighbors is exactly the same. In the slightly disordered structure, which is representative of a liquid, almost every atom still has six neighbors, yet they are not close-packed. Instead, their distances from the central atom vary slightly but are still almost equal to the one in the solid.

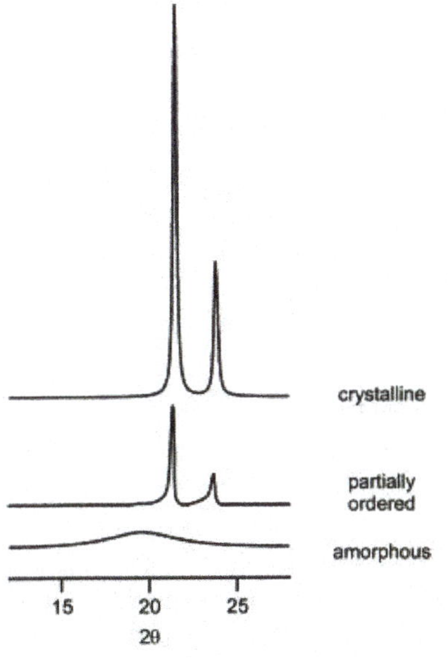

Figure 3.48:
X-ray diffraction of highly crystalline, partially crystalline, and amorphous polyethylene. The sharp peaks are the (110) and (200) reflections.

The fact that neighbor atom-atom distances are almost the same is enough for X-ray diffraction to be useful even for amorphous materials. You can think of it this way: In a liquid, diffraction occurs by simultaneous scattering from two neighbor atoms with an almost fixed distance between them. Of course, the amount of radiation scattered from an atom pair will be minuscule, but at the same time there will be a huge number of atom pairs.

Fig. 3.48 shows a related practical example, namely X-ray diffraction from polyethylene containing various amounts of crystalline and amorphous material. Note the sharp diffraction peaks due to the crystalline portion and the diffuse, broad peak due to the amorphous portion of the material.

From the X-ray diffraction pattern of a simple liquid, it is typically possible to obtain information on the number of nearest neighbors and the average neighbor-neighbor distance.

c) Low energy electron diffraction from crystal surfaces

In many instances it is of interest to know to what extent the physical surface of a crystal has the structure expected for a bulk crystal plane. Subtle differences may show up at the surface because the surface atoms do not have the same number of bonds as the atoms in the bulk in the bulk of the material.

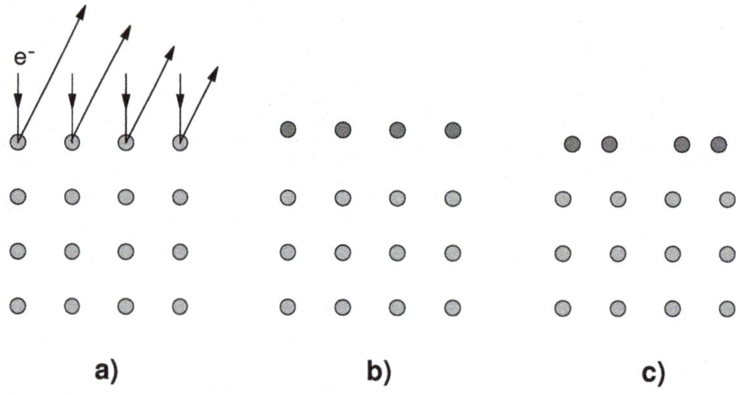

Figure 3.49: a) Low-energy electron diffraction from a perfect surface. b) Surface plane farther out. c) Surface atoms in non-ideal positions.

The surface structure of a material can be investigated using low energy electron diffraction. Low energy electrons (with energy of the order of 100 eV) have the property of penetrating only the first two layers or so of a material, and when they scatter from these near-surface atoms, diffraction can occur.

The usual geometry for low energy electron diffraction is shown in Fig. 3.49. The electrons impinge on the surface in the normal direction and are scattered to the side. Note the path length difference between scattered beams. If this difference has the right value, constructive interference occurs between electron waves, and a diffraction pattern can be observed. Due to the extreme sensitivity of a surface to contamination, such experiments are typically performed with the sample in ultra-high vacuum.

The analysis of low energy electron diffraction is similar to that of X-ray diffraction. It allows to determine the surface structure and to distinguish between the three cases displayed in Fig. 3.49. In addition, it is often possible to ascertain whether foreign atoms have been adsorbed onto a surface.

3.11 Structure and Properties: A Second Glimpse

In the introduction to this chapter on the structure of materials, I mentioned two examples for the close connection between structure and properties. We have a fairly good handle on one of them now, namely ductility vs. structure.

Recall that, for the time being, we understand by ductility the ability of a material do undergo substantial permanent deformation without breaking. By substantial deformation I mean that large-scale motion takes place on the atomic scale. A large number of atoms are displaced over many interatomic distances without the material loosing its integrity.

More specifically, but with some simplification, we said that metals and polymers are ductile, whereas ceramics, covalent crystals, and inorganic glasses are brittle.

As regards metals, let us reiterate that metallic bonding is non-directional, which in itself suggests that a metal can sustain some deformation without really breaking a lot of atom-atom bonds. Furthermore, metals in general are packed quite closely, although they are not always close-packed, and this too should facilitate the motion of atoms relative to each other. The explanation as to why all metals are not ductile will be addressed in the next chapter.

Polymers typically consist of a substantial amount of disordered, loosely entangled carbon-carbon chain molecules, and it would seem reasonable that this type of material can also sustain deformation readily without coming apart.

On the other hand, in ceramics, covalent crystals, and inorganic glasses atomic motion must be seriously restricted. In ceramics, one cannot have equal charges come close to each other; if they did, they would repel each other, and the material would likely fail. Covalent crystals and inorganic glasses have in common that they are essentially extended, rigid networks, held together by directed covalent bonds. Large scale deformation of these materials would involve breaking a large number of bonds, and again the material would be likely to fail.

This is, of course, not the whole story yet. We will fill in many pieces in the next chapter, which is about materials with defects in them, i.e. materials that are not perfect crystals. As it turns out, most materials of interest to an engineer do have defects in them, and often it is precisely those defects that determine the essential properties of the material.

Image sources

Title picture: Field ion microscope image
Courtesy of Prof. K. Hono (National Institute for Materials Science, Japan).

Fig. 3.24:
W. Lu, P. Soukliassian, and J. Boeckl, MRS Bull. 37 (2012) 1119-1124

Fig. 3.43:
A. Kumar, JOM-e 52 (10) (2000)
`http://www.tms.org/pubs/journals/JOM/0010/Kumar/Kumar-0010.html`

Fig. 3.44: reproduced with permission by Elsevier
J. Hayashia, I. Shirotani, K. Hirano, N. Ishimatsu, O. Shimomura, and T. Kikegawa, Sol. State Comm. 125 (2003) 543-546

Fig. 3.45:
Katsuhiko Inoue, Doctoral Thesis, Tokyo University, 1975
`http://www.k2.dion.ne.jp/~inoue_k/inoue_k/Dr-Thesis/68.gif`

Fig. 3.46: reproduced with permission by the American Physical Society
Valeri Petkov et al., Phys. Rev. B 72 (2005) 195402
`http://link.aps.org/doi/10.1103/PhysRevB.72.195402`

Fig. 3.48: reproduced with permission by Elsevier
A.M.E. Baker and A.H. Windle, Polymer 42 (2001) 667-680
`http://www.sciencedirect.com/science/article/pii/S0032386100003645`

References

Note that the values for the ionic radii in Fig. 3.17 are approximate. There are two different sets of ionic radii in use, named after Shannon and Pauling. Also, the radius of an ion depends somewhat on its neighbor ion of the opposite polarity, especially if you compare different N_{coord}.

Both Shackelford, and Callister and Rethwisch provide more data on ionic radii and electronegativities. Shackelford has a complete list of ionic and atomic radii. Both have more examples in ceramics and crystallography.

For a comprehensive treatment of all sorts of crystal structures with many pictures, go to `http://www.chem.ox.ac.uk/icl/heyes/structure_of_solids/Strucsol.html`.

You can find anything you wish to know about quartz at
`http://www.quartzpage.de/gen\_mod.html`

The following book by A. Putnis contains a lot of information on SiO_2, and silicate glasses and minerals:

Putnis, A., *An Introduction to Mineral Sciences,* Cambridge University Press, Cambridge, U.K., 1992.

The packing of hard spheres is an active research area. See Erich's Packing Center, for cute pictures of various packings:
`http://www2.stetson.edu/~efriedma/packing.html`

For the density of random close packing of spheres, see:
G. D. Scott and D. M. Kilgour, J. Phys. D: Appl. Phys. 2 (1969) 863
`http://iopscience.iop.org/0022-3727/2/6/311/pdf/0022-3727_2_6_311.pdf`
The maximum random close-packed density of hard spheres is 0.6366 ± 0.0005.

Exercises

1. What would be the shape of the repulsive part of the `E(r)` curve for atoms as billiard balls?
2. Why is the position of the dashed atom drawn for the cluster labeled 6 in Fig. 3.3 the best position to add the next atom?
3. Based on our discussion, can you formulate a necessary condition for a certain shape to be a unit cell, which would tell you immediately that the smallest square you can draw in Fig. 3.2 is NOT a unit cell?
4. Draw two unit cells for each of the 2D crystals in Fig. 3.1b and Fig. 3.1d.
5. Devise a qualitative argument as to why it is reasonable that the APF of FCC and HCP metal crystals should be the same.
6. What is the number of atoms per small HCP unit cell?
7. Identify the 12 nearest neighbors in an FCC crystal by looking at the atoms in a unit cube, or a suitable number of adjacent unit cubes.
8. What would be the APF for a simple cubic (SC) metal?
9. Are the oranges stacked up at your local supermarket arranged in close-packed fashion, and if so, are they FCC or HCP?
10. Verify that the ions in CsCl obey the stability rule for ionic crystals.
11. What is the structure of RbCl?
12. How many atoms are there per unit cell in the ZnS structure?

13. What is the relationship between the C–C bond length and the lattice constant a for diamond? (Hint: Look straight at a face of the unit cube and view all the atoms inside.)

14. If you assume that SiO_2 is an ionic compound, do the Si^{4+} and O^{2-} ions fulfill the condition for fourfold coordination? The ionic radius of Si^{4+} is 0.041 nm.

15. Find another non-flat hexagon in the Si unit cell in Fig. 3.26. (There are at least two such hexagons visible quite readily).

16. What would be the difference in the way the two network modifiers Na_2 and CaO act when added to silica?

17. What is the number of atoms per unit cell in polyethylene? (Hint: In Fig. 3.33, imagine moving the boundaries of the unit cell to make the counting of atoms easier)?

18. One may be tempted to describe diamond as the super-polymer, a fully connected 3D network of C atoms. Can you discern any polymer-like carbon atom chains in diamond?

19. Practice determining indices of directions.

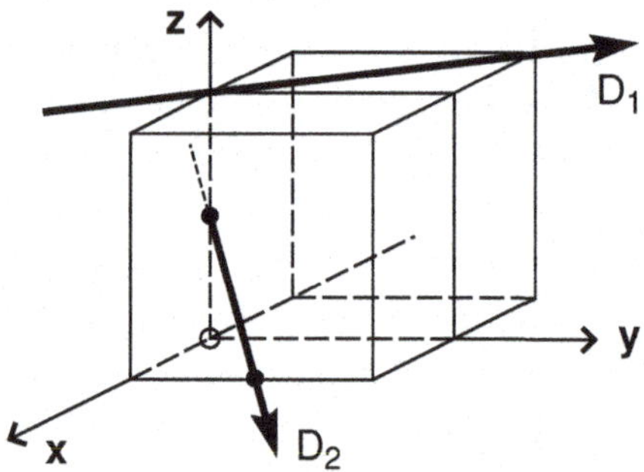

Figure 3.50: Determine the indices of the two directions D_1 and D_2. The small black dots for D_2 are meant to clarify important points of intersection.

20. Draw a coordinate system with as many unit cubes as you need, so that you can draw the [0 3 1] and the $[1\,1\,\bar{1}]$ directions into it. Remember that in a crystal, one unit cube is repeated many times along all three coordinate axes.

21. Determine the Miller indices of the original plane in Fig. 3.37 with the coordinate system moved to the left (i.e. one step in the -y direction). Compare your result to the example in the text. What can you conclude?

22. Draw a coordinate system with a $(1\,\bar{1}\,2)$ and a $(0\,\bar{1}\,2)$ plane.

23. Practice determining Miller indices using Fig. 3.51 below.

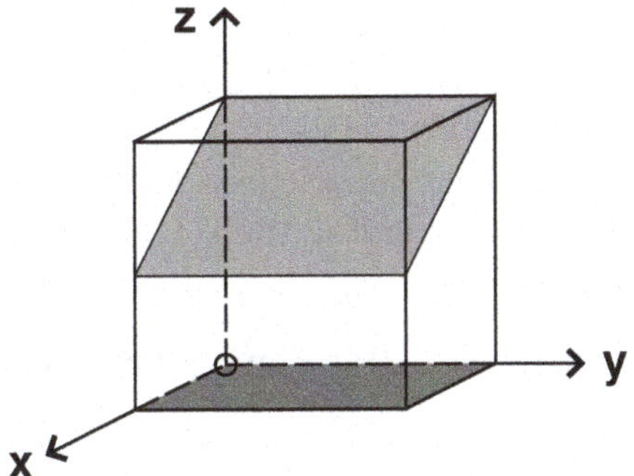

Figure 3.51:
Determine the Miller indices of the two drawn planes.

24. Verify that the peaks at angles 2θ greater than 40° in Fig. 3.43 are indeed consistent with the material analyzed being Au. Remember that $\lambda = 0.154$ nm and $a = 0.408$ nm.

25. From the X-ray diffraction pattern below, determine what the crystal structure of the material is.

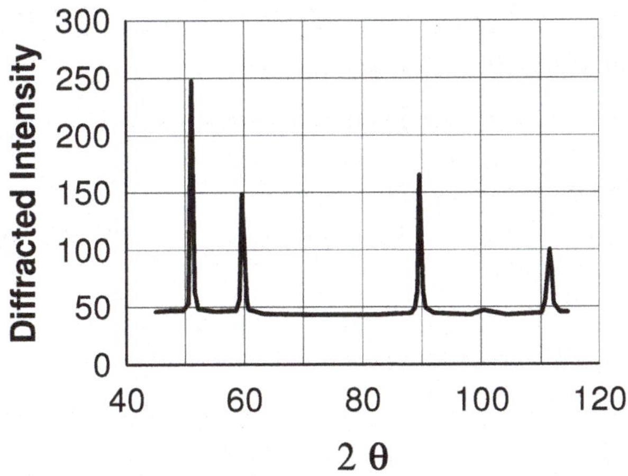

Figure 3.52: X-ray diffraction pattern of an unknown material

26. The sample from Exercise 3.25 had a lattice constant a of 0.370 nm. What was the wavelength λ with which the X-ray data were taken?

27. Calculate the theoretical mass density of molybdenum (Mo).

28. Calculate the lattice constant a for CsCl from the data in Fig. 3.44.

Problems

1. Show that the APF of an HCP crystal is indeed = $\frac{\pi}{3\sqrt{2}}$.
(Hint: Note that c is twice the height of the tetrahedron formed by the three atoms in the bottom A layer and the atom in the B layer touching the three atoms (Fig. 3.12)).

2. In Callister, the ionic radius of H^+ is given as 0.154 nm. What is strange about this number?

3. Positive ion radii are also given in textbooks for the ions of N, P, As. Why did I not list those in Fig. 3.17?

4. Why aren't there any ionic radii listed for the noble gases in Fig. 3.17?

5. Can you explain the trends in the ionic radii for the alkali metals and for the halogens Fig. 3.17?

6. Why are there many elemental metals with FCC or BCC structure, but no elemental SC metals?

7. Is it true that there are no elemental SC metals? How could there be an elemental metal with a SC unit cell?

8. In the text I made the point that the nearest neighbor configuration is the same for HCP and FCC metals, but there is some other difference that gives rise to the two separate structures.
Determine the number and location of the next nearest neighbors in the HCP and FCC structures.

9. Using the ionic radii of Cs^+ and Cl^- given in the text, determine the lattice constant of CsCl. Then look up this lattice constant in the literature and compare it to your own result.

10. What would you expect for the crystal structure of CsF? What is it really, and why?

11. What would be the APF of diamond?

12. Look up the lattice constant of β-cristobalite SiO_2. Is its value consistent with the covalent radii of Si and O, or with the ionic radii of Si^{4+} and O^{2-}? In other words, can you tell whether it is more appropriate to consider SiO_2 a covalent or an ionic compound, or something in-between?

13. In Fig. 3.43, what could the peak at about $2\theta = 27°$ be due to?

14. The alloy Cu_3Au can be in an ordered state (in which it resembles, but is not, FCC) at low temperature and in a disordered state at high temperature (in which the lattice sites are occupied randomly by Cu or Au). What would you expect for the X-ray diffraction patterns of these two forms of the alloy?

15. Compare the X-ray diffraction patterns of RbCl and RbBr. Explain any similarities or differences.

16. In May 2012, it was reported in the press that an Ontario man had exchanged a diamond ring for one with a cubic zirconia stone, and had swallowed the diamond ring in order to hide it. The report went on to say that "An X-ray inspection of the man revealed that he had indeed swallowed a ring, but one with two cubic zirconia stones."
Can this report be true, and if so why? (Hint: Think X-rays, but not diffraction).

17. The idea of modeling crystal structures by stacking spheres is quite old. It has been attributed to the two crystallographers Victor M. Goldschmidt and Fritz H. Laves. It goes back to a conjecture by J. Kepler (1611) that there was no way of packing equivalent spheres more densely than in a face-centered cubic arrangement. T. Hales (U. Michigan) announced a 250 page long proof in 1998.

    In the spirit of packing spheres, here is another question:
    Take the FCC and BCC geometries and figure out what the largest space in-between the atoms is in both cases. If you filled those spaces exactly with a second type of smaller atom, what would be the atomic packing factor?

18. What is the closest packing possible with eggs? This is not a frivolous question. The packing of atoms as ellipsoids has been raised in connection with the fact that many HCP metals are not ideal HCP ($c \neq 1.633a$). The closest random packing of M&Ms has an APF = 0.68, and an M&M crystal can have an APF as high as 0.77. See:
`http://cims.nyu.edu/~donev/Packing/EllipsoidCrystal.pdf`.

19. Design a better egg carton. Is a $2 \times 6 \times 1$ array really the most efficient and cheapest way to pack eggs? Consider a single egg carton, and also the storage and transport of a whole lot of egg cartons together.

Chapter 4
Defects in Solids

Overview

Most solid engineering materials are polycrystalline or amorphous, rather than single crystals. This means that they contain various kinds of imperfections, or defects, which disturb the regular crystalline array of atoms. Different types of defects can be distinguished. Structural imperfections may be confined to individual lattice sites, or they may occur along lines or even planes of atoms. Also, engineering materials often are not pure but may contain isolated impurities or, in the case of an alloy, may be composed of comparable amounts of two or more elements. The proper control of such imperfections in a material allows an engineer to taylor the material properties to a specific purpose or application.

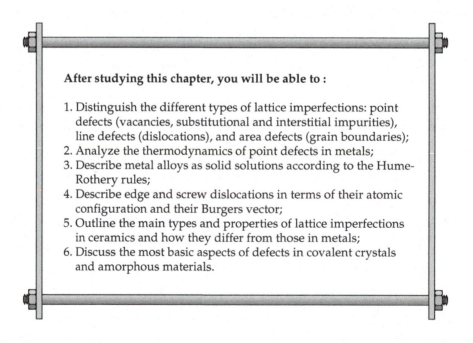

After studying this chapter, you will be able to :

1. Distinguish the different types of lattice imperfections: point defects (vacancies, substitutional and interstitial impurities), line defects (dislocations), and area defects (grain boundaries);
2. Analyze the thermodynamics of point defects in metals;
3. Describe metal alloys as solid solutions according to the Hume-Rothery rules;
4. Describe edge and screw dislocations in terms of their atomic configuration and their Burgers vector;
5. Outline the main types and properties of lattice imperfections in ceramics and how they differ from those in metals;
6. Discuss the most basic aspects of defects in covalent crystals and amorphous materials.

Bubble raft model of a crystal with imperfections

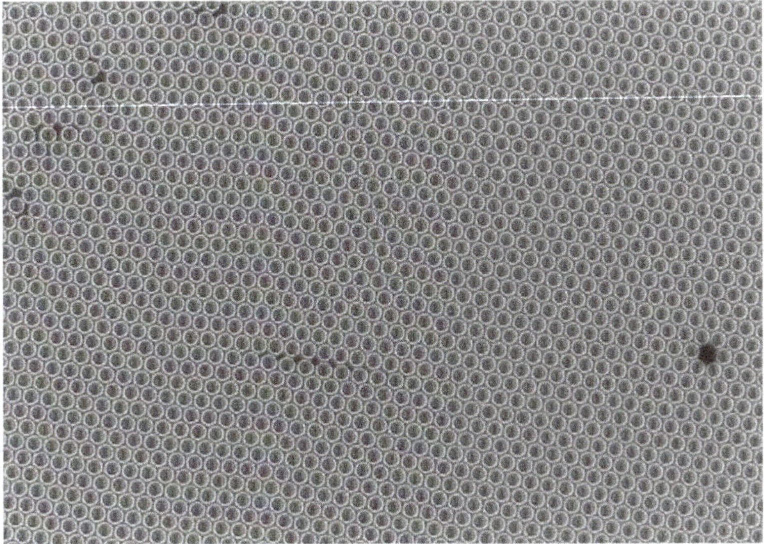

The picture above shows a so-called bubble raft model of a crystal plane with three types of imperfections. (It is an actual layer of soap bubbles.) In the upper right quadrant and elsewhere you can see a perfect hexagonal close-packed layer of bubbles. Near the right edge is a vacancy, a missing bubble. Slightly below the middle is a dislocation; it looks like a small local tear in the hexagonal fabric. Near the upper left hand corner is a grain boundary, a region where two pieces of crystal come together at an angle, i.e. with a slight misorientation.

4 Defects in Solids

In our discussion of the structure of materials we have concentrated so far on crystals and, to a lesser extent, on amorphous materials. With respect to crystals, the tacit assumption was that we were dealing with a **single crystal**, i.e. with a piece of material in which the unit cell repeats itself throughout. It is possible to grow single crystals of many materials, especially most elements, but only with considerable effort. Single crystals are available commercially, but are rather expensive.

Most engineering materials are not single crystals, notable exceptions being silicon and other semiconductor materials used in integrated solid state devices. The reason is that for these devices only single crystals provide the necessary properties, whereas e.g. with structural engineering materials used for their mechanical stability, single crystals would be useless (there is one notable exception here, too, which we will examine later).

Therefore, it is fair to say that most solid engineering materials are polycrystalline or amorphous (or possibly semi-crystalline in the case of polymers). We define a material as **polycrystalline** if it consists of a huge number of tiny crystals of somewhat variable size, oriented fairly randomly relative to each other. (Typical sizes of these little crystals, also referred to as **grains**, might range from 100 nm to 10 μm). It follows that engineering materials are not perfectly ordered, at least not over large distances, but contain various kinds of imperfections, or defects (We will use these two words interchangeably). The study of these defects will be the topic of the present chapter.

4.1 Overview of Crystal Imperfections

The picture at the beginning of this chapter uses a bubble raft model to show three examples of crystal imperfections: a vacancy (i.e. a missing atom) at the right edge, a so-called dislocation a little below the middle of the figure, and a grain boundary (the region between two grains) in the upper left corner. All these defects are important in determining various properties of materials, as we shall see in due course.

It is convenient to classify imperfections according to their dimensionality as follows:

1. **Zero-dimensional (0D), or point defects:**

   These defects involve individual lattice sites. The simplest of them is the **vacancy**, i.e. a missing atom (see the figure on the previous page). Other point defects are illustrated in Fig. 4.1.

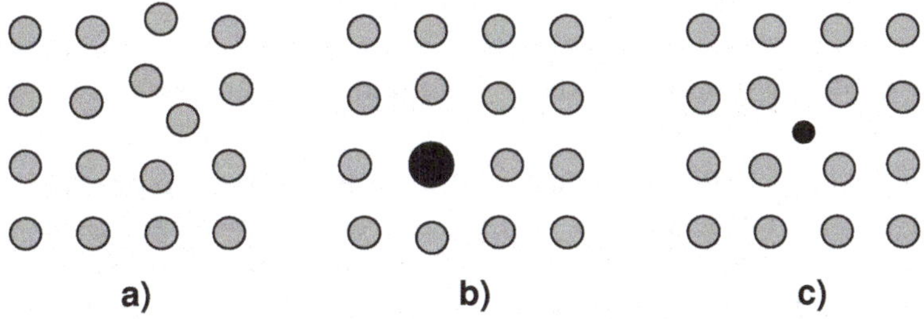

Figure 4.1: Schematic illustration of various types of point defects. a) Self-interstitial; b) substitutional impurity; c) interstitial impurity.

The defect shown in Fig. 4.1a) is called a **self-interstitial**. The word "interstitial" designates the location, namely in-between the regular lattice sites. In a real material, such a defect will also distort the lattice around it because two atoms now have to fit into the space for one. Fig. 4.1b) illustrates a so-called **substitutional** impurity, i.e. an impurity which replaces an original atom at its lattice site. Fig. 4.1c) shows an **interstitial**, i.e an impurity located in-between lattice sites. Both these types of defects generally also result in lattice distortion.

2. **One-dimensional (1D), or line defects: Dislocations**

   This type of defect is easy to spot in the title figure, but hard to visualize properly. In Fig. 4.2 below I have drawn a magnified schematic view of that defect region, with horizontal rows numbered.

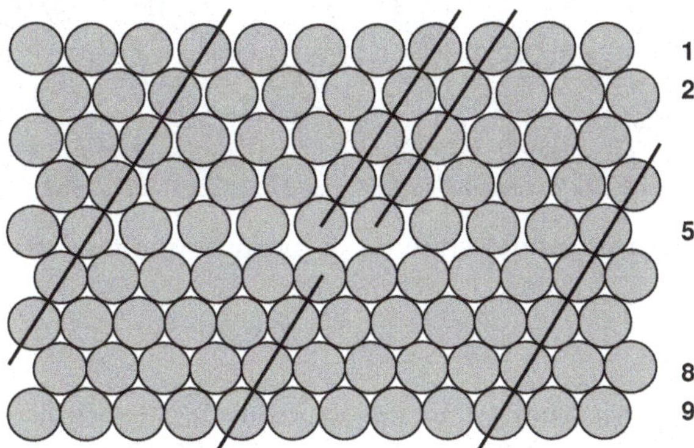

Figure 4.2: Region around the bubble raft dislocation.

The lines on the two sides are meant to align your view along close-packed directions in the hexagonal pattern. If you count atoms in rows

1 and 9, you will realize that row 1 has 11 atoms in it and row 9 has 12. This mismatch shows up clearly in rows 4 and 5, where it appears as if one extra line of atoms in the bottom half has been inserted between two lines of atoms in the top half.

You will note that in row 1 atoms appear to be lined up pretty well again, but you still have 11 atoms fit into the space where 12 atoms are in row 9. So in row 1 the atoms are separated a little farther than they would be in a true equilibrium configuration. It is as if a force has pulled them apart a little, and by way of reaction they are trying to contract. Another way to describe this situation is that the region around a dislocation is stressed.

3. **Two-dimensional (2D), or area defects: Grain boundaries**

   There are two kinds of 2D defects: Grain boundaries (internal surfaces), and the external surfaces of a material.

Figure 4.3: Bubble raft model showing a grain boundary between two grains.

We will not discuss external surfaces in any detail in this book. However, grain boundaries will play an important role for us. An example of a grain boundary can be discerned in the upper left corner of the figure at the very beginning of this chapter, but it is not easy to see. Fig. 4.3, which is another bubble raft image, depicts two grains, with their boundary running diagonally across the picture. Note the difference in the alignment of the "atom" rows above and below the grain boundary. (Also note the dislocation in the upper right quadrant).

From Fig. 4.3 we can conclude that a grain boundary is a rather imperfect region in the sense that atoms at the boundary have unsatisfied bonds and a lower N_{coord} than atoms inside the grains. Still, bonding across the boundary is sufficient to provide strong cohesion for the material.

4. **Three-dimensional (3D), or volume defects**

   Volume defects include pores due to less than optimal materials processing (e.g. with ceramics), or pores formed in the material on purpose, for example if the material is used as support for catalyst particles. We will not discuss volume defects in any detail in this book, even though porous materials are a field of active research in itself.

I should point out that the structure of defects in real materials, with atoms layed out in 3D, is more complex than soap bubbles in a close-packed planar layer. Even so, the figures above illustrate important general features of these defects.

4.2 Point Defects in Metals

4.2.1 Vacancies in Metals

At this point, you may well have doubts about these so-called vacancies: Why should atoms be missing in certain lattice sites? It turns out that vacancies are the one type of defect in a solid which is unavoidable, whereas the other kinds of defects we introduced above are a consequence of the particular processing of a material.

First let me mention that there is good experimental evidence for the existence of vacancies, especially at elevated temperatures. Fig. 4.4 shows a plot of the thermal expansion coefficient of an Ag-3.9Sn sample vs. T. The dashed line delineates the result for a perfect crystal with no vacancies, whose expansion coefficient comes solely from lattice expansion due to the asymmetry of $E(r)$ (Chapter 2, Figs. 2.8 and 2.11). The solid curve describes the actual data. At high temperatures there is a clear difference between the two curves. The effect is also present in a pure Ag sample, but is smaller in magnitude.

This difference between the two curves is attributed to the contribution of vacancies to the volume of the material. Lattice expansion means an increase in the lattice constant, hence an increase in volume. But unoccupied lattice sites add further to the volume because an empty lattice site does not mean that the missing atom has been discarded. Instead, you should think of the removed atom as having been redeposited onto the surface of the sample, so that the total number of atoms remains unchanged.

Another reason why vacancies are important is that they facilitate the motion of atoms through a material. An example is shown in Fig. 4.5. All atoms are identical, but the dark gray atom is tagged so that its motion can be followed more easily.

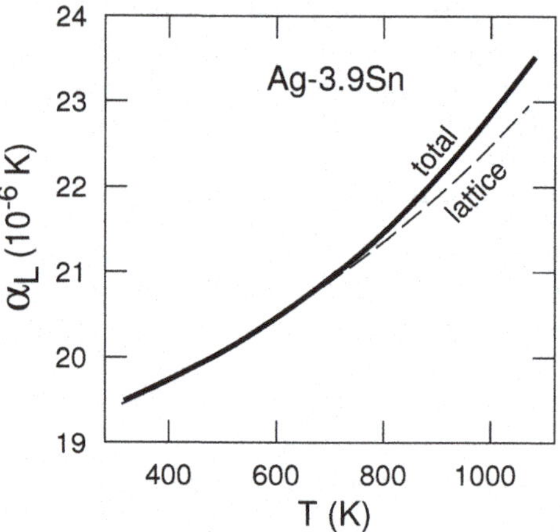

Figure 4.4: Thermal expansion coefficient of Ag-3.9Sn vs temperature T for a perfect crystal, without vacancies (dashed line), and the real crystal, with vacancies (solid line).

We know that at elevated temperature the atoms vibrate more strongly, i.e. with larger amplitude, about their equilibrium positions. If an atom finds itself next to a vacancy, there is a small but finite chance that the atom can jump into the old empty site thus leaving behind a new empty site. The consequence of all these atom jumps is that the vacancy moves, and so will some atoms. An example is provided in Fig. 4.5: After four atom jumps the vacancy has gone around in a circle and the dark gray atom has moved by two lattice spaces.

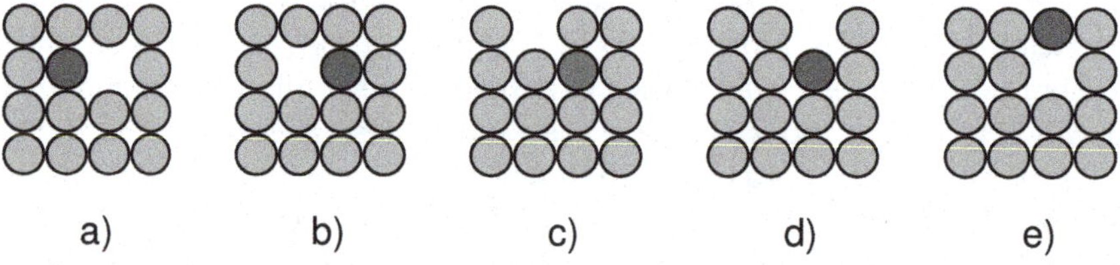

Figure 4.5: Time sequence of pictures showing motion of a vacancy due to atom jumps. Note the motion of the dark gray atom.

So why are vacancies unavoidable? It is true that the absolute lowest energy configuration for a crystal would be with no vacancies, and it costs energy to create a vacancy. Still, thermodynamics dictates that in thermal equilibrium there is a small but finite probability that a configuration with higher energy does exist. Hence there is a small but finite probability that a certain lattice site is empty.

This situation can be described quantitatively as follows:

$$N_v = N\exp(-\frac{Q_v}{kT}) \tag{4.1}$$

or equivalently as

$$N_v = N\exp(-\frac{\Delta H_v}{RT}) \tag{4.2}$$

In these equations N_v is the vacancy concentration (in units of cm^{-3}), and N the concentration of lattice sites (i.e. the number of lattice sites per volume, in units of cm^{-3}). Q_v is the energy of formation for one vacancy, ΔH_v the heat of formation for a mole of vacancies, k the **Boltzmann constant**, R the **universal gas constant**, and T the absolute temperature (in K). You remember, of course, that $k = 1.38 \times 10^{-23}$ J/K, R = 8.31 J/mol K, and $R = kN_{Av}$. The two equations are equivalent: It is just a question of whether you like your energies per atom or per mol.

The equations also have a transparent interpretation: The quantity N_v/N

$$N_v/N = \exp(-\frac{Q_v}{kT}) = \exp(-\frac{\Delta H_v}{RT}) \tag{4.3}$$

represents the probability that a lattice site is not occupied by an atom. Note that Q_v is of the order of 1.6×10^{-19} J, or 1 eV, and ΔH_v is of the order of 100 kJ/mol or 25kcal/mol, so that the exponential is a very strong function of the temperature. Also do not forget that whenever you see the combination kT or RT, the temperature T is in K.

As a numerical example, let us estimate the probability for a lattice site in Al to be vacant at room temperature T_1 = 298 K and at the melting temperature T_2 = 933 K. For Al, the energy of formation Q_v has been reported to be 0.66 eV or 1.06×10^{-19} J. Therefore

$$T_1 = 298\text{K}: \quad N_v(T_1)/N = \exp(-\frac{Q_v}{kT_1}) = 6.39 \times 10^{-12} \tag{4.4}$$

$$T_2 = 933\text{K}: \quad N_v(T_2)/N = \exp(-\frac{Q_v}{kT_2}) = 2.66 \times 10^{-4} \tag{4.5}$$

This clearly demonstrates the strong temperature dependence of the probability of a lattice site to be vacant.

4.2.2 Impurities in Metals

Most metals can be refined to a very high purity, as high as 99.9999%. However, they are rarely used in high purity form. Not only is it expensive to purify them, but highly pure metals generally have one big disadvantage for practical use: they are much too soft. Therefore, engineering metals generally consist at least of a base metal with a small amount of some select impurities. Even more often, they are alloys containing substantial amounts of two or more metals.

By way of example, I list some copper-based engineering materials in the table below:

Table 4.1: Examples of Copper-based engineering materials

Material	Composition	Hardness	Use
Cu (standard)	99.9Cu	40	tubing, wires
Cu-Sn bronze	89.8Cu, 10Sn, 0.2P	240	bells, cymbals, strings, bellows, diaphragms
Yellow brass	70Cu, 30Zn	160	locks, fixtures, door knobs; brass instruments

In this table, the compositions are given in the usual notation: 99.9Cu means 99.9% Cu. You will note the increased hardness of the alloys compared to pure Cu. (The hardness values are approximate according to the Brinnell scale. You should simply use them for a qualitative comparison.) Note that the Cu-Sn bronze is sometimes also called phosphor bronze: It turns out that the small amount of P in it improves the wear resistance and stiffness of the alloy.

For the moment we will confine our discussion to materials which consists mainly of a base metal A, with only a small amount of a second metal B. In this case, we can consider metal A to be the solid **solvent**, and the second metal B will be the solid **solute**. The mixed material represents a **solid solution** of metal B in metal A.

We are interested in answering three questions:

1. Which combinations of metals A and B will form solid solutions?
2. How much of B will it be possible to dissolve in A?
3. Where in the lattice of A will the impurities B be inserted?

With respect to Question 3, what we want to know is whether atoms B will be substitutional or interstitial: Will atoms B replace atoms A in their lattice sites, or will they reside in-between lattice sites.

It is evident that in order for impurities to be interstitial, they have to be small compared to the solvent atoms. A large class of materials of this type, which we will explore later, is known as **steel**, i.e. Fe with a small amount of C impurities in it. But even for those materials, it turns out that the actual space in-between Fe atoms is too small to accommodate C impurities comfortably. So the Fe lattice ends up being stretched a little around a C impurity. This, in turn, limits the amount of C that can be dissolved into Fe (see Fig. 4.1c).

In order for an impurity B to be substitutional, it should be as similar as possible to the host atoms A. This seems to be intuitively clear. But we can make an additional argument that illustrates possible complications when we extend our consideration to cases such as yellow brass: Cu is an FCC metal, and Zn is HCP. So what will a mixture of roughly two thirds of Cu and one third of Zn look like?

The question of what similarity means between solvent and solute, and how it affects **solubility**, has been investigated by William Hume-Rothery, a British metallurgist. He formulated a set of empirical rules, which are named after him. For *unlimited solubility* with respect to composition, atoms A and B should fulfill the following requirements;

Hume-Rothery rules:

1. The difference in the atomic radii should be $\leqslant 8\%$;
2. The crystal structure must be the same;
3. The electronegativities should be similar;
4. The valences should be the same.

The first rule is sometimes given in a slightly different form: Appreciable solubility (>1%) requires that the difference in the atomic radii should be $\leqslant 15\%$. Valence in the fourth rule refers to the positive charge associated with a metal element in a compound (or, equivalently, the number of bonds it can form).

Materials data relevant to the application of the Hume-Rothery rules to the copper-based alloys are listed in the table below:

Table 4.2: Hume-Rothery data for Copper-based alloys

Element	Atomic Radius	Structure	EN	Valence
Cu	0.128 nm	FCC	1.9	+1
Sn	0.151 nm	BCT	1.8	+4
Zn	0.133 nm	HCP	1.6	+2
Ni	0.125 nm	FCC	1.8	+1

The Cu-Ni system is indeed an example for a combination of metals A and B which form a solid solution at any composition, and Sn and Zn do have limited solubility in Cu, as the table suggests.

Note the crystal structure of Sn: It is BCT, which is short for body-centered tetragonal. BCT is similar to BCC, except that for Sn the unit cell is a square in the xy-plane, but is shorter in the z-direction.

In the Cu-Zn system, at low enough Zn content in Cu, a solid solution will form with the FCC structure. On the other hand, at low enough Cu content in Zn, a solid solution will form with the HCP structure. If the Cu and Zn contents are comparable, a mixture of FCC and HCP structures is obtained.

An additional point worth noting is that the formation of a solid solution is generally favored at elevated temperature. What role exactly the composition and the temperature play in the formation of solid solutions or metallic compounds will be examined in greater detail in Chapter 8.

4.3 Point Defects in Ceramics

4.3.1 Vacancies in Ceramics

When it comes to imperfections, we can draw on the same basic ideas, but once again ceramics are more complicated than metals, and once again the reason is that the ion charges place extra restrictions on what is possible.

The most important overriding principle is that overall charge neutrality must be preserved in the ceramic. A ceramic material, being an electrical insulator, cannot sustain a net charge under normal circumstances.

As regards vacancies, this rule implies that one cannot produce a single isolated vacancy, or more generally a single point defect. Rather, point defects must occur in pairs. In the simplest case, there are two possibilities:

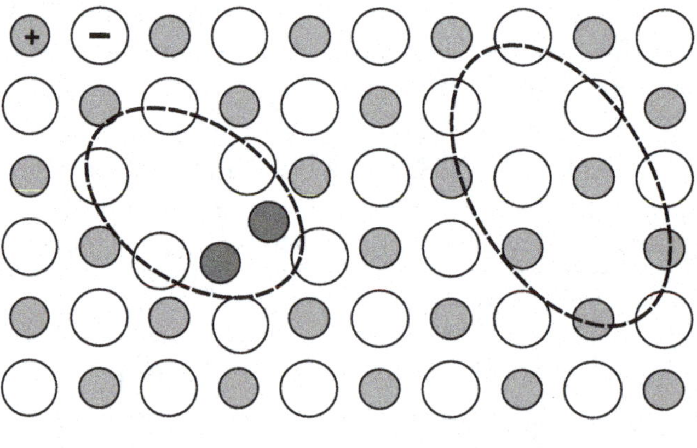

Figure 4.6: Point defects in ceramics. Frenkel defect: Cation vacancy + cation interstitial; Schottky defect: Cation and anion vacancies.

The requirement of charge neutrality can be fulfilled either by removing one type of ion and putting it back in an interstitial position, or by removing one ion each of both types. The two practical alternatives are shown in Fig. 4.6. They are named after the Russian physicist Yakov Frenkel and the American physicist Walter Schottky, respectively. The configuration of one anion vacancy + one anion interstitial, which would also meet charge neutrality, really does not occur. Forming an anion interstitial would be very unfavorable energetically, due to the large size of most anions.

4.3.2 Impurities in Ceramics

Just as with metals, impurities are introduced most easily into a ceramic when they can be substitutional. In order for that to happen, the solute ions should be as similar as possible to the solute ions they are to replace.

A good example for this scenario are the RbCl-KCl and KCl-NaCl systems. All the cations have +1 charge, and the anion is the same. All three compounds assume the NaCl structure (see Section 3.5.2). For Rb^+ and K^+, the size difference is about 7%, and for K^+ and Na^+ it is about 26% (Fig. 3.17). This explains why RbCl and KCl are completely soluble in each other: Rb^+ can be substituted for K^+ in KCl in any proportion, and vice versa. On the other hand, between KCl and NaCl there is only limited solubility.

Interesting new are observed when the impurities have a different charge than the ions they are to replace. In that case, it is still possible to enforce charge neutrality, but with a twist. The first case is for a Ca^{2+} ion replacing a Na^+ ion in NaCl:

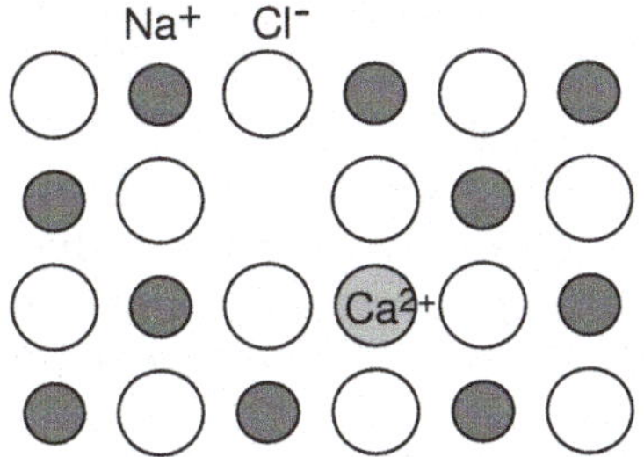

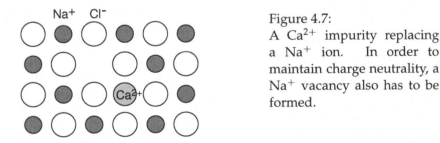

Figure 4.7:
A Ca^{2+} impurity replacing a Na^+ ion. In order to maintain charge neutrality, a Na^+ vacancy also has to be formed.

The second case involves iron oxide, a material in which Fe can appear in two charge states, namely Fe^{2+} and Fe^{3+}.

In a perfect crystal of FeO, all Fe ions are Fe^{2+}, but if FeO were oxidized further, then for every two Fe^{2+} turning into Fe^{3+}, a Fe^{2+} vacancy has to be created. That is why the compound FeO is usually not perfect, and thus has a formula $Fe_{1-x}O$. It is worth noting that FeO is an insulator, whereas iron oxide containing Fe^{3+} ions has semiconductor properties.

For the simplest defects, Frenkel and Schottky, the defect concentration can be obtained in a manner very similar to Eq. 4.3 above. We take as a specific example the Schottky defect. If we let ΔH_{vS} be the heat of formation for the

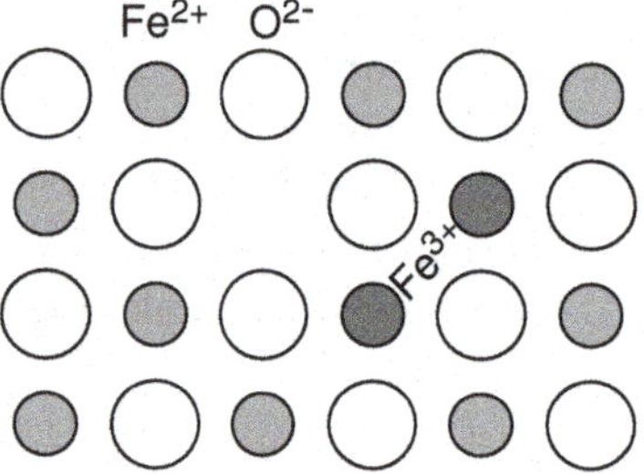

Figure 4.8:
For every two Fe^{2+} turning into Fe^{3+}, a Fe^{2+} vacancy has to be created to maintain charge neutrality.

ENTIRE defect, i.e. for the formation of the cation AND anion vacancy, then it follows from thermodynamics that

$$\frac{N_{vc}}{N}\frac{N_{va}}{N} = \exp(-\frac{\Delta H_{vS}}{RT}) \tag{4.6}$$

where N_{vc}/N and N_{va}/N are the probabilities for a cation site or an anion site to be vacant. Since $N_{vc}/N = N_{va}/N$, we can rewrite Eq. 4.6 as

$$\frac{N_{vc}}{N} = \frac{N_{va}}{N} = \exp(-\frac{\Delta H_{vS}}{2RT}) \tag{4.7}$$

You will note the extra factor 2 in 2RT in Eq. 4.7. This comes from the fact that ΔH_{vS} is the energy to form the complete defect, i.e. two vacancies (cation and anion), and Eq. 4.7 is written for the concentration of one kind of vacancy.

For the Frenkel defect, a similar relationship can be established, with formation energy ΔH_{vF}, once the number and nature of possible interstitial sites is determined. For the more complex types of defects such as those in Figs. 4.7 and 4.8, analogous arguments can be made, and similar relationships derived.

The question as to which of these defects, if any, will occur preferentially in a certain material is difficult to answer in general. All we can say is that this will depend on the relative values of the formation energies ΔH_{vS} and ΔH_{vF}, and these depend sensitively on the actual crystal structure and the details of the possible defect sites.

4.4 Impurities in Covalent Crystals

We will not deal with vacancies in covalent crystals in this book. However, impurities are of great importance for the covalent semiconductors Si and Ge because they allow for the electrical conductivity to be altered over many orders of magnitude. The key to this feature is that the impurities have to be substitutional and from the Groups IIIA or VA, that is from columns in the periodic table on either side of the Group IVA with C, Si, and Ge. We will discuss the details of the properties of Si and Ge with such impurities in Chapter 12.

4.5 Dislocations in Metals

4.5.1 Edge dislocations

We introduced dislocations briefly earlier in this chapter, by noting that they are a line or 1D defect, in the sense that the lattice imperfection is extended in one dimension. The title figure and Fig. 4.2 served to help you visualize such a defect.

We are now ready to examine a few more details about this type of imperfection. What was displayed in Fig. 4.2 above is called more aptly an **edge dislocation**. An alternative schematic is drawn below in Fig. 4.9. Think of it as similar to the situation in Fig. 4.2, but in an SC crystal.

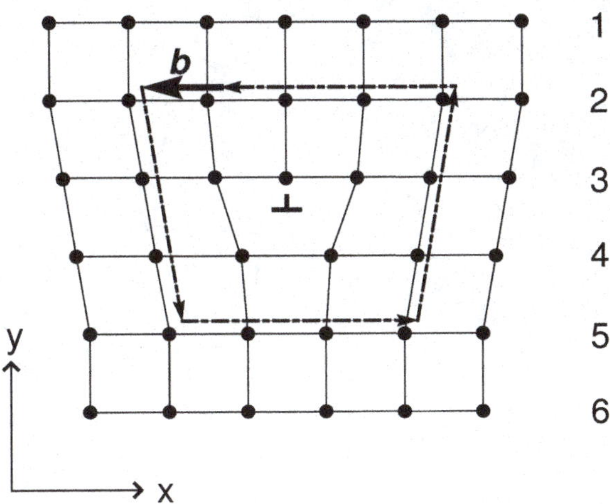

Figure 4.9: Schematic drawing of an edge dislocation. Note the inverted T as a symbol for the dislocation line. Note also the circular path around the dislocation: It takes the Burgers vector **b** to close the path, due to the dislocation inside.

The figure illustrates a number of feature of an edge dislocation. First, note that you can describe this defect as consisting of either a missing half-plane in the bottom part or an extra half-plane in the top part. The inverted T designates the edge of the extra half-plane and indicates the **dislocation line**. An inverted T is often used as a graphic symbol for an edge dislocation.

If you go back to the title figure with the edge dislocation in the bubble raft, it is quite easy to make out this defect. But if you are not sure whether your sample does contain a dislocation, what could you do? You follow a circular path around the suspect area. Such a path is drawn with the dash-dotted line in Fig. 4.9. Start at a certain lattice point (upper left) and go y steps in the -y direction, then x steps in the x-direction, y steps in the y direction, and finally -x steps in the x direction. If the area you enclosed with your path is a perfect crystal, you get back to your point of origin, but if there is a dislocation inside, you need an extra step, indicated by the vector **b** in the figure, to close the path. This vector **b** is characteristic for the dislocation, and is named **Burgers vector** after J. M. Burgers, a Dutch-American scientist.

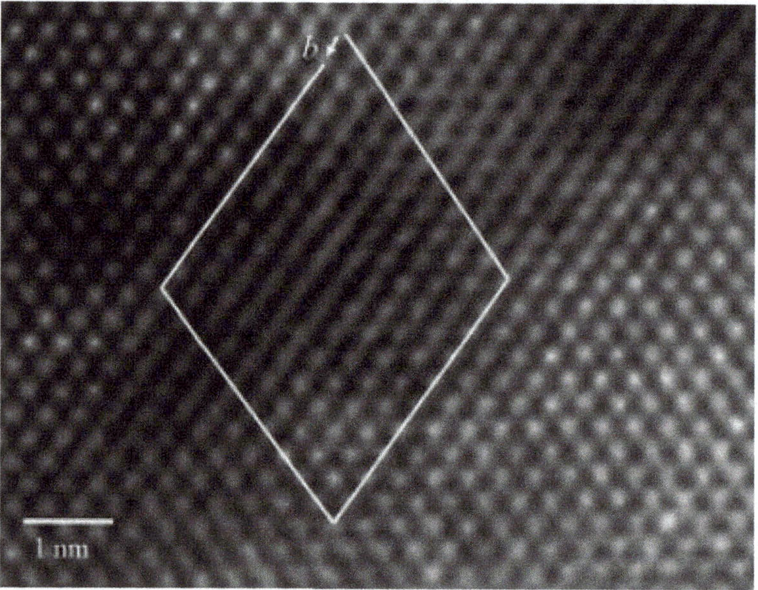

Figure 4.10: Transmission electron microscope image of an edge dislocation in Yttria-Stabilized Zirconia. Note the Burgers vector **b**.

In order to convince you that this is not just a mental exercise, I show here a high-resolution transmission electron microscope picture of an Yttria-Stabilized Zirconia crystal (YSZ: zirconium oxide whose structure is stabilized by the addition of a small amount of yttrium). This material is under investigation for electrodes in solid fuel cells. A circular path is drawn into the atomic-resolution picture, and if you count atom steps, you will see that the Burgers vector shown at the top is necessary to close the path. This proves that there is a dislocation inside the path.

In addition, you will note on Fig. 4.9 that because of the extra half-plane on top, the lattice around the dislocation line is distorted. In Rows 1 and 2, and also 5 and 6, the atom-atom distances are equal to the normal lattice spacing. But in Row 3 the atoms are squeezed together a little laterally, and in Row 4 they are pulled apart. We may conclude that around a dislocation, the crystal lattice is stressed a little.

4.5.2 Screw dislocations

The other important type of dislocation is called screw dislocation. It can be visualized as follows:

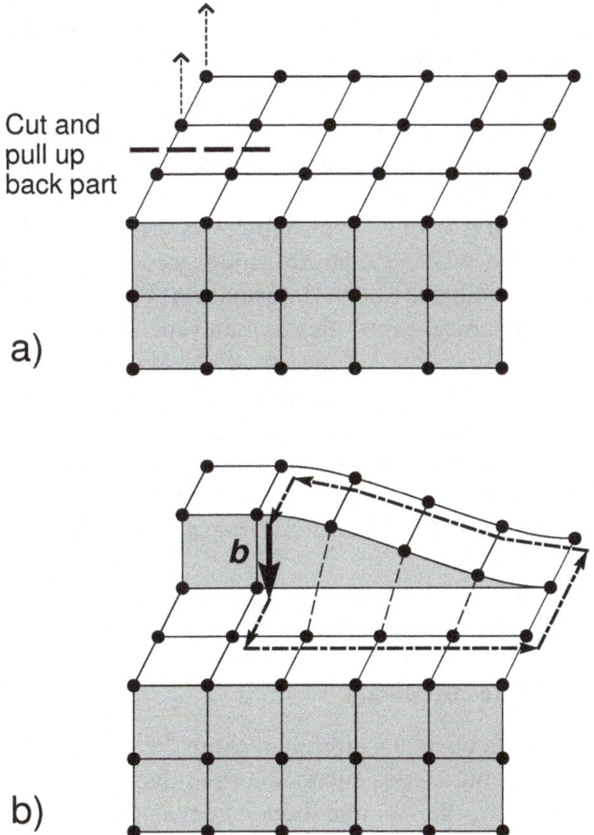

Figure 4.11: Model of a screw dislocation. a) Perspective side and top view of undisturbed crystal with cut; b) Distorted crystal with circular path around dislocation and Burgers vector **b**.

In this case, the dislocation line is given by the end of the cut into the material, and thus extends in the vertical direction. (This definition, i.e. the end of the cut, applies to the edge dislocation as well. You simply say that to make an edge dislocation, you cut the material halfway and insert an extra half-plane into the cut. See Fig. 4.9).

The Burgers vector is not only useful for determining the presence of a dislocation, but also its character. Imagine that you perform your circular path around a suspected dislocation. This path will be approximately planar (Figs. 4.9 and 4.11). If the Burgers vector **b** is in the plane of the path, you will have an edge dislocation, and if it is perpendicular to the plane, you will have a screw dislocation. Equivalently, one can say that for an edge dislocation, the dislocation line and the Burgers vector are perpendicular, and for a screw dislocation they are parallel.

The Burgers vector is useful in yet another way. It should be clear by now that dislocations are high-energy structures, involving substantial lattice deformation (much more so generally than point defects). One can show that the line energy of a dislocation, i.e. the energy per unit length, is proportional to $\mathbf{b}^2$. Therefore, all else being equal, dislocations with a small $\mathbf{b}$ will be favored.

Hence it follows that, in contrast to vacancies, dislocations will not be present in a thermally equilibrated material. Dislocations are usually formed as a consequence of a stress being applied externally or internally to the material. They play an important role when a material undergoes permanent deformation, both by facilitating and limiting such deformation. We will revisit dislocations when we discuss mechanical properties of materials.

The final word in this section concerns dislocations in ceramics. We will not get into any details about this subject, except to note that it must be much more difficult to create dislocations in ceramics than in metals, again because of the electrostatic constraints caused by the ions. Whether we cut a crystal and insert a half-plane for an edge dislocation, or whether we cut and pull up for a screw dislocation, in ceramics ions with the same charge will tend to get in each other's way.

4.6 Area Defects in Metals

4.6.1 Grain Boundaries in Metals

The last types of defects we will examine in some detail are two important area defects, namely grain boundaries and stacking faults.

Grain boundaries are the regions between two grains, i.e. between two small crystals that are not in perfect alignment relative to each other. We saw 2D model examples above in the title figure and in Fig. 4.3. Now we want to focus more closely on this boundary region between two grains in the next figure, Fig. 4.12.

In the graph on the left of Fig. 4.12, the two grains are misaligned by 6°. The dark gray atom at the center of the lowest horizontal row (and also in slanted row 1) touches two atoms on either side, and therefore is common to both grains. The black atom and the two dark gray atoms on top are the first ones that fit into the space created by the misalignment, the grain boundary. The black atom is the first one added on the right which is again in registry with the crystal. This type of grain boundary is called a low-angle boundary because of its small angle of misalignment.

Another way of interpreting the structure of this grain boundary is to focus on the rows of atoms marked with the slanted solid lines. If you count the rows

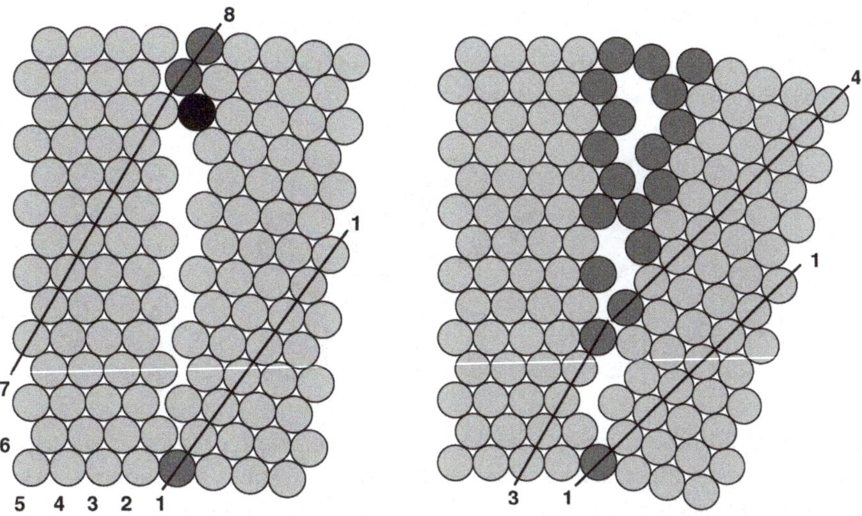

Figure 4.12: Close-ups of two grain boundaries. In the graph on the left, the two grains are misaligned by 6°; on the right the misalignment is 16°. In each graph, the dark gray atom at the center of the lowest horizontal row is common to the two grains. The gray atoms higher up are extra atoms that can be accommodated because of the misalignment.

between the slanted lines, you see that on the left the *seventh line* up is aligned again, except for the tilt angle, but with the *eighth line* up on the right! So the ***upper marked row on the right is extra***, which means that what we have there is an edge dislocation!

Incidentally, if you took away the three darker atoms on top, you could move the two grains together and form a single perfect crystalline region. Furthermore, you could easily extend the grain boundary by repeating the structure, so that row 13 would be just like rows 7 and 1.

In the graph on the right of Fig. 4.12, the two grains are misaligned by 16°. Here a dislocation can also be identified at the bottom of the grain boundary, but this becomes less clear higher up. Also, there is more opportunity to move atoms around inside the boundary, and thus to make the grain boundary more or less disordered. The way it is drawn is fairly ideal, in that all added atoms except one are in registry either on the left or right side of the boundary.

What both grain boundaries in Fig. 4.12 have in common is that they represent a small region of disorder, where the material is less dense. It is for this reason that grain boundaries often facilitate the movement of atoms through a solid. You can appreciate this readily by noting that, especially in the case of the larger misalignment, the exact arrangement of atoms in the boundary region is by no means fixed and unique.

4.6.2 Stacking Faults in Metals

The other area defect worth examining is called a stacking fault. You can visualize it with help of the figure below:

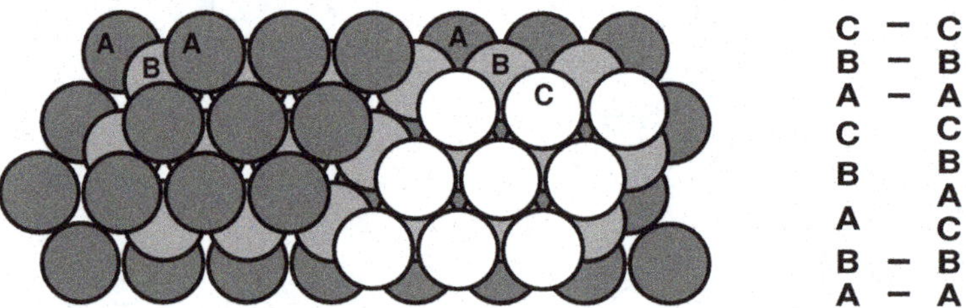

Figure 4.13: Illustration of a stacking fault. The graph on the left is a top-down view of three layers. The graph on the right is a schematic side view showing the sequence and alignment of the stacked layers.

This should remind you of Fig. 3.6. Think of the drawing in Fig. 4.13 as showing the beginning of an FCC crystal with the stacking ABCABC . . . , except that on the left side of the sample an error in the stacking occurred and the third layer ended up being in the A rather than in the C position. Now assume that this is the only fault, and the stacking continues normally on *both* sides of the sample. This would result in the stacking sequence indicated on the right side of Fig. 4.13. But this represents a missing plane in the left part of the sample, at the third plane up from the bottom. In other words, what we have here is an *edge dislocation*! (For the sake of illustration, I assumed that the bottom two layers and the top three layers were in registry all across).

A stacking fault is an example of the close connection between line and area defects. If you study materials science further, you will come across many more such connections.

4.7 Structure and Properties: A Third Glimpse

In concluding this chapter, I want to reiterate that you should take 2D defect models with a grain of salt. A bubble raft cannot replicate the complexity of defect structures in three dimensions. For example, Fig. 4.2 is only a model for a single hexagonal close-packed layer with a line defect in it. It is not obvious how to extend this picture to 3D for a close-packed solid (but you should try). On the other hand, I do think that these models provide useful visualizations, and they point to the factors important for these defects.

You have also seen yet another side of the relationship between structure and properties. Real engineering materials are not perfect crystals, but contain a certain amount of disorder. The effect of this disorder can be manifold. Vacancies may assist in the movement of atoms, especially impurity atoms, through a solid. Dislocations and grain boundaries will have a significant effect on how materials are deformed. Furthermore, they often serve as sinks for impurities, which tend to aggregate at these imperfections.

We can summarize these observations by saying that the properties of engineering materials are determined largely by their **microstructure**, i.e. by details of structure which show up at the scale of micrometers. This includes in particular their defect structure, e.g. the density of dislocations, the grain size, the nature of the grain boundaries, etc.

In addition, when you think of dislocations and grain boundaries as solid regions in which atoms are not in an optimal bonding environment, i.e. atoms have stretched or even unsatisfied bonds, you will not be surprised to learn that defect regions are typically more reactive chemically than crystals. This is another aspect of the structure-property connection. However, we will not pursue this connection to any significant extent in this book.

Image sources

Title picture: Bubble raft image of various types of imperfections, reproduced with permission by D. Stone (Univ. of Wisconsin)
`http://homepages.cae.wisc.edu/~stone/bubble%20raft%20movies.htm`

Fig. 4.3: reproduced with permission by D. Stone (Univ. of Wisconsin)
`http://homepages.cae.wisc.edu/~stone/bubble%20raft%20movies.htm`

Fig. 4.4: Data from:
K Mosig, J Wolff, J -E Kluin and T Hehenkamp, J. Phys.: Condens. Matter 4 (1992) 1447-1458.
`http://dx.doi.org.libproxy.rpi.edu/10.1088/0953-8984/4/6/009`

Fig. 4.10:
`http://npl-web.stanford.edu/archive/energy/micro-fuel-cell/sofc/electrolyte/ion-highway/`

References

Both Shackelford, and Callister and Rethwisch provide more illustrations for dislocations. Guy goes into much more detail on a wider range of topics.

A wonderful resource for defects in solids is at
`http://www.tf.uni-kiel.de/matwis/amat/def_en/index.html`
This is at a more advanced level than this book, but you will be able to appreciate many of the illustrations.

Another site worth exploring is:
`http://www.doitpoms.ac.uk/tlplib/index.php`
This University of Cambridge site has short tutorials on many subjects as well as interesting materials images.

For bubble raft models of grain boundaries at different angles, go to:
W. M. Lomer and J. F. Nye, Proc. Royal Soc. London, 212 (1952) 576-584

The conventional wisdom is that dislocations are not present in ceramics, at least not at low temperature. For a counter-example, see:
`http://composite.about.com/library/PR/2001/blmpi1.htm`
Plasticity and an inverse brittle-to-ductile transition in strontium titanate
P. Gumbsch, S. Taeri-Baghbadrani, D. Brunner, W. Sigle and M. Rühle,
Phys. Rev. Letters 87 (2001) 085505
These researchers found that strontium titanate ($SrTiO_3$) can be deformed at low temperature thanks to dislocations in the material.

Exercises

1. What is the probability of a lattice site to be empty for Cu at room temperature and at the melting point of 1358 K? ΔH_v for Cu is 124 kJ/mol. Compare your result for Cu with the corresponding result for Al.

2. The vacancy concentrations N_v for a certain metal are as follows:

T(°C)	N_v
300	$N \times 1.15 \times 10^{-11}$
900	$N \times 4.53 \times 10^{-6}$

   What is ΔH_v for this material? At 900°C, is it close to its melting point?

3. Fe has the BCC structure at room temperature and a lattice constant a of 0.287 nm. Figure out what the shape and size of the biggest space in-between the atoms is, where an interstitial impurity might reside. By shape I mean something like tetrahedral, or cubic, or something else.

4. Which element is likely to be more soluble in Cu: Sn or Zn?

5. What solubility would you expect between RbCl and NaCl?

6. What is the probability of a cation site to be vacant for NaCl at room temperature and at the melting point of 1074 K? Assume that Schottky defects are prevalent, and ΔH_{vS} = 222 kJ/mol. How does this compare with the numerical example for Al worked out in the text?

7. Determine the theoretical density of rocksalt (NaCl). Use Eq. 3.16, suitably modified for a ceramic. Compare your result to the literature value.

8. When you google, know how to ask the right question! Try a search with the terms "dislocation" and "ceramics". One item coming up is at `http://www.ncbi.nlm.nih.gov/pmc/articles/PMC2504286/` What kind of dislocation are they talking about? What is the relevance of this paper for materials science in general and ceramics in particular.

9. Confirm the angles of misalignment quoted for Fig. 4.12 by a simple geometric argument.

10. Draw yourself a picture like Fig. 4.12, but with a misalignment angle of 30°. Distinguish two cases: a) extra atoms in registry either on the left or on the right side of the boundary. b) disordered extra atoms. In which case can you pack more extra atoms into the grain boundary?

11. In the title picture to this chapter, what can you say about the structure in the upper left corner other than that it is a grain boundary?

Problems

1. Another class of Cu-based bronze materials are those containing Be. Discuss whether you expect Be to be a substitutional or interstitial impurity in Cu, and what its solubility might be.

2. Do the Hume-Rothery rules apply for alloys of HCP metals? Check out Mg and Zr. Make a prediction based on your engineering common sense, and then verify your conclusion in the literature.

3. Look at Exercise 3 above. Assume that you squeeze a C atom into the space in-between Fe atoms as explained, so that all the Fe atoms around the C atom are moved apart a little. By what fraction has the linear dimension of this space been changed in the process?

4. Describe what happens, and explain the consequences, when zirconium oxide is stabilized by adding a small amount of yttrium, as in YSZ.

5. What Burgers vector would you obtain, if there were two edge dislocations inside your circular path?

6. In light of what we said about stresses around an edge dislocation, the Figs. 4.2 and 4.9 are drawn with a subtle simplification. What is it? (Hint: How would the sample respond to these stresses?)

7. Analyze the background pattern on the cover of this book. Is it a crystal? What kind of symmetry does it have? What principle explained in this chapter does it illustrate?

Chapter 5
Mechanical Properties

Overview

Mechanical properties refer to the response of a material to an externally applied loading force. This response is best described in terms of the strain of the material, i.e. its deformation relative to the original dimensions, vs. the applied stress, i.e. the force per unit area. At a small enough stress, a material is deformed elastically, meaning that it will resume its original shape once the stress has been removed. At a larger stress, many materials are deformed plastically, which is to say permanently. At an even larger stress, the point of mechanical failure, or fracture, will be reached. These three stages of behavior can be understood in terms of various deformation mechanisms operating at the atomic level. The mechanical behavior can also be modified to a large extent by controlling the defects present in a material.

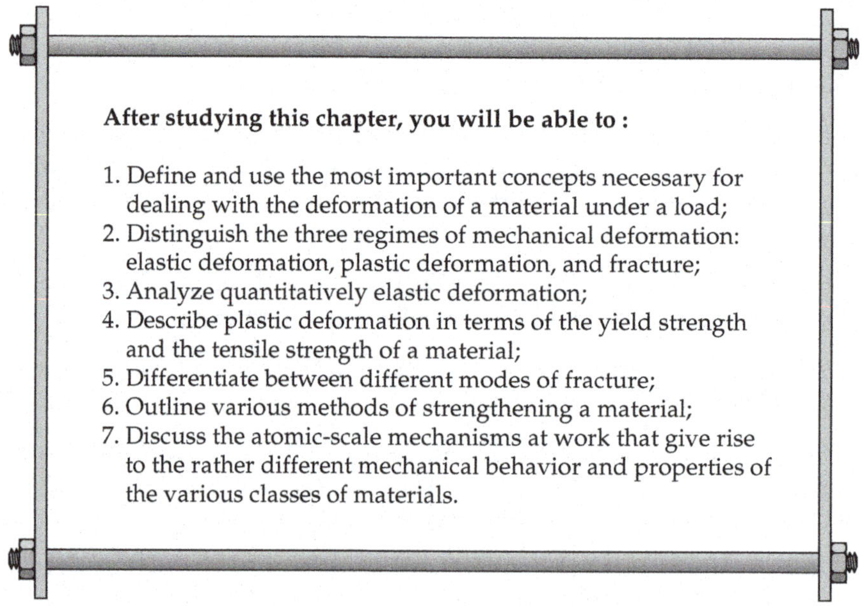

After studying this chapter, you will be able to :

1. Define and use the most important concepts necessary for dealing with the deformation of a material under a load;
2. Distinguish the three regimes of mechanical deformation: elastic deformation, plastic deformation, and fracture;
3. Analyze quantitatively elastic deformation;
4. Describe plastic deformation in terms of the yield strength and the tensile strength of a material;
5. Differentiate between different modes of fracture;
6. Outline various methods of strengthening a material;
7. Discuss the atomic-scale mechanisms at work that give rise to the rather different mechanical behavior and properties of the various classes of materials.

Mechanical materials failure at the Boston "Big Dig"

In the Boston street tunnel known as the "Big Dig", heavy panels were hung on steel bolts fastened with epoxy into holes which had been drilled into the concrete roof. The epoxy failed on some of the bolts. The bolts were pulled out of the holes, and the panels crashed onto the road, and onto a car. One person was killed. The failure was attributed to improper curing of the epoxy, i.e. improper materials processing by the applications engineers.

5 Mechanical Properties

In this chapter, we will examine in considerable detail the mechanical properties of materials in general, and the different classes of materials in particular. By mechanical properties we mean the ways in which a material responds to an externally applied load. This response can go from elastic, reversible deformation to plastic, permanent deformation, and ultimately to fracture. You will see that although there are some universal characteristics to such deformations, different classes of materials exhibit rather different behavior in the details. An issue which we will have a special interest in is how to make a material stronger. You will come to realize that with knowledge of crystalline and defect structures, one is in an excellent position to interpret, and to some extent control, the macroscopic mechanical properties of materials in terms of atomic-level processes.

5.1 Macroscopic Phenomena: Stress and Strain

The main theme of this entire chapter is the response of a material to an applied force, or an applied load. What we are interested in most is to develop a description, and explanations, for the part of the behavior that is characteristic of the material itself, and is not dependent e.g. on its shape or other geometrical features.

We begin our discussion by looking at the two main situations of interest to us (Fig. 5.1). In part a) of Fig. 5.1 a **tensile test** is shown, where a **tensile force** F is applied to the sample. Two equal and opposite forces F act along the axis of the sample, and perpendicular to the initial cross-sectional area A_0. In response to these forces, the sample is stretched from its original length l_0 to its length l when loaded. In the process, the cross-sectional area is reduced from its initial value A_0 to A under load.

Fig. 5.1b) shows a **shear force** F applied to the sample. In this case, the force F is parallel to the area A and the response of the sample to the applied force is reflected in the deformation angle θ, in particular $\tan\theta$.

Experimental observations have shown the following for tensile tests on ***different samples*** (same overall shape, different dimensions) made from the ***same material***:

- For suitably small F and fixed A_0, but variable l_0,
  the change in length $\Delta l = l - l_0$ is proportional to l_0;
- For suitably small F and fixed l_0, but variable A_0,
  the change in length $\Delta l = l - l_0$ is inversely proportional to A_0.

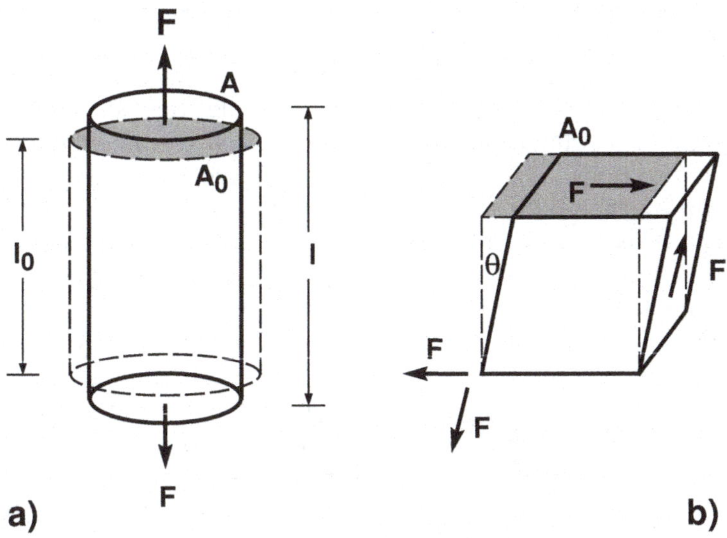

Figure 5.1: The two main experimental setups for studying the mechanical response of a material to an applied force. a) Tensile force; b) Shear force.

This implies that in order to discuss mechanical properties of materials independent of sample shape it is best to use the following variables:

$$\textbf{engineering stress } \sigma = \text{force F per unit area } A_0 \tag{5.1}$$

$$\textbf{engineering strain } \epsilon = (l - l_0)/l_0 = \frac{\Delta l}{l_0} \tag{5.2}$$

The reason why stress in particular is termed engineering stress is because it refers to applied force per unit *initial* area, F/A_0, which is the way stresses are usually measured. For some discussions it may be useful to refer to the **true stress** as the applied force per unit area under load A. Strain is always referenced to the initial length l_0. In the future, our default assumption will be that when we say stress, we mean engineering stress.

If the applied forces are shear forces, one can proceed in similar fashion. **Shear stress** is defined in the same way as tensile stress, namely as force/area, but the force F is applied parallel to the area A_0. The symbol τ is used for shear stress. Also, in a shear stress situation, the quantity $\tan\theta$ illustrated above plays the role of strain, and the symbol γ is used for $\tan\theta$.

At this point, a few comments are in order. First, the illustration in Fig. 5.1a) is for a tensile force. However, we could easily reverse the directions of the two forces, in which case we would call the load **compressive**, and the test a

compression test. What we said above about quantities being proportional to each other will apply to compression as well.

The situation depicted in Fig. 5.1a) is sometimes referred to as **uniaxial loading**, i.e. loading along a single axis. Another simple stress state, which is in a sense the opposite of uniaxial, would be **isotropic stress**, i.e. stress that is the same in all directions. This is achieved for example with a piece of material immersed in water, or with a pressurized tank of gas.

In order to keep signs straight for the various quantities, we will assume that a tensile force F is taken as > 0, so that $\Delta l > 0$ in **tension**, whereas in compression $F < 0$ and $\Delta l < 0$.

Also, you may have noted in Fig. 5.1b) that two shear forces F are shown on either side of the sample. These are necessary so that the sample does not experience any torque. The resulting force from the two pairs of forces F is a force that points along the diagonal of the sample. This is an indication that pure tensile and pure shear stress represent two ideal situations with regard to geometry. In other types of loading, such as torsion or bending, the stress state of the sample is more complex. For example, in bending a bar you will cause tension on one side and compression on the other.

5.2 Stress vs. Strain Curves: Elastic Deformation

The most fundamental information about mechanical properties of materials is obtained from stress-strain curves, i.e. plots of stress σ vs. strain ϵ from a tensile test. Such data are obtained routinely, using commercial tensile testing machines, where the applied stress is measured as a function of elongation imposed on a material . A standard tensile test specimen of a commercial material is cylindrical, with a length of 2", a diameter of 0.505", and with thicker ends that allow for safe attachment to the tensile testing machine.

A tensile test is typically performed until the sample fails. A generic stress-strain curve for a metal alloy sample is shown In Fig. 5.2 below.

The dashed lines delineate three regions in the graph. In the first one, $\sigma(\epsilon)$ is linear; in the second region the curve continues to rise but less rapidly; and in the third region the curve decreases until, at its end, the sample breaks.

The small figures on top of the graph indicate schematically the shape of the sample at different stages of the test. In the first region the sample essentially maintains its shape; in the second region there is a noticeable elongation with a slight reduction in the overall cross-section; and beyond the maximum the sample develops a constriction in the middle with a much reduced cross-section.

For the continuing discussion in this chapter, we will confine ourselves to the first, linear region. In that region, we have that σ is proportional to ϵ, or $\sigma \propto \epsilon$. This means that we can write an equation between σ and ϵ as follows:

$$\sigma = E\epsilon \tag{5.3}$$

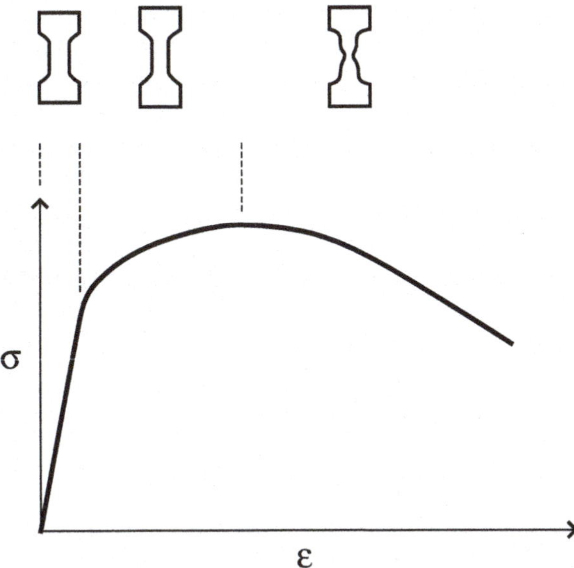

Figure 5.2: Generic stress-strain curve for a metal alloy sample. Note the three regions of the curve and the corresponding sample shapes on top.

The proportionality constant E in Eq. 5.3 is called **elastic modulus**, or **modulus of elasticity**, or **Young's modulus** after the English physicist T. Young, and Eq. 5.3 is known as **Hooke's Law**, after the English physicist R. Hooke.

If we are given a certain stress-strain curve $\sigma(\epsilon)$, then we can also say that the elastic modulus E is the slope of $\sigma(\epsilon)$ at $\epsilon = 0$:

$$E = \left.\frac{d\sigma}{d\epsilon}\right|_{\epsilon=0} \tag{5.4}$$

Physically, the linear part of the curve means that in that region the deformation is **elastic**, i.e. reversible: The sample resumes its original dimensions once the load has been removed. The elastic modulus E is the most important parameter characterizing the mechanical properties of a material.

It needs to be pointed out immediately, however, that a linear $\sigma(\epsilon)$ always implies elastic deformation, but the converse is not necessarily true. In some special cases, material deformation may be elastic (i.e. perfectly recoverable) even a little beyond the linear portion of the curve. Moreover, for polymers the $\sigma(\epsilon)$ may not be linear even near the origin, yet the deformation may still be elastic.

We will examine the atomic basis of mechanical behavior in some detail later. Let it suffice here to note that elastic behavior results from atomic bonds being stretched but not broken. You will recall that this is behavior we would expect based upon the discussion in Chapter 2 on interatomic energy vs. r and interatomic force $F(r)$ curves (cf. Fig. 2.10). The elastic modulus E is related directly to the slope of the curve $F(r)$ at the interatomic equilibrium position.

Note also the units of E: Being the equivalent of a pressure, the unit of E is 1 Pa (Pascal) = 1 N/m^2. We will find that numerical values of E are quite large. So we will often use 1 MPa (1 Megapascal = 10^6 Pa), or even 1 GPa (1 Gigapascal = 10^9 Pa).

As the sample is elongated elastically in the axial direction, one also observes that it contracts slightly in the two directions perpendicular to the sample axis. If we denote the axial direction as the z-axis, Hooke's Law becomes $\sigma = E\epsilon_z$. In order to describe the lateral contraction, we define two strains ϵ_x and ϵ_y in the plane perpendicular to the tensile axis by

$$\epsilon_x = \frac{\Delta x}{x_0}, \quad \epsilon_y = \frac{\Delta y}{y_0} \tag{5.5}$$

If the sample material is isotropic, then $\epsilon_x = \epsilon_y$. In addition, in tension $\epsilon_z > 0$, and at the same time $\epsilon_x = \epsilon_y < 0$. Furthermore, in the elastic region $\epsilon_x \propto \epsilon_z$. Therefore, one can introduce a new equation, and a new parameter ν, by defining

$$\nu = -\frac{\epsilon_x}{\epsilon_z} = -\frac{\epsilon_y}{\epsilon_z} \tag{5.6}$$

where ν is called the **Poisson ratio**. The minus sign in Eq. 5.6 is chosen so that ν is a positive number.

The change in volume $\Delta V = V - V_0$ of the sample due to the tensile stress can be written as

$$\Delta V = V_0(1 + \epsilon_z)(1 + \epsilon_x)(1 + \epsilon_y) - V_0 \tag{5.7}$$

Using the definition of the Poisson ratio, this becomes

$$\Delta V = V_0(1 + \epsilon_z)(1 - \nu\epsilon_z)(1 - \nu\epsilon_z) - V_0 \tag{5.8}$$

For small strains Eq. 5.8 can be simplified to give

$$\frac{\Delta V}{V_0} = \epsilon_z(1 - 2\nu) \tag{5.9}$$

We can conclude that ν is related to the change in volume of the sample. If ν were equal to 0.5, Eq. 5.9 would tell you that there is no change in volume, V_0 = 0. For most engineering metals ν is around 0.32, and for ceramics it is around 0.23, which means that a sample under uniaxial tension will increase slightly in volume.

The principles of elastic deformation also apply to the condition of pure shear stress. In particular, Hooke's Law for shear stress and strain can be formulated in analogy to Eq. 5.3 as:

$$\tau = G \tan\theta = G\gamma \tag{5.10}$$

where τ is the shear stress, G is the shear modulus, and $\tan\theta = \gamma$ is the shear strain (see. Fig. 5.1).

5.3 Inelastic Deformation: Yielding and Ductility

We will now examine $\sigma(\epsilon)$ curves beyond the linear region. For this purpose, I have redrawn Fig. 5.2 with some additional details:

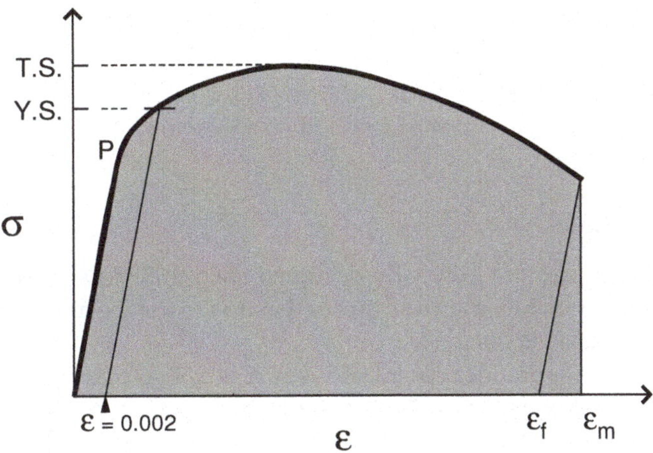

Figure 5.3: Generic stress-strain curve for a metal alloy sample. Note the proportional limit P, i.e. the beginning deviation from linear behavior, the yield strength Y.S., and the ultimate tensile strength T.S.

Fig. 5.3 defines a number of additional mechanical property parameters. The **proportional limit** P designates the point where $\sigma(\epsilon)$ begins to deviate from linear behavior. Next are the **yield strength Y.S.** and the ultimate **tensile strength T.S.**, usually abbreviated to just tensile strength. In addition, the two values ϵ_f and ϵ_m are measures of the strain at fracture, and the gray area under the $\sigma(\epsilon)$ curve is called **toughness**. Each of these quantities will be discussed in further detail below.

First, however, some comments about the general shape of the $\sigma(\epsilon)$ curve in Fig. 5.3 are in order. As I mentioned above, the curve deviating from linear beyond the point P signifies that permanent **inelastic** or, equivalently, **plastic deformation** has begun. The behavior of the curve would certainly appear to be reasonable from there on up to its maximum. But how can it be that beyond the maximum the curve decreases, i.e. the sample continues to become longer, yet the stress is being reduced? This seems to be counter-intuitive.

The explanation can be deduced from the sample shapes in Fig. 5.2. Beyond the maximum of $\sigma(\epsilon)$, the sample starts to develop a "weak area", that is a noticeable constriction or neck. In that region the effective cross-section is much less than the nominal cross-section, and therefore the local effective stress is increased. In other words, σ is not the same as the true stress σ_t in the necked region, which is what really counts and ultimately leads to failure. If you plotted $\sigma_t(\epsilon)$ of the necked region, you would find that it keeps rising up to the instance of fracture.

In our continuing discussion we will focus first on the parameters named yield strength Y.S. and tensile strength T.S. of Fig. 5.3. Their importance, especially to mechanical design, should be evident.

Tensile strength T.S.

The tensile strength is identified as the stress at the maximum of the $\sigma(\epsilon)$ curve. It represents the ultimate limit of applied stress beyond which the material is sure to fail quickly.

Yield strength Y.S.

Mechanical structures are generally designed such that the stresses in the materials used remain below the yield strength, so as to assure that no permanent, inelastic deformation takes place.

However, from a practical point of view, it is not so clear how to ascertain at which point yielding sets in. That point is marked by P on Fig. 5.3, but how do you determine exactly where the slope of the $\sigma(\epsilon)$ curve starts to differ from the value of the elastic modulus E? In order to have a well-defined procedure, people interested in mechanical properties have agreed that they would use as a measure of the onset of yielding the yield strength Y.S. as defined on Fig. 5.3. The value of Y.S. is chosen as the stress at which a straight line with slope E going through the point $\epsilon = 0.002$ on the horizontal axis intersects the actual $\sigma(\epsilon)$ curve. Because of the specific horizontal axis intersect, this is sometimes referred to as the 0.2% rule. As a practical recipe, this can be applied readily to experimental data.

Toughness

You will also note the two strains marked ϵ_m and ϵ_f in Fig. 5.3. Think of these two numbers as the strain just *before* fracture, with the load still applied, and *after* fracture, when the load has been released and the two broken pieces have been put back together. The point is that up to the instance of fracture, energy has been stored in the material both by elastic and inelastic deformation. The elastic part is recovered once the fracture releases the load. (Note that the slanted line going down to ϵ_f in Fig. 5.3 also has a slope of E, i.e. it is parallel to the initial linear portion of the $\sigma(\epsilon)$ curve).

Toughness is defined as the gray area under the $\sigma(\epsilon)$ curve (Fig. 5.3), and we will use the symbol W for it, so that we can write

$$W = \int_0^{\epsilon_m} \sigma(\epsilon) d\epsilon \tag{5.11}$$

The meaning of W becomes clearer if Eq. 5.11 is put into differential form:

$$dW = \sigma(\epsilon) d\epsilon \tag{5.12}$$

From the definitions of the variables σ and ϵ it follows directly that

$$dW = \frac{F(\epsilon)}{A_0} \frac{dl}{l_0} \tag{5.13}$$

and therefore, after a slight rearrangement,

$$dW = \sigma(\epsilon) d\epsilon = \frac{1}{A_0 l_0} F \, dl \tag{5.14}$$

In Eq. 5.14 $A_0 l_0$ is the original volume of the sample. We conclude that dW represents the differential work done per unit volume by the load F on the sample in extending its length by dl. Hence the gray area under the $\sigma(\epsilon)$ curve, the toughness W, is the total work done per unit volume by the load F up to the point of fracture.

Ductility

Ductility is defined as $\epsilon_f = (l - l_0)/l_0$. It denotes the total *plastic* deformation the sample has been able to sustain up to the point of fracture. Figures for the ductility are often expressed in %, i.e. as $100 \times \epsilon_f$. You will note that this definition is in line with how we talked about ductility initially, in Chapter 3, when we related it to the structure of materials.

Hardness

Hardness is a mechanical quantity which is especially useful for quick characterization or quality control, of a material, since it can be tested easily with a small desktop machine. Hardness is defined as the resistance of a material to local plastic deformation. This resistance is quantified by examining a small indent in the material.

Although you will be familiar with the concept of hardness from everyday experience, its use and application in engineering are a little more involved because several different methods exist to measure hardness, as illustrated in the figure.

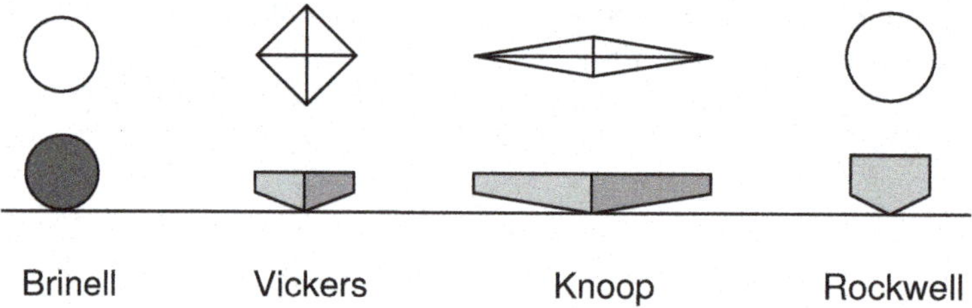

Figure 5.4: Methods for measuring hardness. The bottom row shows a side view of the different indenters, and the top row shows a top view.

The methods differ in the type of indenter and the load they apply. The Brinell method uses a fairly large steel or tungsten carbide sphere, the other methods use a small diamond indenter in the shape of a pyramid (Vickers, Knoop) or a cone (Rockwell). The Rockwell method is complicated in that many different loads are used, depending on the material, and in some cases the indenter may be a sphere rather than a cone. Still, for each type of measurement, hardness is expressed as a single number (actually, a pressure or stress) which is derived from the dimensions of the indentation and the load.

It should be evident that the stress state induced by the indenter in the material is rather complex: It depends on the load and the shape of the indenter. It is clearly three-dimensional but localized. The inelastic strain will vary with distance away from the center of the indenter. So in a sense each method measures something different. However, practice has shown that different hardness scales are correlated with each other and with other measures of inelastic deformation. For example, Eq. 5.15

$$\mathrm{HB} \approx 0.3 \times \mathrm{T.S.}\ \text{(in MPa)} \tag{5.15}$$

provides a surprisingly good relationship between the tensile strength T.S. and the Brinell hardness HB of many types of steel (see homework). A relationship similar to the one in Eq. 5.15 between Vickers hardness and Y.S. holds for Cu and brass as well.

True stress and true strain

We noted above that in principle it is the true stress in the necked region that determines the yielding process. Therefore, true stress would be a useful piece of information to have. Sometimes, data are indeed reported in terms of true stress. However, true stress is not easy to measure, and hence is not measured routinely in a normal tensile test. True stress would require keeping track of the cross-section in the necked region, and true strain would be even more difficult to ascertain as it involves the local elongation in the necked region.

Still it is of interest to elucidate the relationship between engineering stress σ and true stress σ_t. Let us note first that

$$\sigma_t = \frac{F}{A} = \frac{F\,A_0}{A_0\,A} = \sigma\,\frac{A_0}{A} \tag{5.16}$$

Let us now assume that we are below T.S. and the material is isotropic, so that we can write

$$A = A_0(1+\epsilon_x)(1+\epsilon_y) = A_0(1-2\nu\epsilon) \tag{5.17}$$

where ϵ_x and ϵ_x are transverse strains, ϵ the axial strain, and ν the Poisson ratio.

If we combine Eqs. 5.16 and 5.17, we can solve for σ_t to obtain

$$\sigma_t = \frac{\sigma}{1-2\nu\epsilon} = \frac{\sigma}{1-2\nu\frac{\sigma}{E}} \tag{5.18}$$

Sometimes it is assumed that $\Delta V = 0$, i.e. $\nu = 0.5$, in which case Eq. 5.18 reduces to $\sigma_t = \sigma/(1-\epsilon)$. For small ϵ this can be taken as equal to $\sigma_t = \sigma(1+\epsilon)$, the formula given usually for σ_t.

Since for real materials $\nu < 0.5$, it follows from Eq. 5.18 that $\sigma_t > \sigma$. In the elastic region ϵ is in general small, much less than 0.1, so that the difference between σ_t and σ is also small. In the region of uniform yielding, up to T.S., the difference between σ_t and σ continues to increase. (Keep in mind that the Poisson ration ν should be considered to increase gradually towards a value of 0.5 at T.S.). In the neck region, the strain will not be small and at the same time the cross-sectional area will be much reduced. So it is in the neck region that one has $\sigma_t \gg \sigma$.

One can also define the true strain ϵ_t as the total, integrated strain by

$$\epsilon_t = \int_{l_0}^{l} \frac{dz}{z} = \ln(1+\epsilon) \tag{5.19}$$

This reduces to $\epsilon_t = \epsilon$ for $\epsilon \ll 1$, as it should.

5.4 Mechanical Properties of Materials

5.4.1 Generic Properties

As a practical exercise in the analysis of mechanical test data, we will analyze a stress-strain curve for a typical Al-alloy (Fig. 5.5). We will determine first the important mechanical property parameters from the figure step-by-step as follows:

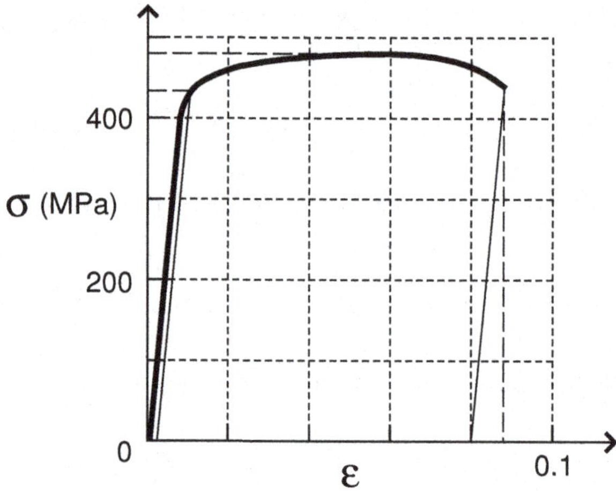

Figure 5.5: Typical stress-strain curve for an Al-alloy.

1. The elastic modulus is the slope of the initial linear portion of the curve:

   $E = \frac{\Delta\sigma}{\Delta\epsilon} = \frac{400 MPa}{0.008} = 50$ GPa

2. You obtain the yield strength Y.S. from the graph by applying the 0.2% offset method. Use the thin line parallel to the initial portion of the $\sigma(\epsilon)$ and the value of σ where the thin line intersects $\sigma(\epsilon)$.
   The result is: Y.S. = 435 MPa.

3. The tensile strength T.S. can be read off of the graph directly:
   T.S. = 480 MPa.

4. The ductility can also be read off of the graph: ϵ_f is 0.08, and ϵ_m is 0.088 (see Fig. 5.3).

Next we illustrate the most important mechanical concepts introduced earlier in this chapter with help of a set of generic $\sigma(\epsilon)$ curves (Fig. 5.6):

Curve a) exemplifies a brittle material with practically zero ductility. This type of curve would be typical for a ceramic. Curve b) is representative of a tough material, with high strength and high ductility. It would be typical for many metal alloys. Curve c) illustrates a weak but highly ductile material, behavior often found in polymers (but some polymers are rather brittle).

Fig. 5.6 points to one of the main challenges for the materials engineer who wants to develop or use a material with both high strength and high ductility. Generally, both these properties are desirable. High T.S. and Y.S. mean that the material will be able to withstand high loads, and high ductility means that the material can be shaped easily during processing because it will be able to

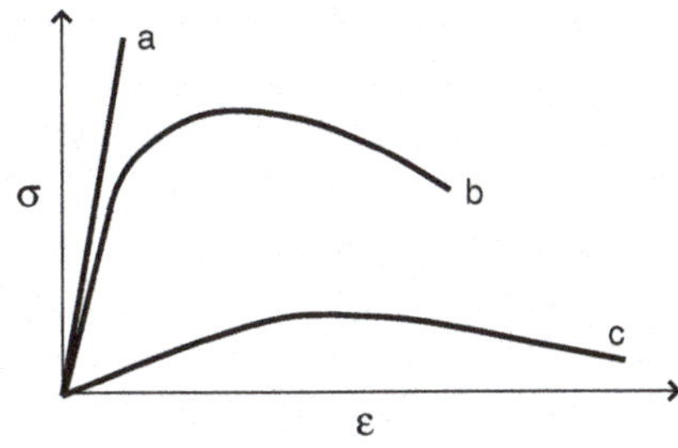

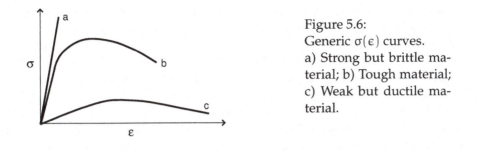

Figure 5.6:
Generic $\sigma(\epsilon)$ curves.
a) Strong but brittle material; b) Tough material; c) Weak but ductile material.

withstand a large amount of plastic deformation. Yet, with most engineering materials, the engineer ends up having to make a trade-off between strength and ductility.

The table below lists data representative of various types of materials, including metal alloys, ceramics, covalent solids, inorganic glasses, and polymers.

Table 5.1: Overview of Mechanical Properties Values

Material	E (GPa)	Y.S. (MPa)	T.S. (MPa)	ϵ_f
Pb-Sn solder	25	-	35	0.40
Mg alloy	45	250	300	0.15
Al alloy	75	250	300	0.18
Cu alloy	120	300	400	0.10
Steel	200	500	850	0.15
Tungsten	400	800	1000	0.02
fused SiO_2	70	-	100	-
Si_3N_4	300	-	500	-
Si	187	-	120	-
SiC	320	-	400	-
soda-lime glass	70	-	70	-
LDPE	0.25	10	15	2.0
PVC	3.1	40	40	0.4
PMMA	2.7	60	60	0.05

As you have undoubtedly realized, the materials in Table 5.1 are arranged according to the classes we named back in Chapter 3. In each class the order is from low to high E. The last three rows list polymers by their abbreviations: LDPE for low-density polyethylene, PVC for polyvinyl chloride, and PMMA for polymethyl-methacrylate (commonly know as plexiglass). We will expand on these properties shortly, especially with respect to polymers.

You should be aware that for each category of materials (Cu alloys, steels, etc.), the parameter values listed are typical, average values. As is the case with most mechanical properties, these values can vary widely within a category because they depend strongly on the processing and the microstructure of the material. This is true in particular for the yield strength Y.S., the tensile strength T.S., and the ductility ϵ_f. For example, the numerical values for the Mg-, Al- and Cu-alloys in Table 5.1 are for materials which have been processed for increased strength. It would be possible to achieve larger ductility, at the cost of reduced strength, by alternative thermal processing. The numbers indicate the trade-off between ductility and strength already mentioned, especially for materials within a certain class, but the correspondence is not perfect.

5.4.2 Mechanical Properties of Metals

Table 5.1 above lists representative values for mechanical properties of broad types of materials. However, even within a given type of material, say Cu alloys or steels, these values vary widely, depending on such parameters as the composition and the processing. In particular the processing determines the microstructure, which in turn determines the detailed properties. You can gain an appreciation of these issues from the two tables below.

Table 5.2: Mechanical Properties Values for Various Types of Steel

Material	Type	Y.S. (MPa)	T.S. (MPa)	ϵ_f	HB
1018	low-C	370	440	0.15	126
1117	low-C	400	480	0.15	137
1045	medium-C	530	630	0.15	180
1095	high-C	460	820	0.10	271
4140	low-C Cr-Mo	620	700	0.18	228
8620	low-C Ni-Cr-Mo	590	700	0.22	210
E8740	med-C Ni-Cr-Mo	660	740	0.17	228

The materials listed here are from an extensive table of a commercial steel manufacturer (http://www.cbsteel.com/pdf/CB-Data.pdf). They are for cold-drawn steels, i.e. steels which have been deformed plastically at a relatively low temperature. The criterion for "low temperature" is, loosely speaking, that atoms will not move around on account of their thermal energy, but only because of the mechanical deformation. Another common process is called hot rolling, which involves deformation at a "high temperature" at which atoms have enough thermal energy to move around easily.

A good general estimate for the temperature that separates "low" from "high" with respect to atom motion turns out to be about $0.5T_m$ where T_m is the melting point of a material on the absolute scale (i.e. K). For steels this is around 900 K, but various temperatures may be used for the treatment at "high temperature".

In Table 5.2, once again representative numbers for the sub-types are shown. Low-C means a carbon concentration of about 0.15 %, and medium-C about 0.4 %. Where nothing else is listed, the steel is a typical carbon steel and the most significant additional alloying element is manganese (Mn) at about 0.5%. Cr-Mo means that the material contains significant amounts (a few tenths of %) of these other alloying elements.

Table 5.3: Mechanical Properties Values for Brass and Bronze

Material	**State**	T.S. (MPa)	ϵ_f	HV
100Cu	annealed	210	0.35	55
90Cu 10Zn		245	0.35	75
80Cu 20Zn		265	0.40	80
90Cu 10Zn	half hard	310	0.07	95
80Cu 20Zn		340	0.10	95
96Cu 4Sn	annealed	295	0.40	80
94Cu 6Sn		340	0.50	90
96Cu 4Sn	half hard	430	0.08	150
94Cu 6Sn		525	0.12	170

Table 5.3 shows data for different Cu alloys, both annealed and with hardness half hard (there are several different grades of hardness). The hardness is given here on the Vickers scale HV. The data are from a commercial producer (http://www.indiamart.com/arihanta-metal-company/brass.html). You will

note an increase in T.S. and hardness HV with an increase in alloying and between the annealed and the hardened grade of material. Also keep in mind in comparing Tables 5.2 and 5.3 that to a first approximation, Brinnell hardness HB and Vickers hardness HV in their most common versions are numerically equal (specifically, HV $= 1.05 \times$ HB).

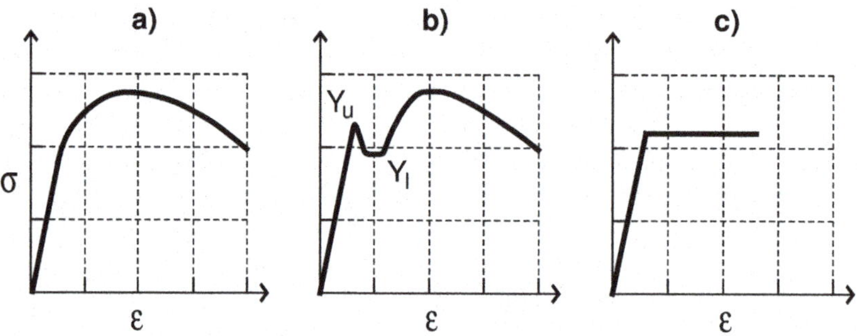

Figure 5.7: Types of stress-strain curves. a) Normal curve; b) Curve with upper yield point Y_u and lower yield point Y_l; c) Idealized curve with elastic deformation followed by ideal plastic flow.

Another issue in mechanical properties is illustrated in Fig. 5.7. Most metals, especially after being processed mechanically, will exhibit a stress-strain curve of type a). However, certain carefully **annealed**, so-called mild steels may exhibit a curve of type b). Annealing is done for the purpose of making the material softer. It involves bringing the material up to what we called above a "high temperature", keeping it there for some time, and cooling it down again very gradually.

Curve b) is characterized by an **upper yield point** Y_u and a **lower yield point** Y_l. It is only after the lower yield point that the normal yielding, with the gradual rise of the stress-strain curve, is observed. Curve c) is a hypothetical curve, in that it shows a normal elastic range, but this is followed by "ideal" plastic deformation in which the material essentially flows, i.e. deforms plastically at constant stress.

In order to understand the behavior of curve b), we need to expand a little first on what happens in curve a). Then you can think of b) as being like a) but combined with a certain amount of c).

You know that the normal curve a) begins with a range of elastic deformation, followed by inelastic, permanent deformation as it goes past the yield strength. It does so requiring an increasing stress in order to increase the strain further. This means that as the material is yielding, *it is in effect getting harder* ! In other words, the energy that is expended to cause permanent deformation results in increased hardness of the material, and thus increased strength.

Curve c) represents the opposite case: After a certain threshold stress is reached, at the end of the elastic range, the material behaves as an ideal plastic material which flows under the applied stress without an increase in this stress.

What happens in curve b) is that at the upper yield point Y_u behavior very similar to ideal plastic flow sets in briefly, which is followed after a short delay by normal yielding starting at the lower yield point Y_l. The exact mechanism that gives rise to this behavior is still being debated, and we will get a somewhat better idea about it in the next chapter on the behavior at the atom scale underlying mechanical properties.

5.4.3 Mechanical Properties of Ceramics

Fig. 5.6 above illustrated the main feature of the mechanical properties of a ceramic: Deformation is elastic up to the point of fracture. Ceramics exhibit essentially no plastic deformation or, in other words, they have zero ductility. The ultimate tensile strength is also the stress at fracture.

The fundamental reason for this behavior is not difficult to deduce in light of what we said in Chapter 3 about the crystal structures of ceramics. Plastic deformation would require substantial movement of a large number of ions relative to their equilibrium positions. Such movement is severely restricted because of the electrostatic repulsion between ions with the same charge.

Another difference in the mechanical behavior of metals and ceramics is that metals behave similarly in tension and in compression: A ductile metal will show elastic deformation and yielding either way, and the same definition of the yield strength in terms of the 0.2% offset is used. Of course, in compression no neck will form, and thus fracture occurs in a different manner. Ceramics, on the other hand, are typically about 10 times stronger in compression than in tension. This asymmetry in the mechanical behavior is related to the presence of flaws, an issue to be explored further below.

A practical problem with mechanical testing of ceramics is that it is difficult to machine them into a test specimen for a tensile test. Furthermore, even if machining is feasible, the process may introduce hard-to-control variations because of surface flaws. Hence ceramics are often tested in a four-point bending configuration: The specimen is supported at one point at each end and is loaded between two points in the middle. The complication in that case is that the stress state in the sample is more complex, since bending causes compression on the side where the load is applied, and tension on the opposite side. Still, a figure of merit called modulus of rupture can be extracted from bending tests, which is comparable to the tensile strength.

Hardness testing, as an indirect measure of strength, is also applied to ceramics and has a long history going back to the Mohs hardness scale for minerals. This is an approximate relative scale devised by the geologist Friedrich Mohs in 1812, and is based on which material can scratch another one. Talc rates at a hardness of 1 on the Mohs scale, Al_2O_3 at 9, and diamond at 10.

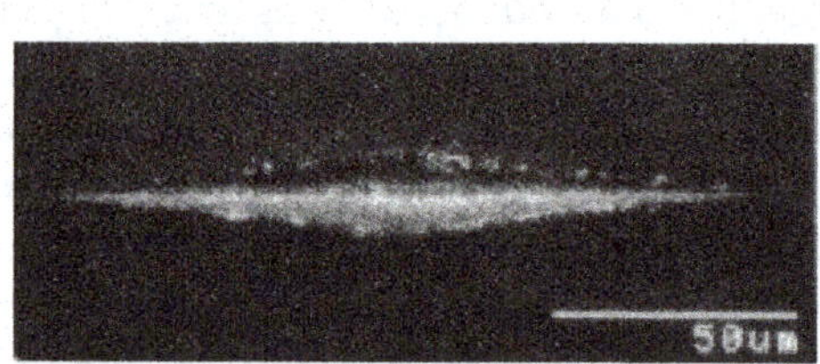

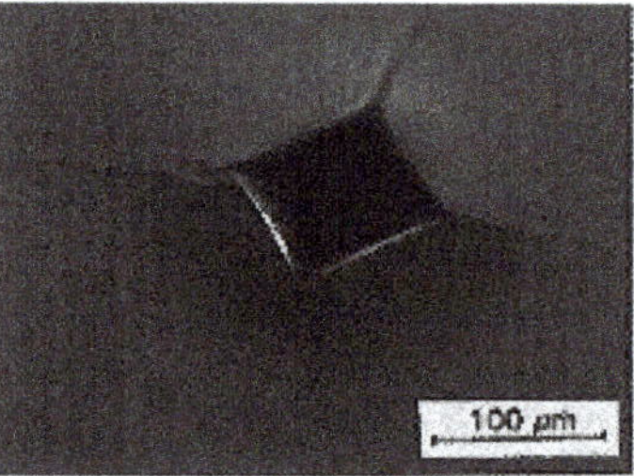

Figure 5.8: Optical microscope images of the ceramic Si_3N_4 subject to hardness tests. Left figure: Knoop test. Right figure: Vickers test.

From the materials science point of view, as ceramics are brittle, their hardness cannot be framed in terms of resistance to plastic deformation. Instead, for ceramics hardness is defined as resistance to abrasion. Of course, abrasion and scratching are somewhat destructive processes, involving the application of concentrated stress to cause local fracture and removal of material.

You will notice, on the right-hand picture of Fig. 5.8, a small amount of inelastic deformation from the gently bent-up edges of the pyramidal indentation. This appears to be required in order to accommodate material displaced by the indenter. At the same time, in the corners of the indent small cracks extend outward, indicating the onset of fracture.

5.4.4 Mechanical Properties of Polymers

The mechanical properties of polymers in general can be summarized by noting that, compared to metals and ceramics:

a) The same mechanical concepts can be used.

b) Some concepts need to be modified slightly.

c) The mechanical parameters vary over a much wider range.

More explicitly, the phenomena listed below are crucial for the mechanical behavior of polymers, and are tied closely to the structure of a material:

1. Stress-strain curves $\sigma(\epsilon)$ vary widely in shape;
2. The elastic modulus E and the tensile strength T.S. vary over a wide range;
3. $\sigma(\epsilon)$, E, T.S., etc. depend strongly on *temperature*;
4. $\sigma(\epsilon)$, E, T.S., etc. depend strongly on *time* in the sense that the *strain rate*, i.e. the time interval over which a strain is induced, or a stress is applied, is very important.

We will begin our discussion of polymers by revisiting the issue of structure, adding a few more details to what we know from Chapter 3. Then we will proceed to demonstrate the four principles enunciated above with specific examples.

More on Polymer Structure

When we discussed the structure of polymers initially in Chapter 3, I pointed out that the natural building blocks of a polymer are long chain molecules with a -C-C- backbone, and that to a first approximation you should think of a polymer as a partially crystalline material. This means that small crystalline regions, with parts of chains aligned in parallel, are embedded in an amorphous matrix of coiled-up, back-folded chains.

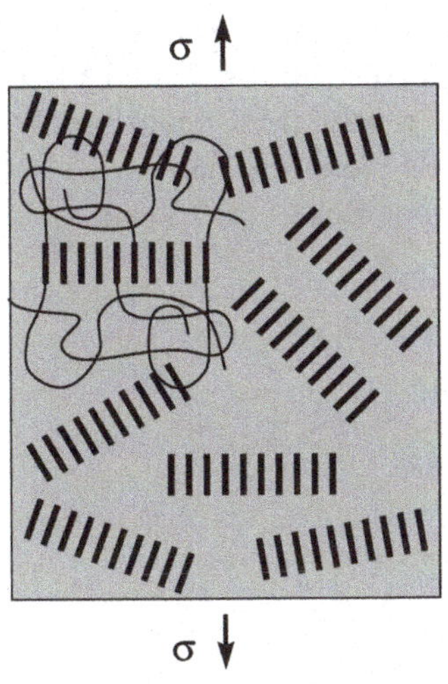

Figure 5.9: Schematic picture of a polymer subject to a tensile stress. Note the small crystals embedded in the amorphous gray region. Only a few coiled-up chains from one crystal are shown, but all crystals contribute many such chains to the amorphous region.

The question will be how such a material responds to an applied stress. Two things are clear immediately: First, a polymer is in essence a composite consisting of crystalline and amorphous regions, and these two components of the material will respond differently to a stress. And second, we should expect the crystalline parts to be stronger and harder than the amorphous parts.

Before we get into specific aspects of the mechanical properties, we have to add a few details with respect to the structure of this type of material, in particular in its amorphous regions. As we noted before, "amorphous" does not mean "without any structure". How polymer chains come together in the amorphous regions depends on what kind of chains they are, particularly what side groups are attached to the backbone, and on what connections, if any, there are between chains.

First let us focus on individual polymer chains. Three examples are displayed in Fig. 5.10. At each corner of a chain are either two H atoms or one H atom and another side group (the H atoms are not drawn). The side groups are

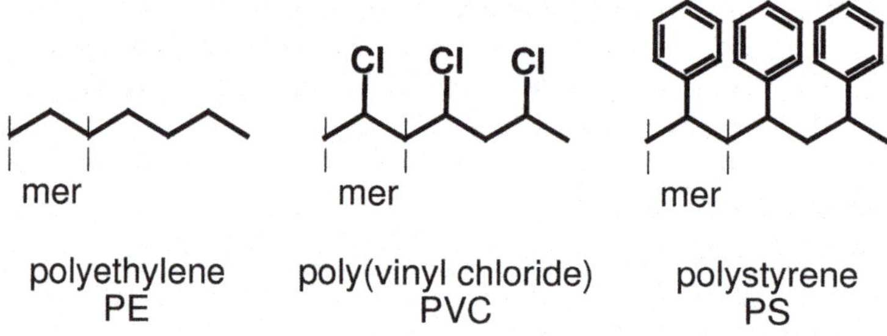

Figure 5.10: Examples of polymer chains with different side groups.

shown approximately to scale relative to the -C-C- bonds in the chain.

The two dashed vertical lines in Fig. 5.10 delineate what is called the **mer** for each molecule. The mer is the fundamental repeat unit *along* the chain, and a molecule consists of many mers. For polyethylene the mer is C_2H_4, for poly(vinyl chloride) it is C_2H_3Cl, and for polystyrene it is $C_2H_3(C_6H_5)$.

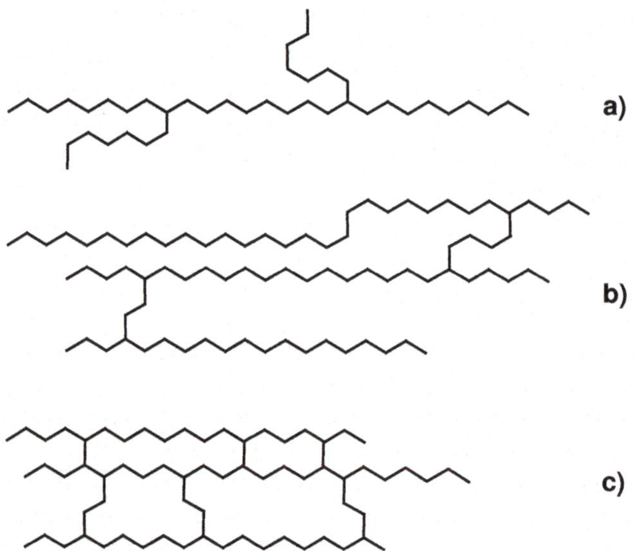

Figure 5.11: Polymers with various degrees of cross-linking.
a) Branched polymer; b) lightly cross-linked elastomer; c) heavily cross-linked network.

The three examples in Fig. 5.10 are called linear polymers, since the -C-C- chains are linear in their simplest geometry. But you know that the true physical conformation is not linear but involves coiled-up and intertwined chains.

The second aspect of structure that turns out to be important for the mechanical properties is to what extent individual chains are linked together in various places by covalent bonds. (You recall that separate linear chains only interact with each other via van der Waals forces). Case a) in Fig. 5.11 shows a so-called branched polymer. The branches of one chain are not connected to other chains but in effect act as rather complex side groups. Case b) is representative of rubber-like materials, also referred to as **elastomers**. Its key feature is light **cross-linking**, where light means something of the order of one cross-link for every 25 mers. Case c) illustrates heavy cross-linking, to the point where one can consider the material to be an amorphous network. A class of materials with this type of structure are epoxies.

The third structural feature important for the mechanical properties is the **degree of crystallinity** in the material. This again depends on the type of chain present or, we could say now, on the type of mer making up the chain. It should be evident that the larger and the more complex the side groups attached to the mer, the less likely the material is to form crystalline regions. A brief illustration of this effect is provided by the fact that polyethylene can be made with a very high degree of crystallinity, up to around 90%, commercial PVC has about 10% crystallinity, and polystyrene always stays entirely amorphous. (In Fig. 5.11, apparently parallel segments of chains are not meant to imply that these are crystalline regions. This is just to simplify the drawing).

Stress-Strain Curves for Polymers

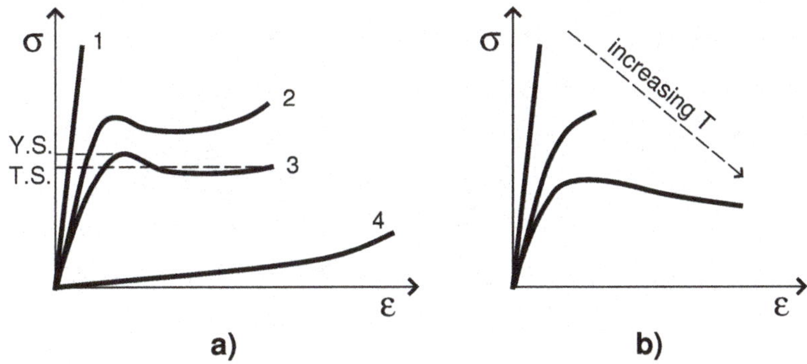

Figure 5.12: Generic $\sigma(\epsilon)$ curves for polymers. a) Curves for different types of polymer materials at room temperature. b) Temperature dependence for a type 1 polymer.

Sets of typical stress-strain curves for polymers are shown below in Fig. 5.12. Compare these curves to those in Fig. 5.6 and also keep in mind the parameters in Table 5.1. Curve 1 in Fig. 5.12a represents a brittle material, curves 2 and 3 ductile materials, and curve 4 an elastomer. Note that for the elastomer the

ϵ scale is not really the same as for the other materials, since elastomers can sustain much higher strains than other polymers, and such strain are elastic and fully recoverable. Fig. 5.12b displays a typical temperature dependence of $\sigma(\epsilon)$ curves for many polymers.

The curves in Fig. 5.12 are interpreted in terms of mechanical parameters as before. The elastic modulus E is defined as the initial slope of $\sigma(\epsilon)$ even though there may be hardly any linear portion to the curve. For polymers, the local maximum after the initial increase is used to designate the yield strength Y.S. (see curve 3). Both curves 2 and 3 show upper and lower yield points. In contrast to metals and ceramics, it is the end point of the $\sigma(\epsilon)$ curve that is identified as the tensile strength T.S.. Note that T.S. may be higher or lower than Y.S. for a polymer.

Fig. 5.12b also illustrates the strong dependence of the elastic modulus on temperature. E may vary by a factor of 3 or more over a temperature range of typically 50°C, whereas for a metal E might decrease by something like 20% with a temperature increase of several hundred °C.

Moreover, an entirely new set of phenomena is observed with polymers, in that all the mechanical parameters are basically time-dependent. For example, the elastic modulus of a material may depend not only on the magnitude of the applied stress, but also on how quickly the stress is applied. This is expressed by saying that the ***mechanical properties of polymers***, such as the elastic modulus, the ductility, or the toughness, ***depend on the strain rate***, not just on the magnitude of the strain itself.

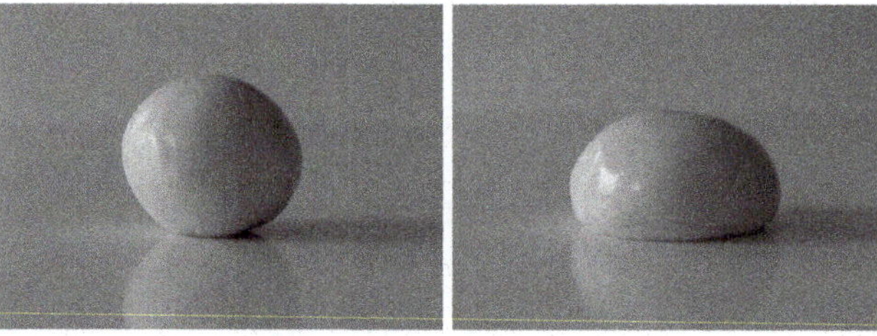

Figure 5.13: A ball of silly putty, freshly rolled on the left, and after lying on the table for 30 minutes on the right.

A good example of strain rate dependence is in evidence with the children's' toy known as "silly putty". (It is over 60 years old but still available. I got it in the Smithsonian Institution gift shop). If you roll the material gently in your hands, you can form it easily into a ball (low strain rate). If you take this ball and throw it to the ground, it bounces back elastically as if it were rubber (high strain rate). Yet if it is left lying on a table for 30 minutes, it flows noticeably under the tiny stress of its own weight (very low strain rate). One can summarize this behavior by saying that the higher the strain rate, the higher is the stiffness of the material.

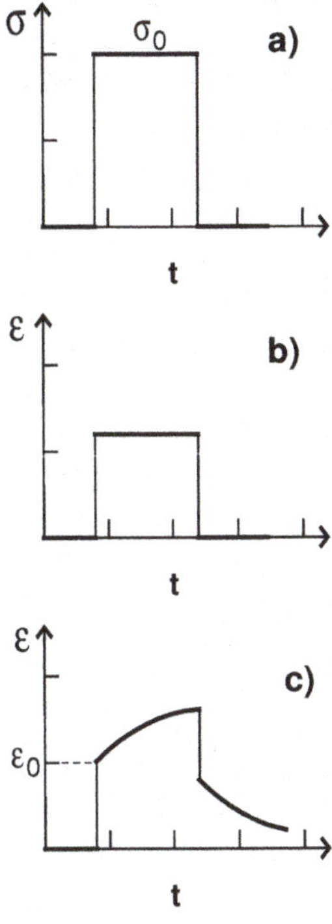

Figure 5.14: Time-dependent stress-strain curves for metals (b) and polymers (c).

Let us now compare in a little more detail the stress-strain behavior of "silly putty" and other polymers to that of a typical metal or ceramic. If a constant, sufficiently small stress σ_0 is applied for a certain period of time to a metal sample as in Fig. 5.14, curve a), then the elastic strain ϵ will be given by curve b). That is, the material response will be instantaneous and elastic, involving the stretching of primary atom-atom bonds.

For a polymer subject to the same stress σ_0, there is also an instantaneous elastic response, but this is typically followed by a slow additional deformation (curve c)), due to gradual changes in the conformations of the polymer chains.

Upon unloading, for a metal or ceramic the material response is again instantaneous, but for a polymer is a function of time. This type of time-dependent deformation is called **viscoelastic**. (Note that for certain polymers some plastic deformation may remain, as the material may not recover fully after the load has been removed.

In order to characterize polymer mechanical behavior more quantitatively, we define a time-dependent modulus of elasticity $E_{ve}(t)$ by setting

$$E_{ve}(t) = \frac{\sigma_0}{\epsilon(t)} \tag{5.20}$$

in analogy to Hooke's Law. The subscript ve, for viscoelastic, is to remind you that this modulus does not describe simple elastic behavior except at very short times.

It is possible to obtain equivalent information from a slightly different experimental setup in which the strain ϵ is kept constant after the initial elastic rise to ϵ_0, and the stress $\sigma(t)$ is continually reduced in time so as to maintain a constant ϵ_0. The ratio $\sigma(t)/\epsilon_0$ acts again as a time-dependent modulus of elasticity, which is sometimes referred to as relaxation modulus.

For further discussion of the viscoelastic effect we will use values of $E_{ve}(t)$ at a fixed but arbitrary time of 10s. An illustration of $E_{ve}(t)$ for a viscoelastic material is given in Fig. 5.15, where data similar to Fig. 5.14 have been used.

Note that here $\sigma(t)$ is applied as a step function in time. This is in a sense the simplest case. If the change in σ occurred over a longer time, more complex behavior would result for $\epsilon(t)$ and $E_{ve}(t)$, depending on the relationship between the time scale for application of σ and the viscoelastic response time.

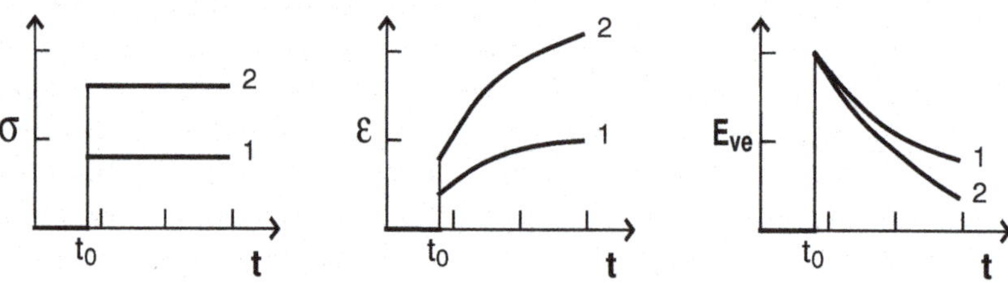

Figure 5.15: Schematic representation of $E_{ve}(t)$.

$E_{ve}(t)$ in general, and $E_{ve}(10)$ in particular, are also strong functions of temperature. This is pictured for a generic polymer in Fig. 5.16. Curve 1 refers to a highly crystalline form of the polymer, curve 2 to a fully amorphous form, and curve 3 to a lightly cross-linked form. Note the log scale on the y-axis.

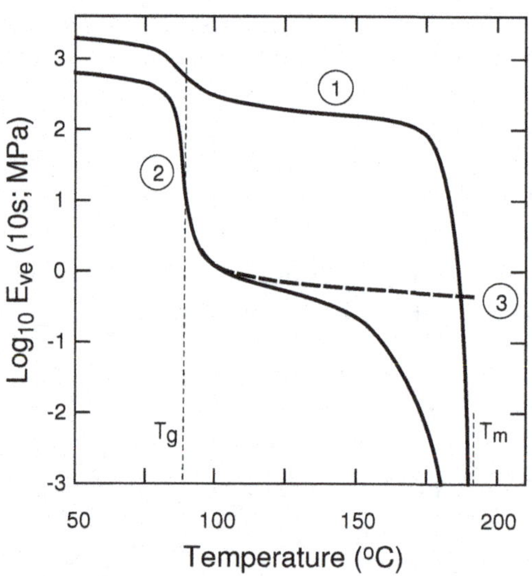

Figure 5.16: Dependence of $E_{ve}(10)$ on temperature.

In all three curves in Fig. 5.16 three temperature ranges can be identified: below T_g, between roughly T_g and T_m, and very near T_m. Below T_g, the materials are stiff and brittle. T_g is called the **glass transition temperature** (see Chapter 7 for more details), although T_g really denotes a temperature range over which the material becomes much softer and weaker but also tougher. T_m is the **melting temperature**, at which the material turns into liquid.

A brief explanation for the phenomena shown in Fig. 5.16 follows from Fig. 5.9. The key point is that many polymers are essentially composites of hard, strong crystalline regions embedded in a much softer, weaker amorphous matrix. (Some polymers may be entirely amorphous).

The drop-off in E_{ve} at around T_g is due to the temperature becoming high enough to activate the relative motion of polymer molecular chains in the amorphous parts of the material (see Fig. 5.9). On the other hand, the crystalline parts remain structurally unaffected up to T_g and beyond. The difference between curves 1 and 2 in Fig. 5.16 arises from the different amounts of crystalline material contained in the materials (say 80% vs. 0%).

At T_m the crystalline parts melt abruptly, while the amorphous parts approach the melted liquid state gradually as T approaches T_m. For cross-linked materials (curve 3 in Fig. 5.16), there is generally no real melting observed because the cross-links limit the relative mobility of the polymer chains. Instead, the material decomposes at a high enough temperature.

5.5 Atomic Models of Mechanical Behavior

In this section we will explore how much of the mechanical behavior of materials we can interpret using our knowledge of the structure and interatomic forces in solids. We will focus on elastic and inelastic deformation of crystalline solids, in particular metals, but will outline the extension of these arguments to ceramics as well. At the end of this section, we will comment briefly on atomic processes in polymers. As always, these will be rather different than for crystalline materials.

5.5.1 Elastic Deformation: Mainly Metals

The principal definition of elastic deformation is that it must be recovered fully when the applied stress is removed. As a secondary characteristic we noted that elastic deformation usually, but not always, goes along with a linear relationship between the applied stress σ and the resulting strain ϵ.

For deformation to be elastic, it seems evident that it must involve atom-atom bonds that are only stretched but not broken. If breaking of bonds occurred on a large scale, there would be no reason to expect that all the broken bonds would be reformed as in the original state once the load is removed.

Fig. 5.17 illustrates this situation with a schematic drawing of a crystalline rod subject to a tensile stress. Clearly, the forces between neighboring atoms resist the deformation and will restore the original shape when the load is removed.

Although this picture appears to be qualitatively reasonable, it is worth asking whether it is in agreement with the notions put forth in Chapter 2 on atom-atom bonding in solids (see Figs. 2.8-2.10). For the sake of this discussion I have redrawn Fig. 2.8 as the new Fig. 5.18 below.

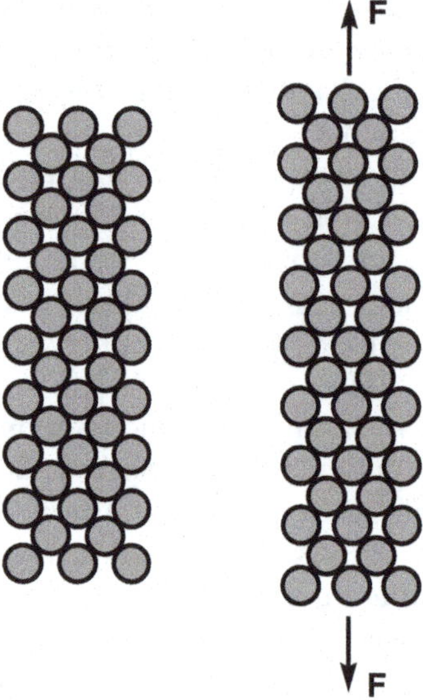

Figure 5.17: Atomic model of elastic deformation. Crystalline rod of FCC metal subject to load F in z-direction. Note axial stretching and lateral contraction.

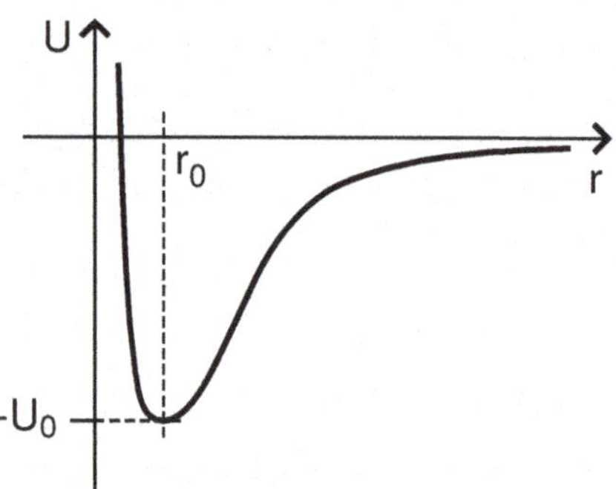

Figure 5.18: Potential energy $U(r)$ as a function of interatomic distance r.

This time we are using $U(r)$ for the potential energy because E is the elastic modulus. $U_0 = U(r = r_0)$ is a measure of the interatomic bond energy and r_0 represents the equilibrium atom-atom separation. You also recall that $F(r) = \frac{dU}{dr}$ is the force on an atom, and the slope of $F(r)$ at r_0 is correlated with the stiffness, or what we now call the elastic modulus. Moreover, the slope of $F(r)$ at r_0 is essentially the curvature of the $U(r)$ curve at the minimum.

When you examine the properties of metals across the periodic table, you will make the interesting observation that the equilibrium distances r_0 (or equivalently the lattice constants a) vary by less than a factor 1.2. In fact, for the common cubic metals, it would be an excellent approximation to say that $a = 0.35 \pm 0.06$ nm. So if you plotted $U(r)$ curves for different metals on the same graph, they would lie practically on top of each other and vary mostly with respect to their depth rather than their width or their position of r_0.

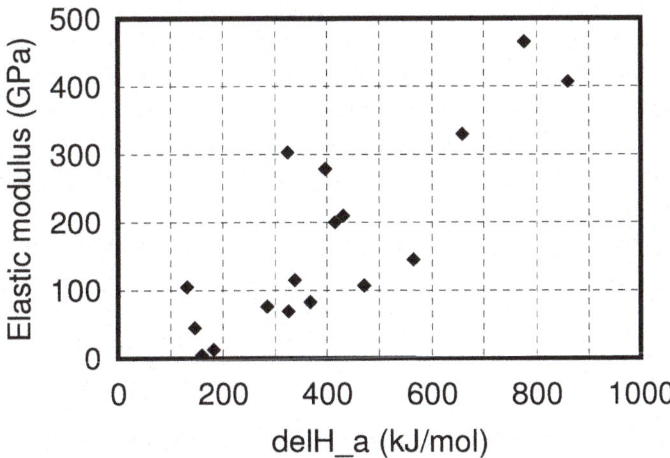

Figure 5.19: Plot of elastic modulus E versus heat of atomization ΔH_a for a sample of mostlv common metals.

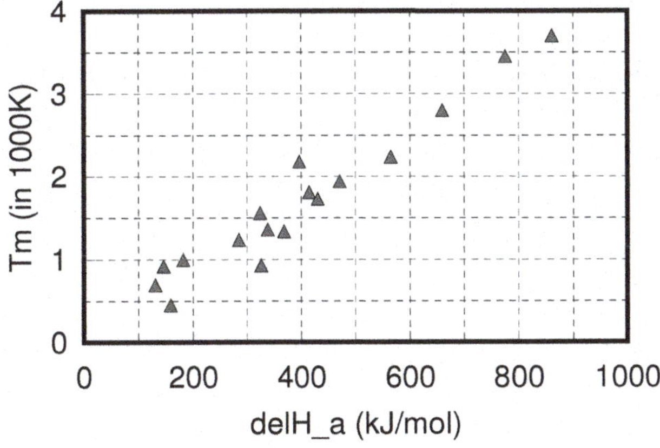

Figure 5.20: Plot of melting temperature T_m versus heat of atomization ΔH_a for the same metals as in Fig. 5.19.

These observations suggest that the elastic modulus of a metal should be correlated with the depth U_0 of the $U(r)$ curve. Now, what experimental quantity should be used for U_0? Remember that $U(r)$ is a short-hand description of what really goes in a solid. $U(r)$ in a sense represents the interactions between a certain given atom and all its neighbors as if it were just between two atoms. Therefore, let us choose the so-called heat of atomization ΔH_a as the measure of U_0. ΔH_a is a thermodynamic quantity and designates the energy per atom it takes to atomize the solid, i.e. to separate all the atoms completely.

In Fig. 5.19, the elastic moduli of a number of metals are plotted as a function of the heat of atomization ΔH_a. Being thermodynamic quantities, the values for ΔH_a are listed per mole. The metals used for the plot are, in order of increasing E (from 5 to 466 GPa): Li, Ba, Mg, Al, Ag, Au, Zn, Ti, Cu, Pt, Fe, Ni, Cr, Be, Mo, W , Re. This list should represent a good sample, as it contains about equal numbers of BCC, FCC, and HCP metals. You will agree that the plot shows a clear correlation between E and ΔH_a, although there is also a fair amount of scatter between the data points.

An even better correlation exists between the melting temperature T_m and ΔH_a, as you can see in Fig. 5.20. Even though in the molten state atoms are not separated much farther than in the crystalline solid state, the quantity ΔH_a, as an average atom-atom bonding energy, is apparently an excellent predictor of T_m. It follows from these two plots that a very good correlation also exists between T_m and E.

An explicit function with the general form of $U(r)$ in Fig. 5.18 is often used to model the interactions between atoms in a solid in general, i.e. between a specific atom and *all* its neighbors. The function is known as the Lennard-Jones potential. Applied to our situation, it can be written as

$$U(r) = U_0 \, [\, (\frac{r_0}{r})^{12} - 2 \, (\frac{r_0}{r})^{6} \,] \tag{5.21}$$

You can easily convince yourself that it has a minimum of $-U_0$ at $r = r_0$ by taking the first derivative. The attractive part is negative and there is some physical justification for it having the form r^{-6}, but the r^{-12} term is just a mathematically convenient form for the steep repulsive part.

Fig. 5.21 displays the Lennard-Jones potential in a reduced form, $u(x)$, together with the associated reduced force $f(x)$, by letting $U_0 = 1$ and $x = r/r_0$. That is, the equations plotted are:

$$u(x) = \frac{1}{x^{12}} - \frac{2}{x^{6}} \tag{5.22}$$

$$f(x) = \frac{du}{dx} = -\frac{12}{x^{13}} + \frac{12}{x^{7}} \tag{5.23}$$

The main thing you should take away from this graph is that the potential is very steep indeed in its repulsive part, and that on the attractive side it quickly approaches the value 0. You can also see that $U(r)$ is clearly asymmetric near its minimum.

If we accept this potential energy function as reasonable, then it is straightforward to derive a mathematical expression for the elastic modulus from it.

Have another look at Fig. 5.17: Assume that at the ends where the force F is applied, the cross-sectional area per atom is A_a. Now, let the force per atom F_a be given by the derivative of $U(r)$ in Eq. 5.20. Then by the definition of F_a, Hooke's Law can be written as:

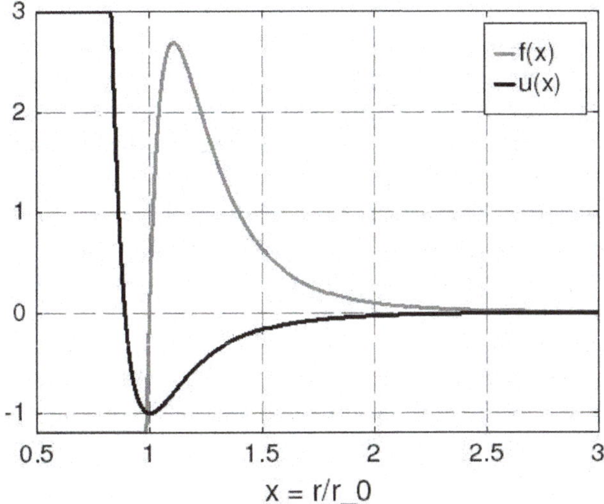

Figure 5.21: Plots of Lennard-Jones potential energy (black curve) and associated atom-atom force (light gray curve).

$$\sigma = \frac{F_a}{A_a} = \frac{\frac{dU}{dr}}{A_a} = E\,\frac{\Delta r}{r_0} \tag{5.24}$$

Think of $\frac{\Delta r}{r_0}$ as the strain of an atom-atom bond and Eq. 5.24 as Hooke's Law for an atom-atom bond.

From here it is just a matter of taking the derivative of Eq. 5.21 twice and doing some algebra to arrive at an expression for the *macroscopic* quantity E in terms of *atomic* parameters. First, solve Eq. 5.24 for E:

$$E = \frac{1}{A_a}\frac{F_a}{\frac{\Delta r}{r_0}} = \frac{r_0}{A_a}\frac{dF_a}{dr}\bigg|_{r=r_0} \tag{5.25}$$

This can be evaluated explicitly using $\frac{dF_a}{dr} = \frac{d^2U}{dr^2}$, with the result

$$E = 72\,\frac{U_0}{r_0\,A_a} \tag{5.26}$$

At the simplest level, Eq. 5.26 says that the elastic modulus E should be proportional to the atom-atom bond energy U_0. This is, of course, exactly what the data demonstrated in Fig. 5.19 above. To what extent Eq. 5.26 is quantitatively accurate will be explored in the homework.

5.5.2 Plastic Deformation in Metals

Whereas in the previous section it appeared that our simple atomic solid model gave a reasonable description of elastic deformation, it may likely be a different matter with inelastic, or plastic, deformation. Remember: The latter involves large-scale, permanent deformation with the breaking and remaking of atom-atom bonds, not just the stretching of them.

Moreover, when you revisit Fig. 5.17, it is hard to see how, under the influence of a tensile stress, bonds can be broken and then quickly reformed. It would seem more plausible to expect that once the stress is large enough, the sample will simply break straight across.

A clue to what is really going on can be gleaned from what happens to the shape of a tensile test sample beyond T.S. (Fig. 5.2). When a neck is formed, clearly this must involve shear forces inside the sample, although at the ends it is loaded in tension. But where do such shear forces come from?

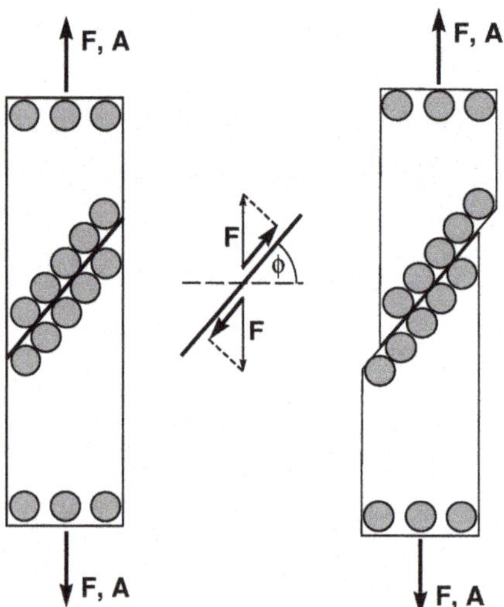

Figure 5.22: Tensile test sample with resolved shear stress along a plane not normal to the externally applied force F.

You should realize that Fig. 5.17, with F applied perpendicularly to the end faces, which themselves are perpendicular to the sample axis, suggests a geometry that looks simpler than it really is. In Fig. 5.22, with the same sample as Fig. 5.17, two adjacent internal crystal planes are drawn which are at an angle other than 90° relative to the axis. Now it is obvious that F has a component parallel to the inclined planes, and hence these planes experience a shear stress.

Furthermore, this internal shear stress may be able to cause a deformation as indicated by the graph on the right of Fig. 5.22, where a certain plane has slipped by one atomic step over the same type of plane underneath it. The result would be a permanent deformation that came about by a minimal disruption of the crystal lattice, i.e. by a gentle breaking, or perhaps just loosening, and then reforming of atomic bonds.

The internal shear stress produced by the external tensile force F is called **resolved shear stress** τ_r and is defined by

$$\tau_r = \frac{\text{resolved force}}{\text{inclined area}} \tag{5.27}$$

From Fig. 5.22, τ_r can be expressed in terms of the applied stress $\sigma = F/A$ as

$$\tau_r = \frac{F\sin(\phi)}{A/\cos(\phi)} = \sigma\sin(\phi)\cos(\phi) = \frac{1}{2}\sigma\sin(2\phi) \tag{5.28}$$

where I used the trigonometric identity $2\sin(\phi)\cos(\phi) = \sin(2\phi)$. You can see from this equation that τ_r reaches its maximum value of 0.5σ for $\phi = 45°$.

The process illustrated in the right-hand graph of Fig. 5.22, with an entire crystal plane slipping over another one, is called **homogeneous slip**. The question is whether this is really the process giving rise to inelastic deformation of metals in general. For an answer to this question, we will examine again a simple atomic model to see whether it reproduces the correct inelastic behavior, especially the key parameter of the yield strength Y.S.

Plastic Deformation by Homogeneous Slip

The model is illustrated in Fig. 5.23. In the top part of the figure, you are looking in the y-axis direction at the side of an FCC crystal subject to a shear stress τ in the x-direction. The potential energy $U(x)$ is what the atoms in the planes on top see when being displaced laterally by the shear force. U_0 is the bond energy as before, but in the lateral direction $U(x)$ is now a periodic function with a period equal to the lattice constant a.

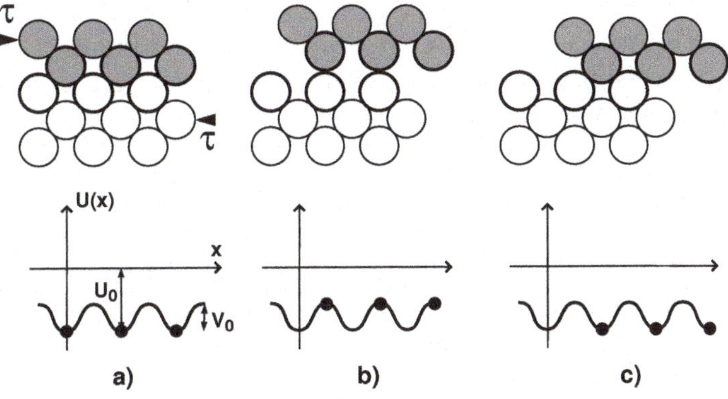

Figure 5.23: Atomic model for homogeneous shear.

We want to estimate the shear stress it will take for the lowest of the gray planes of atoms to overcome the local potential maximum V_0 in order to be able to slip from one lattice position to the next. For a quantitative analysis, let us focus in on the bottom part of Fig. 5.23:

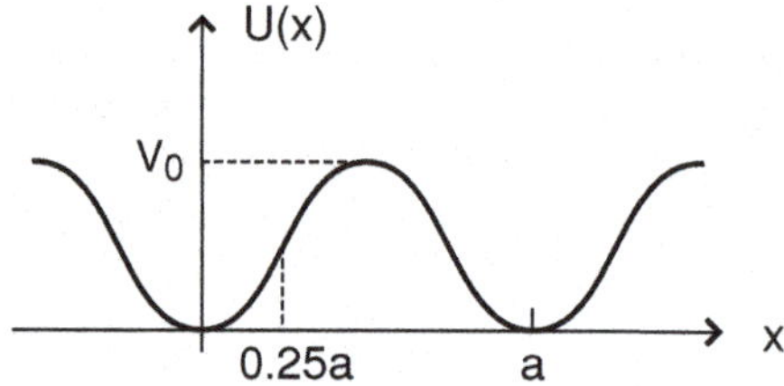

Figure 5.24: Local potential for lateral motion of atom planes.

We do not know the exact shape of $U(x)$ nor the value of V_0, only that $U(x)$ repeats itself over a lattice spacing a. To make things simple, we assume that $U(x)$ consists of parabola pieces, pointing up and down. The condition for slip to occur is that the shear force move the atoms along $U(x)$ up to the inflection point at $x = 0.25a$, where the force needed to push against $U(x)$ is largest.

With these assumptions we can write for $U(x)$ near $x = 0$:

$$U(x) = \frac{8V_0}{a^2} x^2 \tag{5.29}$$

You can check that this makes $U(x = 0.25a) = 0.5V_0$. The force $F(x)$ is the derivative of $U(x)$ as usual, so that for the maximum force $F_{max}(x) = F(x = 0.25a)$ we obtain

$$F_{max} = \frac{4V_0}{a} \tag{5.30}$$

A sensible guess for V_0 would seem to be $V_0 = 0.25U_0$. Now all that is left to do is to convert the force F_{max} into a shear stress τ. We accomplish this by dividing the force by the area per atom, which we take to be $0.5a^2$, for a (100) plane of an FCC crystal (this value will depend on the kind of plane we are looking at). The final result for the maximum stress τ represents the yield strength Y.S. in our model:

$$\text{Y.S.} = \tau_{max} = \frac{F_{max}}{0.5a^2} = \frac{2U_0}{a^3} \tag{5.31}$$

This is a nice, compact result that can be checked numerically. Again let us identify U_0 with the heat of atomization, ΔH_a. As an example, for Al ΔH_a = 326 kJ/mol and a = 0.405 nm. Then

$$\text{Y.S.(Al)} = \frac{2 \times 326 \times 10^3}{6 \times 10^{23} \times (0.405 \times 10^{-9})^3} = 18\,\text{GPa} \tag{5.32}$$

A typical experimental value for Y.S. of an Al alloy is about 250 MPa (Table 5.1); for pure Al it would be lower. We are forced to conclude that our model predicts a yield strength which is too high by about two orders of magnitude! The corollary is that homogeneous slip cannot be the mechanism by which metals yield and deform inelastically.

Plastic Deformation by Dislocation Motion

The correct mechanism for inelastic deformation turns out not to involve homogeneous slip, but slip by the motion of dislocations. To see how this comes about, consider the following model:

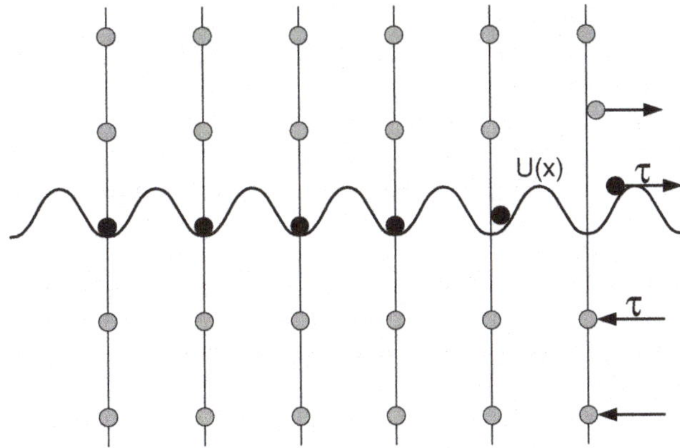

Figure 5.25: Lattice model with applied shear stress.

Imagine that the applied shear stress τ pulls the row of black atoms out of their equilibrium positions (Fig. 5.25). Some of this deformation is felt also in a few gray rows above the black row, as shown. The potential energy curve $U(x)$ is the same as in Fig. 5.23, indicating the potential set up by the gray atoms below for the black atoms, as these are forced to move laterally.

For the further discussion, we will consider just the black atoms moving in the potential $U(x)$ under the shear stress τ, and we will assume that the interaction of the black atoms with the gray atoms above them can be represented by spring forces between the black atoms (Fig. 5.26).

As we increase τ gradually, there will come a point when the first black atom on the right will go over the potential hump, thereby leaving the potential well it came from originally. At the same time, the second black atom is pulled out of its equilibrium position and moves up its potential well. In a way, the second row $U(x)$ in Fig. 5.26 displays a temporary "equilibrium" for the black atoms while the shear stress is on.

Now you can see how the situation will proceed: With just a little more τ the second black atom flips over and pulls up the third black atom, and so forth. So with only a limited amount of shear stress, and with only a couple of atoms

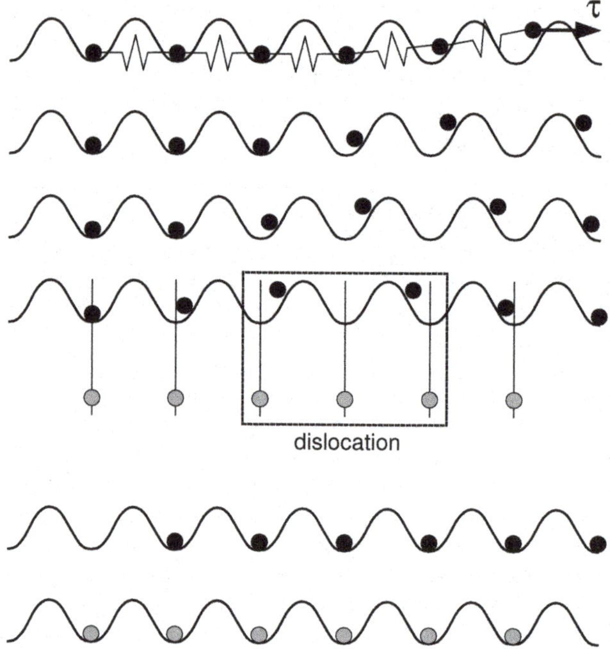

Figure 5.26: Lattice disturbance moving across the crystal due to the applied shear stress.

displaced at a time, the initial disturbance in the black row of atoms moves left, all across the sample. The bottom two panels in Fig. 5.26 show the end result: The ***entire*** top part of the crystal has been moved to the right by one lattice spacing. There is your permanent deformation!

If you focus on the little frame in Fig. 5.26, you will realize that you have seen this so-called ***disturbance*** before: It is an ***edge dislocation***! It looks like an extra half-plane of gray atoms at the bottom inserted between the black atoms on top.

It seems evident that this process of moving a dislocation across the sample should require much less shear stress than homogeneous slip. In the case of homogeneous slip, all atoms must move up their local potential energy wells at the same time, whereas with a dislocation only a few atoms have to do so at a time. You could even say that what really moves in the case of a dislocation is the empty potential well.

Keeping in mind the usual caveat with respect to extending two-dimensional models to real three-dimensional solids, we can conclude from this discussion that the way to envision plastic deformation is that, first, shear stress will create a dislocation, and second, the shear stress will cause the dislocation to move across the material. This type of dislocation-assisted slip can proceed at a much lower shear stress than would be required for homogeneous slip. Furthermore, even if the externally applied stress is tensile, shear stress is present along any internal crystal plane not normal to the tensile stress.

If you accept this overall picture, then it follows that the easiest way for dislocations to move is along close-packed directions in a close-packed plane. The reason is that these types of directions are smoothest, and thus have the lowest potential well V_0 to overcome (Figs. 5.23 and 5.24). Remember that the close-packed directions are those along which atoms touch. In the cubic system these are the following:

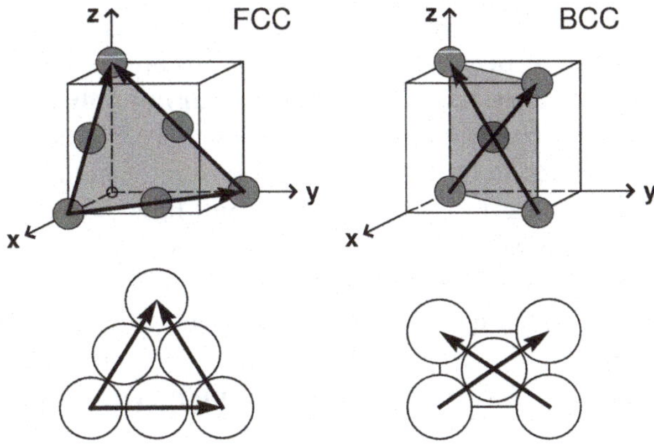

Figure 5.27: Planes and directions along which dislocations will be most favored to move.

In FCC crystals, the close-packed planes are like the (1 1 1) plane shown in Fig. 5.27, with the three slip directions $[\bar{1}\,0\,1]$, $[0\,\bar{1}\,1]$, and $[\bar{1}\,1\,0]$. (Make sure to refresh your memory with respect to crystallographic conventions). A close-packed plane together with a close-packed direction is called a **slip system**. In an FCC crystal there are four independent close-packed planes: (1 1 1), $(\bar{1}\,1\,1)$, $(\bar{1}\,\bar{1}\,1)$, and $(1\,\bar{1}\,1)$. Each has three slip directions like the ones shown in Fig. 5.27. Hence we can say that FCC crystals have 12 slip systems designated by the families of slip planes and directions involved as $\{1\,1\,1\}\,\langle 1\,1\,0\rangle$.

In BCC crystals, the close-packed planes are of the $(\bar{1}\,1\,0)$ type (cf. Fig. 5.27) with two slip directions, $[\bar{1}\,\bar{1}\,1]$ and [1 1 1]. There are a total of six independent planes like $(\bar{1}\,1\,0)$ in a cube, each with two slip directions. Hence we can say that BCC crystals also have 12 systems. (See Table 5.4 and Problem 5.11 for a slight modification to this last statement).

On the other hand, you can convince yourself easily that HCP crystals have only three slip systems, because the basal plane is the only close-packed plane (with hexagonal symmetry). Our conclusions about slip systems are summarized in Table 5.4 below.

Table 5.4: Crystal Types and Slip Systems

Crystal type	Nr. of slip systems
FCC	12
BCC	> 12
HCP	3

The table sums up neatly why certain metals are ductile and others are brittle. The ductile ones are FCC and BCC, and the brittle ones are HCP. Ductility requires a large number of slip systems. BCC is listed as having more than 12 slip systems in Table 5.4. because in BCC, the $[\bar{1}\,\bar{1}\,1]$ and [1 1 1] type crystal directions, along which atoms touch each other, occur not only in the family {110} of planes, but also in some other low-index planes (see Problem 5.11).

To conclude this section on inelastic deformation of metals, you should be aware that most of our arguments up to this point have been based on considering a 2D model of one individual crystal. However, as you know, engineering materials are polycrystalline in 3D: Each grain is its own little crystal, but the orientation of these grains is pretty much random. If shear stress appears in such a material, you should envision dislocations being created, and slip occurring, in each grain. The deformation of the entire material is a grand average over what happens in all the individual grains.

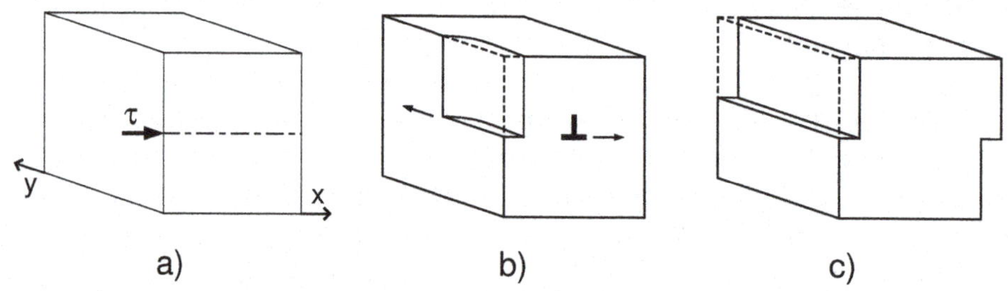

Figure 5.28: Motion of a mixed dislocation across a grain.

We have also said very little about the dynamics of dislocations, except to point out that they can move fairly easily at a suitable shear stress. The motion of dislocations can become very complex in three dimensions, especially when not just one, but many dislocations move at the same time and interact with each other. For a simple illustration of dislocation motion in 3D, examine Fig. 5.28. If a shear stress τ is applied locally, in Panel a), the entire plane of atoms perpendicular to the x-axis may not move together. This causes the dislocation to be mixed: It has edge character in the x-direction, but looks like a screw dislocation in the y-direction. As this structure moves under the influence of τ, an arc in the disturbed lattice sweeps across the sample until, in Panel c), the same final configuration is reached as in Fig. 5.26.

5.5.3 Elastic and Plastic Deformation in Ceramics

The picture we envisioned above for elastic deformation of metals carries over directly to ceramics. Again, elastic deformation means recoverable deformation after removal of the load. This comes about because in the elastic regime atom-atoms bonds, or in this case ion-ion bonds, are only stretched, not broken.

Our arguments about ductility and slip systems also extend directly to ceramics. Basically, ceramics are brittle because they have few slip systems. Given the complexity of ceramic crystal structures, we will not look into their slip systems in any detail. However, a figure will illustrate the issue.

Not only are there few slip systems in ceramics, but it would also be very difficult to create dislocations. Once again, this is because of the additional constraints brought about by the mutual repulsion of like ions.

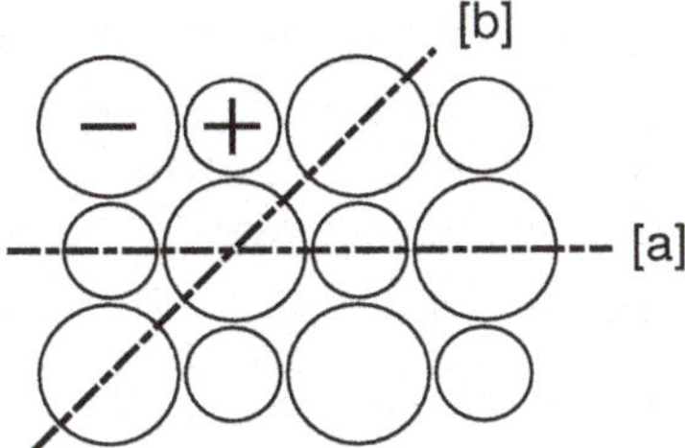

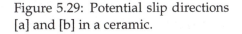

Figure 5.29: Potential slip directions [a] and [b] in a ceramic.

You can see in Fig. 5.29 that direction [a] is not a feasible slip direction. If you were to move the upper part of the crystal along direction [a], ions with like charge would come in contact, and that is not allowed. On the other hand, direction [b] would avoid that problem. The figure should also make it clear why creating dislocations in a ceramic would be difficult. For example, if you try to imagine an edge dislocation, where you need to insert an extra half-plane of ions into the existing crystal, it is hard to see how this could be accomplished without having like ions in conflict with each other.

5.5.4 Elastic and Plastic Deformation in Polymers

The mechanisms giving rise to plastic deformation in polymers are different than in crystalline metals, since a polymer is at best partially crystalline. In that case, the applied stress in effect operates on a composite material consisting of two components with different mechanical characteristics. The crystalline parts will be stronger, harder, and have a higher elastic modulus, than the amorphous parts.

If the polymer is entirely amorphous, then its properties depend on the extent to which it is cross-linked. The degree of cross-linking can range from none at all to highly cross-linked (meaning that the number of cross-links is of the same order as the number of mers in a polymer chain).

If a small enough stress is applied to a semi-crystalline polymer, such that the deformation is elastic, it is primarily the amorphous parts that are strained. Some uncoiling of the highly coiled parts of the polymer chain molecules between the crystals takes place, to the extent it is allowed by the surrounding chains. This uncoiling involves mostly rotation of -C-C- bonds and is reversible upon release of the stress.

At a high enough stress, chains begin to slide by each other and inelastic deformation sets in. Beyond the yield strength Y.S. (curves 2 and 3 in Fig. 5.12), one can say that the material has begun to flow. Eventually, this will result in some alignment of the crystalline regions in the direction of the applied stress (see Fig. 5.9) and even movement of parts of chains relative to each other within the crystals.

In an entirely amorphous polymer, the motion of individual chains relative to each other is constrained by the cross-links, if there are any. If the density of cross-links is high, virtually no such motion will be allowed (Fig. 5.11c). In that case the material is a rigid amorphous network, something we know from the inorganic glasses. It will be brittle, i.e. elastic up to the point of fracture.

A particularly interesting case is the one with a low enough density of cross-links so that it allows for substantial uncoiling/extension of the coiled chain molecules without chain sliding (Fig. 5.11b). This is the principle behind elastomeric, rubber-like materials. They can undergo very large strains (several 100%) which are entirely elastic (curve 4 in Fig. 5.12). Note that even though the stress-strain curves are non-linear, the large strains are reversible.

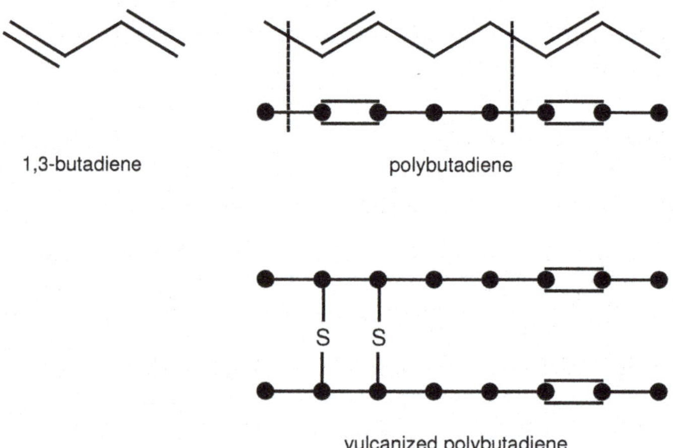

Figure 5.30: Vulcanization of polybutadiene by cross-linking with sulfur to form polybutadiene rubber . Black circles represent C atoms. The vertical dotted lines indicate the mer.

Low density cross-linking has been implemented in a process know as vulcanization of **rubber**. The principle is illustrated schematically in Fig. 5.30. The figure shows on the upper left a side view of the monomer 1,3-butadiene. Only bonds between C atoms are displayed. You have to complete the picture with the necessary C-H bonds: 2 at each of the two end C atoms, and 1 at each of the two inner C atoms. One of the double bonds per monomer is used to form a polymer chain molecule. The panel on the upper right shows a side and a top view of a small section of such a chain.

In the vulcanization process (patented by Charles Goodyear in 1844) sulfur is used to form a pair of cross-links at about every tenth remaining double bond. The unvulcanized polybutadiene is a rather soft material at room temperature and would not be stable enough for use in inflated car tires. The cross-linking provides the material with strength and stability as well as large, elastically recoverable strain. The cross-links allow for uncoiling and extension of polymer chains while keeping them from sliding past each other.

Real materials used as rubbers are typically more complex than the example based on butadiene from above. For instance, natural rubber is polyisoprene, a polymer of isoprene (2-methyl-1,3-butadiene). In it one of the H atoms attached to an inner C atom of butadiene (Fig. 5.30) is replaced with a CH_3 group. In neoprene (polychloroprene), a common synthetic elastomeric material, a Cl atom is used in place of the CH_3 group (Fig. 5.31).

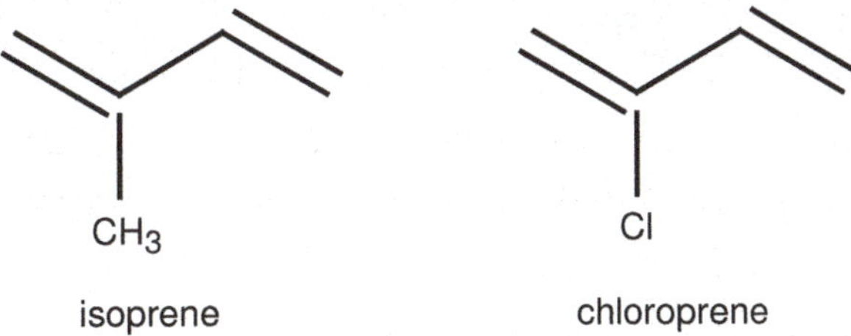

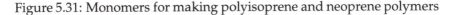

Figure 5.31: Monomers for making polyisoprene and neoprene polymers

The mechanical properties of an elastomer, of course, depend on the number of cross-links and on the type of side group. The larger the density of cross-links, the stronger and harder the material will be.

Let me emphasize again that Fig. 5.30 shows an idealized, linear geometry of the two polymer chains. At every C atom with two single bonds to neighboring C atoms, a chain can rotate about a -C-C- axis as explained in Fig. 3.32. On the other hand a C=C double bond is rigid and thus provides some stiffness to the backbone of the molecule.

5.6 Fracture

Fracture represents the last stage of mechanical deformation, the point at which the material looses its integrity and breaks apart. As with the other mechanical phenomena, fracture depends mostly on whether a material is ductile or brittle, and again polymers are in a class by themselves.

5.6.1 Ductile Fracture

In a tensile test, fracture of a ductile metal occurs after considerable plastic deformation has taken place in the material and, specifically, in the region where a neck has formed. The fractured sample of a tensile test has the typical shape shown in the next figure.

This sample geometry is referred as cup-and-cone, for obvious reasons. The necked region, and the break within it, are apparent. Also, in the central part of the break you can see very extensive plastic deformation.

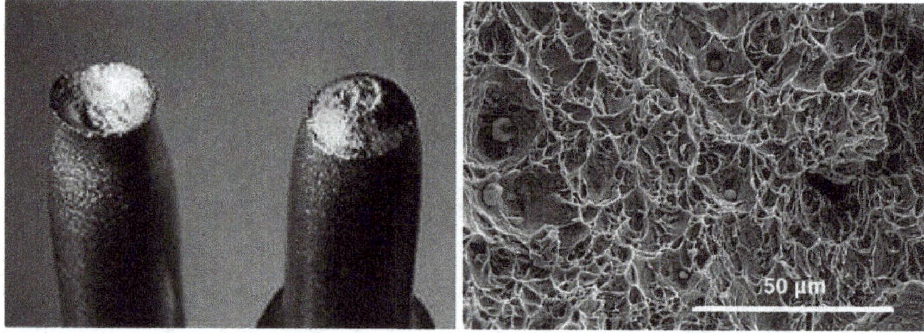

Figure 5.32: Left picture: Cup-and-cone geometry of a fractured ductile metal. Right picture: Central part of the fracture surface of a ductile metal sample.

A magnified view of this central part (from a different sample) is displayed in the scanning electron micrograph on the right in Fig. 5.32. Note the scale bar of 50 μm at the bottom. This pictures indicates the presence of numerous voids, some fairly large. Hence it appears that ductile fracture originates around small voids which, once formed, lead to locally increased stresses that eventually cause voids to coalesce and the remaining solid to be pulled apart.

The other interesting aspect of the fractured pieces in Fig. 5.32 is that the sidewalls of both the outside of the cone and the inside of the cup are at an angle that looks like it is close to 45°. This is exactly what we would expect to happen due to the maximum resolved shear stress at that angle.

5.6.2 Brittle Fracture

Brittle fracture is in a way worse than ductile fracture because it occurs abruptly and with no forewarning from plastic deformation. In a tensile test or a four-point bending test, it typically produces a break straight across the sample with a surface which is much smoother than in ductile fracture.

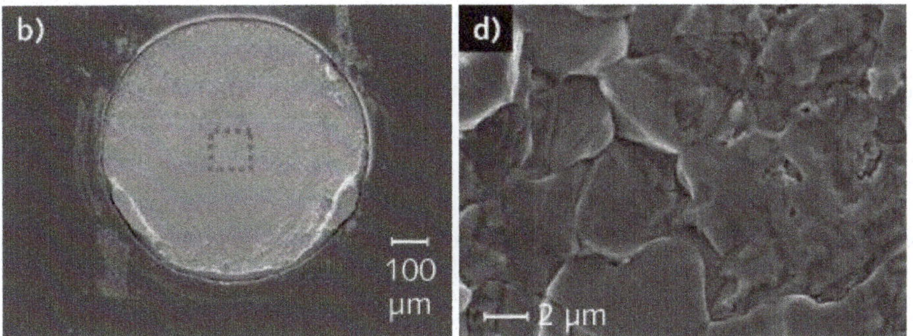

Figure 5.33: Two views at different magnification of a soldering pad from which a soldered-on ball has been pulled off.

Fig. 5.33 shows an example of brittle metal fracture from the world of electronics. It is from tests of lead-free solder with which two components had been soldered together. One component has an array of small metal balls on it, which are attached to an array of matching metal pads on the other component. At the lower magnification, the fractured surface looks very smooth. In the marked central area, at the higher magnification, individual grains of the solder are visible.

Theoretical Tensile Strength

We will pursue an approach here which is similar to the one with plastic deformation of metals. We will estimate the tensile, i.e. fracture, strength of an ideal brittle material based on the atom-atom interaction potential in Fig. 5.18 and its representation in Eqs. 5.21 to 5.23. This model implies that fracture involves breaking atom-atom bonds simultaneously across the entire sample.

Looking at $f(x)$ in Eq. 5.23, we assume that fracture occurs at the interatomic distance x_f where $f(x)$ reaches its maximum. Remember that $f(x)$ is a measure of the tensile stress applied to the sample. Therefore, taking the derivative of Eq. 5.23 we obtain

$$\frac{df}{dx} = -12\,(-\frac{13}{x^{14}}) + (\frac{7}{x^{8}}) \tag{5.33}$$

This expression will be equal to 0 at $x = x_f$ if $7x_f^6 = 13$, that is if

$$x_f^6 = \frac{13}{7} = 1.857 \quad \text{or if} \quad x_f = 1.108683 \tag{5.34}$$

At that point the force $f(x_f)$ is equal to

$$f(x_f) = -12 \times (\frac{1}{x_f^{13}} - \frac{1}{x_f^{7}}) = 2.7 \tag{5.35}$$

Eq. 5.35 denotes the reduced force. The real force $F(x = r/r_0)$ is obtained as $F = f/r_0$, and the stress σ is equal to F/A_0 as before. Hence, the final result for the tensile strength according to the Lennard=Jones potential is:

$$T.S. = 2.7 \frac{U_0}{r_0 A_0} \tag{5.36}$$

If you compare this result to the one for the elastic modulus E in Eq. 5.26, you see that the model predicts

$$T.S. = E \frac{2.7}{72} = 0.037 E \tag{5.37}$$

Earlier, we left the question open whether the model predicts E quantitatively (It does not, and you will have figured that out doing Homework Problem 5.7). Here, comparing Eq. 5.37 to the data for the brittle materials in Table 5.1., it is evident that the model overestimates T.S. by about a factor 40. The data indicate that, for the brittle materials, typically, $T.S. \approx 10^{-3}E$.

Moreover, from Hooke's Law applied to a brittle material, it follows by the same argument that at fracture $\epsilon_f \approx 10^{-3}$, which is much less than the model prediction of 0.109 (see Eq. 5.34). This should not surprise you because brittle materials exhibit elastic deformation right up to the point of fracture, with an essentially constant elastic modulus E, whereas according to the model E decreases continually up to fracture, and in fact is equal to 0 at $x = x_f$ or, equivalently, at $\epsilon = \epsilon_f$. In addition, a strain of 0.109 would be unrealistically high for a ceramic material.

Griffith Theory of Brittle Fracture

The English engineer A. A. Griffith developed a different approach to the fracture of brittle materials in the early 1920's. He observed the behavior of glass fibers and noted that

- the tensile strength of freshly drawn glass fibers was much greater than that of fibers which had been exposed to the environment for a while;
- the tensile strength of fibers decreased with increasing length;
- there was a great deal of variability between fibers of nominally the same material.

Griffith attributed these observations to the fact that glass fibers contained flaws of various sizes, and that these flaws were responsible for the tensile strength being much lower than the ideal value.

You may be aware of this issue from your own practical experience, if you ever had to cut a glass tube to size in the chemistry laboratory, or cut a pane of glass for repair of a window. Of course, you cannot cut the glass, literally, but it is possible to break it in a controlled fashion by scoring it, i.e. by producing a small scratch on its surface. With enough stress applied to the material at the scratch in the right direction, and with some luck, it will break cleanly.

The scratch in the glass underscores the dual role of a flaw: It produces an initiating crack which concentrates the applied stress, and at sufficiently high stress it causes the crack to propagate across the material. This is the basic mechanism of brittle fracture envisioned by Griffith.

We investigate this process in two steps. First we will look at the case of an ideal brittle continuum material which, although being macroscopic, turns out to be related closely to the atom-atom bond breaking model from above. Then we will consider the role of flaws, especially thin cracks, on a stressed ceramic in some detail.

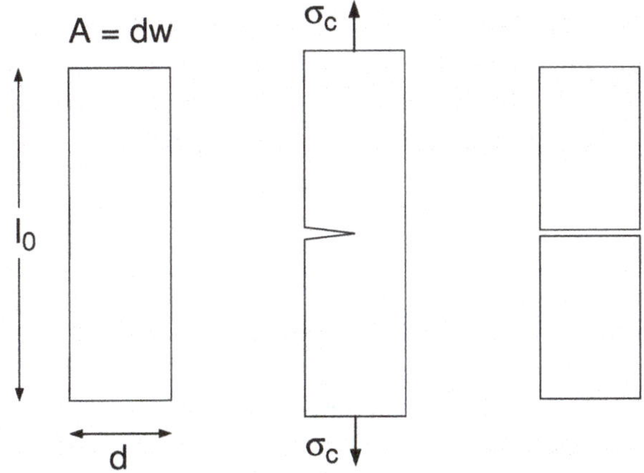

Figure 5.34: Schematic for ideal fracture of a brittle material.

Let us imagine first a flawless sample of material with initial length l_0 and cross-sectional area $A = dw$ where d and w are the thickness and width, respectively (Fig. 5.34). As we increase the applied tensile stress, there comes a point, at the critical stress σ_c, where a crack starts on one side and propagates rapidly across the sample. The end result is two pieces of material.

Griffith's point of view is to examine this situation in terms of energies involved. What happens when fracture occurs is that the elastic energy stored in the material is released, and thereby two smaller samples, *each with a new surface*, are created. Creating these two new surfaces costs energy, since atom-atom bonds have been broken and the atoms at the new surfaces form fewer bonds than they did before the fracture.

Let us look at the energy balance of this process in detail. We will assume that the $\sigma(\epsilon)$ curve for the material is linear up to the stress σ_c where fracture occurs. Now, using Eqs. 5.11 and 5.14 and Hooke's Law, the total elastic energy W_{tot} stored in the material at the point of fracture (which we called toughness earlier) can be written as

$$W_{tot} = A\, l_0 \int_0^{\epsilon_f} \sigma(\epsilon) d\epsilon = V_0 \int_0^{\epsilon_f} E\epsilon d\epsilon = \frac{V_0}{2} E\epsilon_f^2 = \frac{V_0}{2E} \sigma_c^2 \qquad (5.38)$$

where ϵ_f is the strain at which fracture occurs and we have used $V_0 = A\, l_0$.

The energy of the two newly created surfaces is written as

$$E_{surf} = 2\gamma_s A \qquad (5.39)$$

where γ_s is called specific surface energy. γ_s is a measure of the higher energy state the atoms at the new surfaces are in, due to the fact that some of their bonds have been broken and their coordination number is smaller than for bulk atoms.

Now we define W_f as the part of the total stored W_{tot} which is used for creating the two new surfaces. Most importantly, we assume that W_f is the energy stored in the single layer of atom-atom bonds broken when the material is fractured. That is, we assume, in analogy to Eq. 5.38:

$$W_f = A \frac{a}{2E} \sigma_c^2 \qquad (5.40)$$

where a is the interatomic distance in the broken layer. This is certainly debatable: Perhaps we should use $2a$ or $3a$. However, we shall see shortly that this assumption does not introduce an important limiting factor.

The threshold condition for σ_c, when the stored energy in the breaking layer is just enough to compensate for the two new surfaces, is now given by setting equal W_f and E_{surf}. From Eqs. 5.39 and 5.40

$$A \frac{a}{2E} \sigma_c^2 = 2\gamma_s A \qquad (5.41)$$

and solving for σ_c

$$\sigma_c = 2\sqrt{\frac{E\gamma_s}{a}} \qquad (5.42)$$

In Table 5.5 below is a brief list of experimental literature values for γ_s, E, and T.S., and a comparison between T.S. and σ_c calculated from Eq. 5.42. (I assumed $a = 0.3$ nm for the calculations).

Table 5.5: Comparison of Experimental and Calculated Values for the Tensile Strength of Select Ceramics

Material	γ_s (J/m^2)	E (GPa)	T.S. (MPa)	σ_c (GPa)
SiC	2.8	480	310	130
Al_2O_3	1.1	400	300	73
MgO	2.4	320	60	97
TiO_2	0.75	205	100	43
SiO_2	0.87	120	50	57

In interpreting these figures, keep in mind that experimental data for engineering ceramics are variable precisely because they depend on unspecified flaws in the material. So the data in Table 5.5 represent reasonable average values. In addition, data on T.S. are often inferred from bending tests rather than direct tensile tests. Still, in light of the fact that the T.S. figures are given in MPa and the σ_c figures in GPa, it is clear that our calculated values overestimate the real strengths by at least two orders of magnitude (cf. Table 5.1).

Thus we arrive at the conclusion that the assumption of all bonds in the fracture plane being broken at once must be in error. Moreover, both the present continuum model and the earlier, atomic Lennard-Jones model have pointed to this conclusion. Hence we must look for a different mechanism. (Remember also that we came to a similar conclusion earlier with regard to the mechanism of plastic deformation and slip).

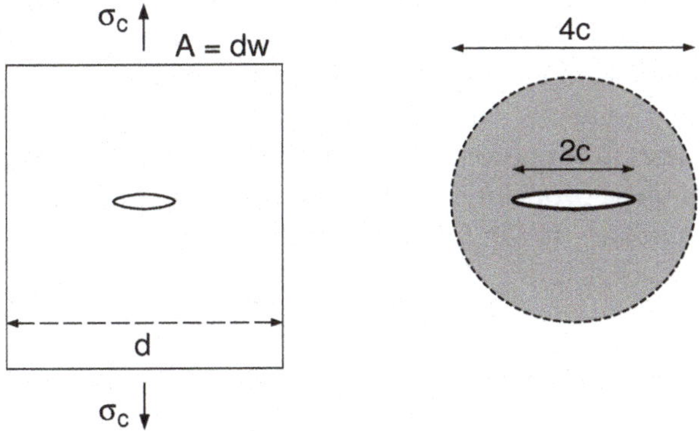

Figure 5.35: Ceramic sample with a sharp elliptical crack. The panel on the right shows an enlarged view, with the affected area around the crack indicated schematically in gray.

Let us now assume that the material in question is not perfect but has a small interior flaw in the form of a thin crack (Fig. 5.35). The shape of the crack is shown as elliptical for illustration purposes, but you should really think of it as a very thin "atomic cut" with atomically sharp corners. The total length of the crack is 2c, and we assume that it extends along the entire width w of the sample (i.e. the direction perpendicular to the plane of the figure).

The crack serves two roles. First, it acts as a locus of raised stress, especially at its corners, as one can show by some rather complex analysis of the stress distribution around such a crack. However, the critical thing is whether or not the applied stress is large enough to make the crack grow. If it is not, then nothing bad happens, yet. But if it is, the crack will propagate rapidly through the sample and lead to failure.

In what follows below, I will sketch a simplified version of Griffith's argument determining the condition for crack propagation. The idea here is to come up with the correct dependence of the phenomenon on experimentally relevant parameters. Griffith proposed that a crack will grow when the strain energy released during the process of its growth is at least as large as the energy required to produce new surface area for the crack.

If you look at Fig. 5.35, I hope you will agree that: 1) no stress can be transmitted across the crack, and therefore the areas just above and below the crack must have had their strain energy released, and 2) the stress field around the crack must vary with position, but this variation should be of limited range. That is, at a large enough distance, the presence of the crack will not be felt.

Now we will make our *main assumptions*: First, for the crack to grow, *the released energy from the volume affected by the crack has to be at least as large as the surface energy of the crack.* A related assumption has to do with how large that affected volume should be. We will make an educated engineering guess (Fig. 5.35): We take it as a cylinder with diameter 4c around the crack, i.e. with a diameter twice the length of the crack, and extending the width of the sample, as does the crack itself. We could have chosen another value, but 2c (really the minimum feasible) seems too small, and much larger than 4c looks too large. (This is the major part of our simplified argument following Griffith). The cylindrical area in question is marked in gray in the right panel of Fig. 5.35.

If you accept these propositions, the condition for crack growth can be written down very similarly to Eq. 5.41 as follows:

$$\pi\,(2c)^2 w\,\frac{{\sigma_c}^2}{2E} = 2\gamma_s\,2cw \tag{5.43}$$

Remember that in Eq. 5.41 the volume whose stored energy is released was Aa, but here we need to use the volume of the gray cylinder, $\pi\,(2c)^2 w$. Now we can solve Eq. 5.43 for ${\sigma_c}^2$:

$${\sigma_c}^2 = \frac{2\gamma_s E}{\pi c} \tag{5.44}$$

and hence the critical stress σ_c at which the crack continues to grow is

$$\sigma_c = \sqrt{\frac{2\gamma_s E}{\pi c}} \tag{5.45}$$

Eq. 5.45 is known as the Griffith equation for the propagation of a crack in a brittle material. It shows immediately the importance of cracks, or flaws, for the strength of a brittle material: The strength will be limited by the largest flaws. Moreover, since the size and size distribution of flaws may vary from sample to sample, depending on the processing, variations in the observed strength should be expected.

You should also note the similarity of Eqs. 5.42 and 5.45. The difference is that in Griffith's model, the distance determining the condition for fracture is not an atom-atom separation a but the length of a crack 2c. Since $2c \gg a$, the critical stress σ_c for a material with flaws will be much smaller than in the ideal case.

The **Griffith criterion** Eq. 5.45 is very important for the mechanical design with ceramic materials, but it is generally used in a somewhat different form. One can say that E and γ_s are inherent, primary properties of a given type of material regardless of any flaws. On the other hand, flaws and their sizes, including in particular the parameter 2c, are secondary in the sense that they depend on the processing of the material. Therefore, Eq. 5.45 can be written in a form where these two effects are separated:

$$\sigma_c \sqrt{\pi c} = \sqrt{2\gamma_s E} \tag{5.46}$$

or equivalently

$$\sigma_c \sqrt{\pi c} = K_{Ic} \tag{5.47}$$

The parameter K_{Ic} is called **fracture toughness** and represents the material-inherent information embodied by γ_s and E in the expression $\sqrt{2\gamma_s E}$.

K_{Ic} is pronounced as "kay one cee". The I in K_{Ic} refers to the type of loading of the sample and essentially denotes a wide sample (i.e. a sample with large w in Fig. 5.34) stressed so that the crack is pulled apart as in Fig. 5.35. Other modes of loading are possible, e.g. involving shear along a crack.

Furthermore, in mechanical design Eq. 5.47 is commonly used in a slightly modified form, namely with an additional dimensionless parameter Y:

$$K_{Ic} = Y\sigma_c \sqrt{\pi c} \tag{5.48}$$

The parameter Y takes into account empirically certain geometrical factors such as the shape of the sample. To a first approximation, for a wide sample one can take $Y = 1$. For a narrow sample of the same material, Y becomes larger than 1.

Values of K_{Ic} are tabulated or provided by the manufacturers of ceramic engineering materials for use by the mechanical designer. The role of the materials engineers is to make sure that the flaws in the material are as small as possible, and that the size of the largest flaws is known. With knowledge of K_{Ic} and c, Eq. 5.48 enables the determination of the critical tensile stress σ_c at which fracture should be expected to occur.

When dealing with K_{Ic}, make sure you use consistent units for c and σ_c, which really means S.I. units. From Eq. 5.47 you can see that if c is in m and σ_c is in N/m^2, then K_{Ic} would have to be in units of $N/m^{3/2}$, but it is given more commonly in MPa $m^{1/2}$.

5.6.3 Comments on Fracture

With the discussion above in mind, it is now apparent why brittle materials are much stronger in compression than in tension. It is because tension tends to promote the growth of cracks, whereas compression tends to prevent cracks from opening up further.

In the previous two sections, we separated materials neatly into two categories, ductile and brittle. Whereas this distinction is quite appropriate when it comes to ceramics, in other cases it is an idealization. This is true even for metals, and especially for polymers.

In the case of metals, ductility is really the critical concept. Even though we categorized metals as ductile or brittle, based upon their inelastic deformation behavior, i.e. based upon the number of slip systems, it would be better to consider them as more or less ductile.

The question then arises whether Griffith's point of view can be applied to metals as well, at least to those with high strength but low ductility (i.e. low toughness). The answer is yes, with some modifications. One of the key assumptions in the Griffith theory is that ***all*** of the energy stored around a crack is used for the growth of the crack and no plastic deformation takes place. For a material with some ductility, this cannot be true. The material will undergo some yielding, including around cracks, before fracture, which means some of the stored energy is converted into plastic deformation rather than crack propagation. As a result, a ductile material requires a larger amount of stored energy, hence a larger stress, for a crack to grow.

This effect can be captured in Griffith's formalism by modifying Eq. 5.46 as follows:

$$\sigma_c \sqrt{\pi c} = \sqrt{2(\gamma_s + \gamma_p)E} = K_{Ic} \tag{5.49}$$

where the additional parameter γ_p is formulated as if it were a second type of surface energy, but is really just the energy absorbed into plastic deformation localized around the flaw.

This means that Eq. 5.47 can be used as well for metals with low ductility, if K_{Ic} is interpreted in accordance with Eq. 5.49. Indeed, values of K_{Ic} are provided for metals with low toughness just as they are for ceramics.

As regards the ***fracture of polymers***, this again depends on all the structural details you are familiar with: degree of crystallinity, linear vs. branched vs. cross-linked chains, etc. Thus, in analogy to metals, the fracture of polymers may range from ductile to brittle, including anything in-between, or a combination of both. The details depend on what deformation mechanisms are relevant.

5.7 Strengthening of Materials

It should be evident to you by now that the strength of a material is one of its major figures of merit. Therefore, methods for increasing the strength of a material are of great interest. It is also clear that these methods will vary, depending on the type of the material in question. However, there is a common idea underlying all these methods, and it can be formulated simply as follows:

> **Strengthening a material:**
> Inhibit the deformation mechanism leading to failure

A second consideration follows from the general observation that in most instances increased strength can only be achieved at the expense of a reduction in ductility. Of course, this will only apply to those materials which have ductility in the first place. So when it comes to strengthening, ceramics and glasses are in a class of their own. For metals, increasing strength without loss of ductility, or increasing ductility without loss of strength, are the holy grail of materials design.

5.7.1 Strengthening of Metals

For metals, the deformation mechanism that needs to be inhibited is the motion of dislocations, since this is how plastic deformation occurs, and plastic deformation precedes ductile fracture. As it turns out, all the different defects you have become familiar with impede dislocation motion. These defects include grain boundaries, impurities, and other dislocations.

Strengthening by Grain Size Refinement

A grain boundary presents a natural obstacle for the motion of dislocations. Remember that dislocations, although being defects, need slip planes and slip directions in order to move across material. That is, they need a continuous crystalline structure, and this structure is interrupted at a grain boundary.

In order for a dislocation to cross a grain boundary, a higher shear stress is required than for slip just within a grain. Furthermore, the wider a grain boundary, the more difficult this crossing will be.

One factor determining the width of a grain boundary is the misorientation between two adjacent grains (see Fig. 4.12). Another factor to consider is that, as dislocations move to a grain boundary, they will pile up there. In doing so, they will make it more difficult for additional dislocations to move. At the same time, the pile-up will increase the disorder at the grain boundary and increase its width.

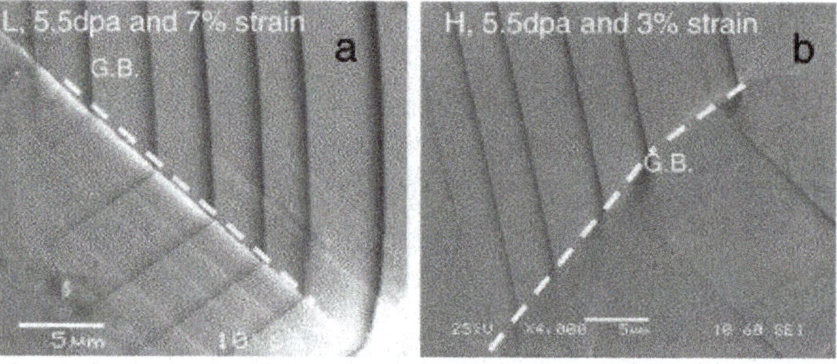

Figure 5.36: Slip lines at grain boundaries of samples of proton-irradiated stainless steel. At the higher strain (i.e. the higher stress) in a) dislocations can cross the boundary. In b), at a lower strain, this does not take place.

Fig. 5.36 shows scanning electron micrographs of two stainless steel samples in which stress had been induced by irradiation with energetic protons in a nuclear reactor environment. In Fig. 5.36a) the stress was large enough for dislocations to be transmitted across a grain boundary. At the smaller stress in Fig. 5.36b) no such transmission took place.

Based upon these arguments, we expect strength to be a function of grain size. More specifically, the yield strength Y.S. should increase with decreasing grain size. The smaller the grains, on average, the more space is taken up by the boundaries relative to the interior of the grains. Hence, the smaller the grains, the larger is the volume of material through which it is difficult for dislocations to move.

An empirical relationship has been identified to describe this effect:

$$\text{Y.S.} = \sigma_0 + k_y d^{-1/2} \tag{5.50}$$

σ_0 and k_y are experimentally determined constants and d is the average diameter of a grain.

Eq. 5.50 is known as the Hall-Petch equation. It usually applies only over a certain range of grain sizes, and breaks down at very large or extremely small grain sizes. Fig. 5.37 shows an example for copper.

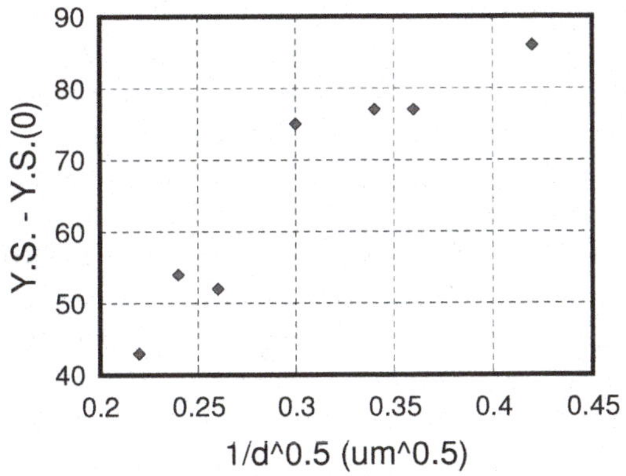

Figure 5.37: Data for yield strength vs. grain size of copper, showing Hall-Petch relationship.

Note that other mechanical properties (those correlated with yield strength) also depend on grain size and tend to follow a Hall-Petch type of relationship. That is, tensile strength, hardness, and toughness improve similarly with a reduction in grain size.

Strengthening by Dislocations: Cold Work

Dislocations themselves, of course, represent major disruptions of a crystalline structure. Therefore, it is not surprising that a higher density of dislocations will go along with reduced dislocation mobility, i.e. with increased strength.

The way to increase the dislocation density is to subject the material to plastic deformation. That is, continued plastic deformation beyond the yield strength creates more and more dislocations, thereby making the metal stronger and harder. This method is known as **work hardening** or **strain hardening**. Additionally, doing **cold work** on a material refers to deforming it plastically around room temperature (i.e. at a temperature much lower than when individual atoms can move between lattice sites due to their thermal energy).

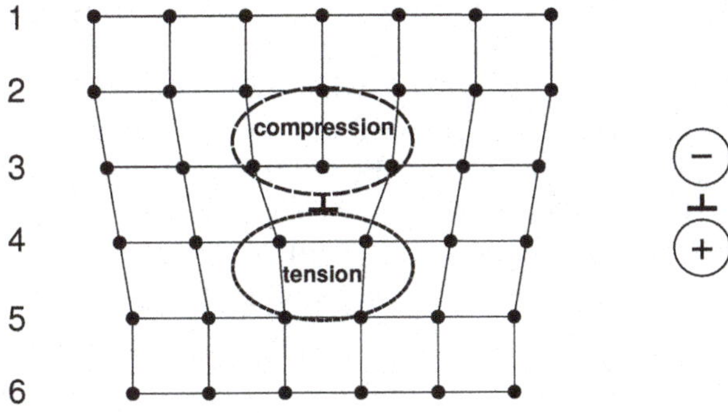

Figure 5.38: Stress field around an edge dislocation

The interaction of dislocations can be visualized as follows. First, note that above the dislocation line there is a region which is under compression, and below the dislocation line there is a region under tension. This is indicated in Fig. 5.38 by the fact that in Rows 1 and 2, and also in Rows 5 and 6, the atoms are drawn at their equilibrium separation. However, in order to reduce the number of vertical planes from seven above to six below, the atoms in Row 3 must be squeezed together slightly, and the atoms in Row 4 must be pulled apart a little.

The small graph on the right of Fig. 5.38 is a shorthand notation for an edge dislocation with its stress field. The plus sign denotes tension (positive strain, by convention), and the minus sign denotes compression.

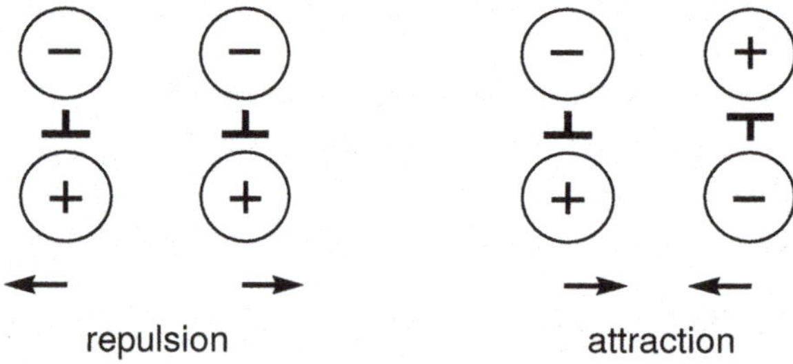

Figure 5.39: Interactions of two edge dislocations

Because of these stress fields, two edge dislocations near each other experience an interaction (Fig. 5.39). If the dislocations have the same orientation, they repel each other (Try drawing a picture like Fig. 5.38 with two dislocation very close to each other, and you will see the difficulty). If they have opposite orientation, they attract each other. In the former case, one dislocation cannot move past the other one. In the latter case, if they are aligned just right, they will annihilate each other, and if they are not aligned perfectly, they end up trapping each other in a dislocation dipole.

These are some of the possibilities how two dislocations can "get into each other's way", making plastic deformation more difficult. But the issue is magnified if you consider a large number of dislocations together. Remember that in an edge dislocation, the dislocation line is perpendicular to the slip plane. Furthermore, in each crystalline grain there will be many slip planes, with various orientations relative to each other, and each of these many planes may have many dislocations in it. What you end up with is sometimes called a dislocation forest, in which the mobility of dislocations is greatly reduced.

The effect of cold work on the mechanical properties is shown in Fig. 5.40, with data for copper and two types of brass. Similar results are obtained with other metals. Note that the amount of cold work, CW, is defined as

$$CW = 1 - \frac{A}{A_0} \tag{5.51}$$

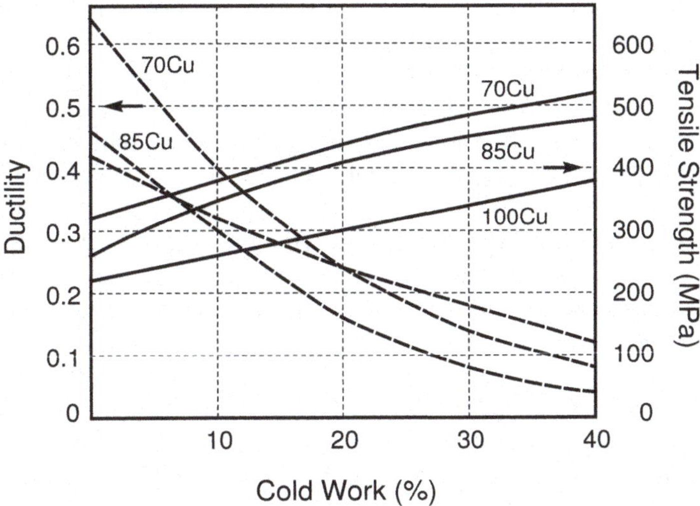

Figure 5.40: The effect of cold work on the tensile strength and the ductility of copper and two types of brass.

where A_0 is the initial cross-section of the sample, and A is the final cross-section after cold work has been performed on it. The numerical value of *CW* is usually given in percent. The area ratio is used because cold work is most often performed either by rolling a flat metal plate or by drawing a metal rod or wire. However, if we assume, to a first approximation, that the total volume of a sample is constant during the plastic deformation of cold work, *CW* can be considered about equal to the strain ϵ caused by the cold work.

An additional consequence is apparent from Fig. 5.40: All materials see an increase in the tensile strength due to the cold work, but also a decrease in the ductility. This illustrates the typical trade-off mentioned earlier that accompanies the strengthening of materials.

Strengthening by Alloying

On several previous occasions I have made the point that engineering metals are most often used as alloys rather than in very pure form, the reason being that alloys are stronger and harder than their base metals. Our previous examples included copper, brass, and bronze, among others. The data of Fig. 5.40 for Cu with various amount of Zn in it demonstrate the effect. (Just focus on T.S. at 0% cold work.)

Another example is noble metals used in jewelery, such as Ag and Au . If these were pure, they would be be too soft and not durable enough for everyday wear. Sterling silver is 92.5Ag-7.5Cu (or 7.5Ni), and the purest form of gold commonly used is 22 karat yellow gold, meaning it contains 91.67% Au, the rest being 5% Ag, 2% Cu and 1.33% Zn. You will notice that in both cases, the alloying elements are cheaper than the base metals, too.

An exception to the rule of commercial metals usually being alloys is copper. The copper used as seamless tubing in home plumbing applications is typically 99.9% pure. This makes the tubes soft and easy to bend, which is desirable in this case, but the material hardens after it has been bent (i.e. after it has been cold-worked).

As we will explain shortly, impurities have the same effect on dislocations as the other defects: They make dislocation motion more difficult. But before we examine this effect in more detail, we need to revisit some generic issues about impurities in metals. You recall from our discussion in Chapter 4 that impurities in metals may be interstitial or substitutional, i.e. they may reside in-between lattice sites or on lattice sites. For obvious reasons, only very small impurities, compared to the solvent atoms, can be interstitial. On the other hand, for impurities to be substitutional, they should be as similar as possible, including in size, to the solvent atoms.

In either case, impurities will induce strain in the lattice (Fig. 4.1), and this causes the allowed impurity concentration to be limited. With respect to the induced strain, one can say that interstitial impurities will always cause local compression. In addition, if the impurities are substitutional and small, they will cause local tension, and if they are substitutional and large, they will cause local compression, as the lattice has to accommodate the mismatch in size.

So in order to avoid complications, we will confine our discussion here to alloys with suitably small impurity concentrations. "Suitably small" is rather ill-defined but means roughly that impurities are isolated from each other. (At a later point in the book, we will be able to relax this assumptions).

Fig. 5.41 presents two additional sets of data. Cu with added Ni is of special interest because in this case the assumption of a small impurity concentration is not necessary. The Cu/Ni system turns out to be a combination of elements where *any* mixture of the two components produces a perfect solid solution. In other words, Cu and Ni are so similar, including being both FCC, that any amount of Ni can substitute for the corresponding amount of Cu, and vice versa. Yet despite this similarity, Ni atoms in Cu disturb the lattice enough so that a Cu-Ni alloy is somewhat stronger than pure Cu.

For the Fe-based alloys, solute and solvent atoms are not very similar. Therefore, the solubility is limited for all alloying elements. (It would certainly not extend to 10% and more, as with Ni in Cu). In addition, keep in mind that N is interstitial, whereas the other elements are substitutional. Note also the different behavior for Cr in Fe.

It goes without saying that in order for the effect of impurities to show unambiguously, the alloy must not have been cold-worked. For the time being, that is what the specification of "annealed" for Cu in Fig. 5.41 indicates (more on annealing below). The same should be assumed for the steel alloys.

So how do impurities affect the mobility of dislocations? It turns out that

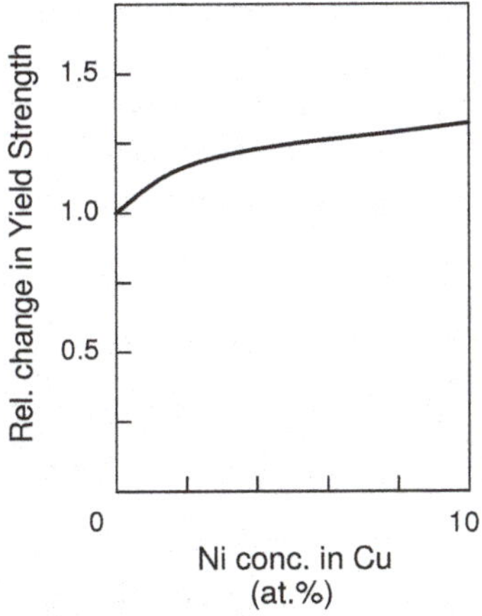

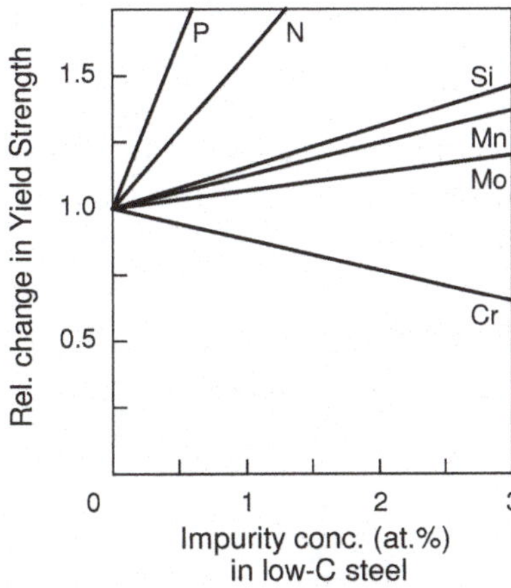

Figure 5.41: Left: Increase in Y.S. of annealed Cu by adding Ni. Right: Increase in Y.S. of low-C steel by adding various alloying elements.

the interaction of impurities with a dislocation is a complex phenomenon. An important part of it is related to where an impurity prefers to locate itself when it is near a dislocation. This is a function of the stress field around the dislocation on the one hand, and the stress the impurity itself creates in the lattice on the other hand.

Above we argued that around a dislocation, a state of local compression exists above the dislocation line, and a state of local tension below the dislocation line (Fig. 5.38). It follows that the most ***favorable position for a small impurity is above the dislocation line*** because there it is able to relieve some of the stress caused by the dislocation. This, in turn, makes the shear stress necessary to move the dislocation larger, thus increasing Y.S.. With a large impurity, it is the other way around, for the same reason: A large impurity will tend to reside below the dislocation line, because that is where it can reduce the stress from the dislocation itself. This again makes the shear stress necessary to move the dislocation larger, thus increasing the yield strength.

Annealing to Reverse Strengthening

The strengthening induced by grain size refinement or cold work can be reversed to a large extent by a process called thermal **annealing**. This refers to a heat treatment in which the sample is kept at an elevated temperature of typically between $\frac{1}{3}T_m$ and $\frac{1}{2}T_m$, where T_m is the absolute melting temperature, for a suitable period of time. It is in this range of temperatures that individual atoms acquire enough mobility to undo some of the distortions in material microstructure brought on by cold work. In addition, at these kinds of temperatures it is also possible for small grains to grow in size, thereby increasing the average grain size but decreasing the number of grains.

Both cold-worked and fine-grained specimens constitute non-equilibrium structures. As you know, cold work creates a large number of dislocations. It also introduces severe deformation of the grains present. Grain boundaries are high-energy regions as well, and the smaller the grains on average, the larger is the volume occupied by these disordered regions compared to the crystalline grains themselves.

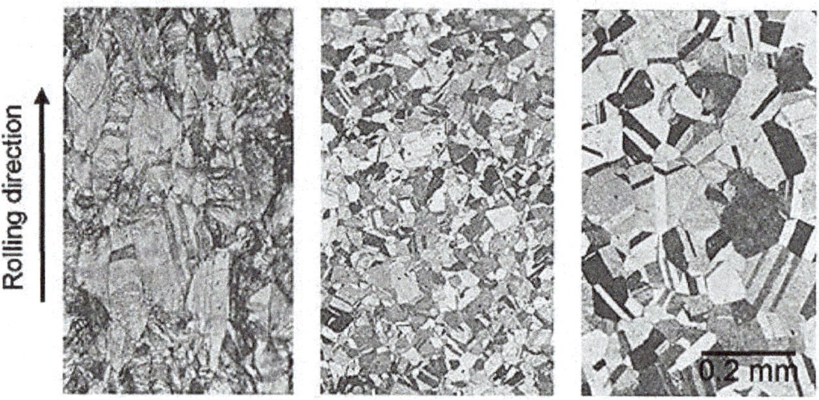

Figure 5.42: Effects of annealing on cold-worked brass. Left panel: Sample after cold-rolling. Middle panel: After annealing at 550°C for 1 hour. Right panel: After annealing at 650°C for 1 hour.

Note that in the left panel of Fig. 5.42, all grains are elongated in the rolling direction. In the original state, they were roughly equiaxial, i.e. with similar dimensions in all directions. So cold working produced a great deal of mechanical distortion, together with dislocations and their built-in stresses. The other two panels show a complete rebuilding of the grain structure as a consequence of the annealing process. At the higher temperature, with its greater atom mobility, grains have grown substantially larger than during the same period at the lower temperature.

The entire process of annealing is sometimes considered as occurring in three stages: **recovery**, followed by **recrystallization**, and then **grain growth**. During recovery, stresses internal to the metal are relieved, some dislocations are able to assume lower-energy configurations, and some point defects are healed. During recrystallization, entirely new grains are nucleated continually and begin to grow. Eventually, during grain growth, a new microstructure with large, stress-free, equilibrated grains is achieved.

From our prior discussion of strengthening it should be evident that annealing will lower a material's tensile strength, yield strength and hardness, but the ductility will increase. For this reason, annealing is sometimes used to fine-tune the mechanical properties of a previously strengthened material to the desired combination of strength and ductility.

5.7.2 Strengthening of Ceramics and Glasses

How can one implement the generic strengthening principle formulated at the beginning of Section 5.7. for ceramics and inorganic glasses? These being brittle materials, the real question is how to reduce the likelihood of the occurrence of brittle fracture.

The remedy for the problem is embodied in Eqs. 5.46 and 5.47: Your goal is to achieve the highest possible critical stress σ_c necessary for crack propagation. Therefore, you need to select a material with a high fracture toughness K_{Ic}. At the same time, you should process and handle the material so that the concentration of flaws is as low as possible and the length c of the largest flaws is as small as possible. You should also try to use a brittle material in such a fashion that the influence of inherent flaws is mitigated. Specifically, keep the following issues in mind:

1. **Optimize fracture toughness**
   In general you do not have control over K_{Ic} of a material; it is a parameter which the manufacturer tries to optimize. However, K_{Ic} is affected by the specific surface energy γ_s and the elastic modulus E (cf. Eq. 5.46). So if you are in the business of developing a ceramic material or a glass, it may be possible to tweak its composition primarily for the purpose of increasing γ_s.

   Also, ceramic materials in particular are often synthesized from fine powders by mechanical compacting and by **sintering** (a high-temperature consolidation process that fuses particles together). Because of the nature of these processes, ceramics are often not fully dense. That is, they do not achieve the highest possible density conforming with their composition and crystal structure. Rather, they end up containing a certain amount of pores. Even though pores may not be stress raisers to the extent sharp cracks are, pores will still reduce the density of the material, and thus its effective elastic modulus and fracture toughness. Therefore, it would seem to be a good strategy to try to minimize the porosity in a ceramic.

   It is also interesting to note that, in comparison to metals, it appears that in ceramics there is no direct correlation between grain size and strength or fracture toughness. Presumably, this is because other factors, such as the presence of pores and other flaws, are much more important.

2. **Reduce flaws in the material**
   We pointed out that ceramics and glasses may contain flaws in the form of small cracks, either internal or external. Such flaws act to concentrate an externally applied stress, and they may cause internal stresses as well. Most importantly, they are potential points of origin for cracks to propagate.

The first conclusion we can draw is that one should always design processing and handling of ceramics and glasses so as to minimize the presence of flaws in general, and their size in particular. Additionally, since processing often involves some kind of thermal treatment, such treatment should be chosen so as to minimize internal stresses caused by heating and subsequent cooling of the material.

With respect to alleviating the danger of crack propagation, a general approach often taken is to modify the microstructure of a material in such a way that cracks are likely to encounter obstacles that prevent them from propagating all across the sample. This might include incorporating small particles or fibers of a different material into the base material, such as in metal-reinforced ceramics, or using a multilayered material.

3. **Use material in compression**
   The relative lack of mechanical strength in ceramics and glasses is not only caused by their inherent structure, but also by the practical fact that mechanical structures are often used under tensile stress. This is, of course, the very condition which promotes crack propagation and thus mechanical failure. It is also the reason why these materials are typically ten times stronger in compression than in tension. It follows that if as a designer you have a choice, you should try to use a structural ceramic in compression rather than in tension.

An interesting example for these principles in action can be seen in ***tempered glass***. The key feature of this material is that by a clever manufacturing process, the two surfaces of the glass pane have been put into compression. This is achieved by cooling the two surfaces of the still hot glass pane very rapidly by streams of air. During the process, the outermost surface layers contract and become rigid while the interior is still soft and in a mechanically relaxed state. As the interior itself cools down, it contracts and thus puts a compressive stress on the already rigid outer layers. This treatment not only makes the material several times stronger than untempered glass, but it also causes the tempered glass to break into a large number of small pieces if indeed it does break.

A similar result can be achieved by chemical means in some glasses containing Na^+ ions. If such a glass is exposed to a solution containing K^+ ions at a suitably high temperature, some Na^+ ions in the near surface regions are exchanged with K^+ ions. The size difference between these ions again causes compressive surface stress, which strengthens the glass.

A third example along these is represented by the material making up the dishware introduced originally as Corelle ® by Corning Glass Works. It consists of a glassy inner layer, with a glaze on both sides that has a lower thermal expansion coefficient than the center material. By the mechanism described above, this puts the surfaces into compression when the plate or cup cools down, thus making the material more resistant to scratching and chipping.

5.7.3 Strengthening of Polymers

Polymers are the most diverse class of materials when it comes to mechanical properties: They can vary from completely amorphous to highly crystalline, from linear to highly cross-linked, and from highly ductile to brittle. By the same token, the methods for strengthening them are just as varied. However, the general principle still applies that one should inhibit the deformation processes that ultimately lead to fracture.

The main approaches to the strengthening of polymers are listed below. The mechanisms by which they improve strength should be quite evident in light of the general description of mechanical deformation in Section 5.5.4 above. The main idea is to make the uncoiling, untangling, and sliding-by of polymer chains more difficult.

1. **Molecular weight**
   All other things being equal, the longer the polymer molecules, the stronger in general the material tends to be.

2. **Crystallinity**
   For a given type of polymer, say polyethylene, increasing the degree of crystallinity will increase the strength.

3. **Side groups**
   Bulky side groups, such as e.g. in poly(vinyl chloride) or polystyrene, tend to make a material stronger.

4. **Secondary bonds**
   This refers to bonds between different chains. These may be in the form of actual cross-links or hydrogen bonds. The more of these bonds there are, and the stronger they are, the stronger will be the material.

5. **Chain orientation**
   Since in plastic deformation chains are being uncoiled, a polymer can sometimes be strengthened by straightening out the chains to begin with, for example when polymer fibers are made by drawing the material through an orifice. This would correspond to cold working of a metal, in the sense that plastic deformation strengthens a material.

6. **Adding particles**
   A strengthening effect similar to bulky side groups can be achieved by including small particles in the polymer matrix. This is implemented, for example, by adding Carbon Black (very small carbon particles) to rubber used for car tires.

Relevant mechanical data for a few common polymers are given in the following table:

Table 5.6: Mechanical property data for some common polymers

Polymer	Density ρ (g/cm^3)	E (GPa)	T.S. (MPa)	ϵ_f
LDPE	0.92	0.25	10	4
HDPE	0.96	0.4	30	1.5
PP	0.905	1.5	36	1.5
PVC	1.33	3.4	50	2
PS	1.05	2.5	44	0.03

For the materials and abbreviations, see Fig. 5.10. Also, note that LDPE and HDPE refer to low-density and high-density polyethylene, respectively, and PP means polypropylene. LDPE is branched and with shorter chains, whereas HDPE has longer chains and is not branched. This causes LDPE to have a much lower degree of crystallinity, and thus a lower density, than HDPE.

The data in the table illustrate the methods listed earlier. For example, the difference between LDPE and HDPE is due mainly to the difference in crystallinity. The higher strength of PP, PVC, and PC can be ascribed to their side groups.

Two additional comments are worth making, especially with regard to polymers, as we finish up the discussion of mechanical properties. First, keep in mind that when you make a material stronger, you may not improve the toughness at the same time. In many applications, toughness may well be more important than absolute strength. And second, we emphasized earlier that, among other things, the mechanical properties of polymers depend on temperature, but we did not go into much detail about it. Suffice it to say that the discussion in this chapter applies by and large around room temperature. We will pay much more attention to effects of temperature on materials properties in general in the next several chapters.

Image sources

Title picture: Mechanical materials failure at the Boston "Big Dig"
`http://www.ims.uconn.edu/associate/images/publications/Ezrin-ANTEC2008-Big-Dig.pdf`

Fig. 5.8: reproduced with permission by ASM International
"Hardness Testing of Ceramics" by George D. Quinn, in *Advanced Materials and Processes* Vol. 154, No. 2 (1998), ASM International, Materials Park, OH.
`http://www.metallography.com/ceramics/ceramics.htm`

Fig. 5.32:
`http://hsc.csu.edu.au/engineering_studies/application/lift/3210/index.html`
`http://www.tescan-usa.com/gallery/ductileFracture.htm`

Fig. 5.33:
`http://images.pennnet.com/articles/ap/cap/cap_0711apdetecting03.gif`

Fig. 5.36: reproduced with permission by Elsevier
Z. Jiao, J.T. Busby, G.S. Was, J. Nucl. Mat. 361 (2007) 218-227

Fig. 5.37: Data from:
A.H. Chokshi, A. Rosen, J. Karch, and H. Gleiter, Scripta Metall. 23 (1989) 1679

Fig. 5.42: reproduced with permission by M. Pfeifer (Imet LLC)
`http://www.imetllc.com/industrial-metallurgists-llc-newsletters/5-minute-metallurgy-lesson-july-2012.html`

References

Both Shackelford, and Callister and Rethwisch discuss further aspects of mechanical properties such as creep and fatigue. Guy goes more in depth on some mechanical properties topics.

For developing a better idea about dislocations, visit
`http://www.tf.uni-kiel.de/matwis/amat/def_en/kap_5/backbone/r5_2_3.html` to learn about the energy of a dislocation, and

`http://www.tf.uni-kiel.de/matwis/amat/def_en/kap_5/backbone/r5_4_1.html` to learn about partial dislocations and stacking faults.

Two wonderful sites with animations of moving dislocations and other mechanical properties phenomena are
`http://simap.grenoble-inp.fr/groupes-de-recherche/simulations-par-dynamique-des-dislocations-discretes-`

219401.kjsp?RH=SIMAP_GPM2

http://zig.onera.fr/DisGallery/index.html

For more information on fracture toughness, consult http://www.ndt-ed.org/EducationResources/CommunityCollege/Materials/Mechanical/FractureToughness.htm

http://www.substech.com/dokuwiki/doku.php?id=fracture_toughness

Exercises

1. An aluminum alloy rod with 1 cm diameter is put under a tensile load of 1000 N.
   a) What is the stress in the rod?
   b) If the elastic modulus of the material is E = 75 GPa, what will be the strain?

2. A 1 m long piece of high-strength-steel piano wire has the following mechanical property values: T.S. = 1100 MPa, Y.S. = 1000 MPa, E = 200 GPa, diameter = 1.0 mm.
   a) If the maximum stress is not to exceed 0.9 Y.S., what is the maximum load that can be applied to the wire?
   b) What will be the length of the wire under that load?

3. Determine E, Y.S., T.S., and ϵ_f from the diagram below:

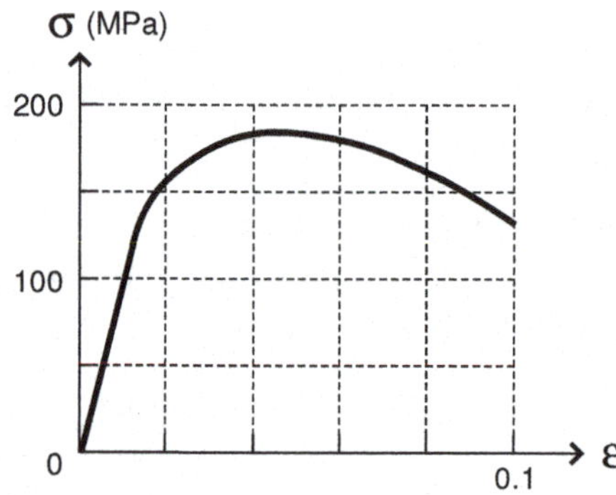

4. Draw a stress-strain diagram for a material with the following parameters (these are quite realistic for a metal):

   $$E = 100 \text{ GPa}, \text{ Y.S.} = 70 \text{ MPa}, \text{ T.S.} = 200 \text{ MPa}, \epsilon_f = 0.40$$

   Be as quantitative as you can and note when you have to make an educated guess. What is the main difference between your diagram and the one drawn above in Exercise 3 ?

5. Draw a stress-strain diagram with curves for a brittle metal, a ductile polymer, and a brittle polymer. Show the important parameters in the proper qualitative proportion.

6. The table below shows data for a tensile test on a Mg rod with diameter 0.25 cm.

load (N)	0	1×10^3	2×10^3	3×10^3	3.5×10^3
l (mm)	50.000	50.086	50.171	50.322	50.450

   a) Can you determine E just from the given data? If so, why and how?
   b) Plot the data and determine Y.S..
   c) From your plot, estimate T.S. if you can.
   d) From your plot, estimate ϵ_m if you can.

7. Determine the true stress and true strain at the maximum of the $\epsilon(\sigma)$ curve in Fig. 5.5. Assume a Poisson ratio of 0.3. Compare the true values to T.S. and $\epsilon(\sigma = \text{T.S.})$.

8. For the material of Fig. 5.5, give a worst case estimate of the amount by which the true stress could exceed the engineering stress at T.S..

9. Are the data in Table 5.2 in agreement with Eq. 5.15 ?

10. The figure below shows the result of a tensile test on a commercial polyester film with the following dimension: Length = 20 mm, width = 15 mm, thickness = 0.038 mm. The x-axis displays the elongation Δl.

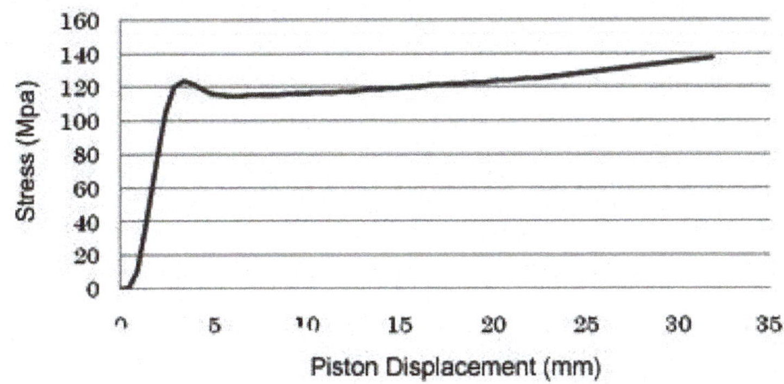

   a) Determine the load at a stress of 80 MPa.
   b) Determine the elastic modulus of the material.
   c) Determine the yield strength and the tensile strength.

11. Does the argument outlined in Fig. 5.23 apply to compression? If so, how? If not, why not?

12. In Figs. 5.25 and 5.26 we examined the case where the rightmost black atom was pulled outward. Consider here the alternative situation where the leftmost black atom is pushed inward by the shear stress τ (see Figure below) to the point where it has been displaced by one lattice spacing.

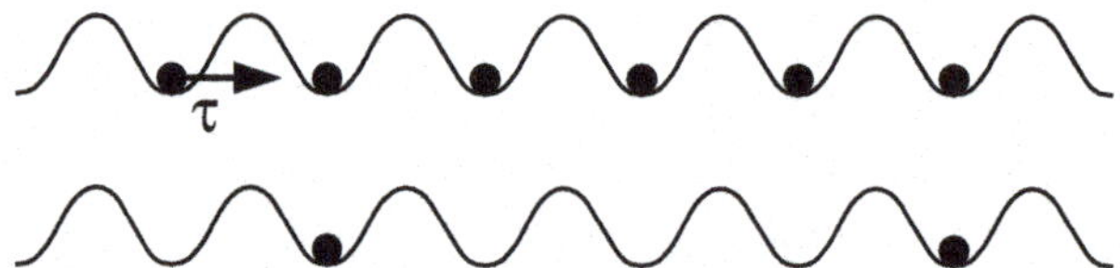

Draw the other black atoms into the lower panel and show that this creates a similar type of disturbance that can also move easily across the sample. Discuss how this case is related to the one shown in Fig. 5.26.

13. Draw a short chain with a couple of mers for polyisoprene. How do you think the mechanical properties of polyisoprene would compare with those of polybutadiene? What if the two polymers were vulcanized?

14. What do you expect to happen to "silly putty" if it suffers a VERY quick blow?

15. Why is the picture of the crack in Fig. 5.35 unrealistic for a brittle material?

16. For a certain Al alloy, it has been found that Y.S. = 345 MPa and K_{Ic} = 44 MPa $m^{1/2}$. If this material contained cracks, what would the crack length have to be so that Y.S. equals σ_c ?

17. A certain Ti alloy with Y.S. = 910 MPa and K_{Ic} = 55 MPa $m^{1/2}$ is being used in a structural application where it may be exposed to tensile stresses as large as one third of Y.S.. If the largest cracks in the Ti alloy have a length 2c of at most 0.1 mm, is the material likely to undergo fracture ?

18. What are the largest and the smallest grain size in the plot of Fig. 5.37?

19. Plot Y.S. vs. composition for Ni in Cu and for 100% Ni. How should the curve be completed so as to make it continuous ?

20. Look up the compositions of the penny, the nickel, and the dime. Why are these compositions what they are? What are the values of the coins in terms of actual metal prices?

21. How would you explain the following in Table 5.6:
    a) The difference in strength between LDPE and PP?
    b) The similarity in strength but difference in density between PP and HDPE ?

22. In what way is PS in Table 5.6 different from all other materials listed ?

23. The PS referred to in Table 5.6 is known as general purpose polystyrene, but there exists also high impact polystyrene with much improved toughness. By what method of strengthening is this achieved ?

Problems

1. There are some rather special materials which have a negative Poisson ratio ν, meaning that they expand upon application of a uniaxial stress. Find some examples and describe the structure that leads to such behavior.

2. Calculate the toughness for the material in Fig. 5.5 by doing the integral given in Eq. 5.11.

3. Using Fig. 5.5, devise an approximate mathematical expression for the toughness that does not require doing an integral. Compare its accuracy to your result of Problem 2.

4. Assume a certain type of steel has a ductility of 0.18. Make a prediction about the amount by which the true stress σ_t at $\epsilon(\sigma = \text{T.S.})$ is likely to exceed T.S.. What is the uncertainty in your estimate?

5. In Table 5.3 two types of brass are compared: 90Cu 10Zn and 80Cu 20Zn. Do you think this is a valid comparison, or has something important been left out? (Hint: Think solid solution).

6. Go to the reference noted in the text (http://www.cbsteel.com/pdf/CB-Data.pdf) and gather data for some hot-rolled steels, preferably the ones listed in Table 5.2. Determine what difference this treatment makes, in comparison to cold drawing, for the mechanical properties of the materials.

7. Take 3 BCC metals and 3 FCC metals of your choice, with a decent range of elastic moduli E, and check whether Eq. 5.25 is able to describe these moduli quantitatively.

8. Assume that the force curve in Fig. 5.21 is applicable to an "atomic tensile test"and that it provides a valid description of elastic deformation up to a point where the slope of the $F(r)$ curve deviates by 10% from its limiting value at $r = 0$. At what strain is this 10% deviation reached ?

9. Assume a Lennard-Jones potential of the form in Fig. 5.21 describes the interaction between each pair of atoms in a BCC crystal. Focus on a particular atom and locate and count all nearest, second-nearest, and third-nearest neighbors. Determine how much each set of neighbors contributes to the total potential energy at the central atom.

10. In BCC crystals, the close-packed planes are the {1 1 0 } family. Show that the close-packed direction $[\bar{1}\,\bar{1}\,1]$ of Fig. 5.27 is also in the (1 1 2) plane, even though (1 1 2) is not part of {1 1 0 }. Therefore (1 1 2) $[\bar{1}\,\bar{1}\,1]$ is another slip system in BCC. Draw a picture of (1 1 2) that shows its important atomic features.

11. It can be shown that the elastic energy stored in the stress field around a dislocation is equal to cGb^2, where G is the shear modulus, b the magnitude of the Burgers vector, and c a constant of the order of 1.
In the figure, two edge dislocations with the same orientation are displayed schematically. In a) the dislocations are separated, and in b) they have moved together.

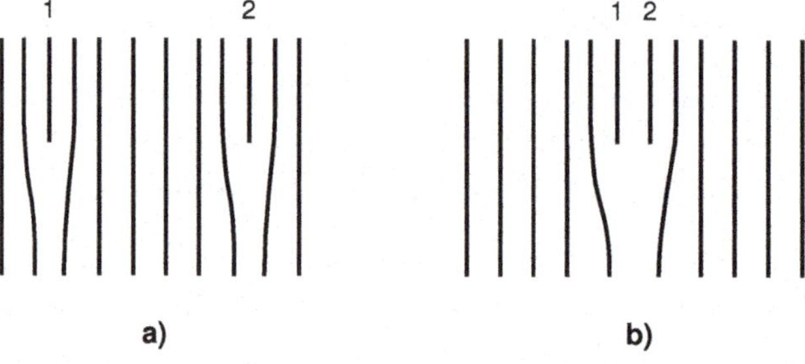

Show that the stored energy in the configuration b) is higher than in a), thus proving that two edge dislocations of the same orientation repel each other.

12. Find a commercial Cu/Ni alloy containing more Ni than Cu. Describe its properties and indicate what is special about it, and why?

Chapter 6
Diffusion

Overview

Diffusion is a high-temperature process by which atoms in a solid can move about and redistribute themselves. At sufficiently high temperature, atoms vibrate so strongly about their equilibrium positions that occasionally they are able to jump to a neighboring site. This type of step-by-step atom motion, when combined with local differences in atom concentrations, gives rise to mass transport over macroscopic distances. Two situations are of particular importance: 1) Steady-state or, equivalently, time-independent diffusion, where atoms move without changing the overall concentrations; and 2) Non-steady-state diffusion, where atom concentrations in the sample change with time. The diffusion process depends strongly on temperature and on the material and its microstructure. It is especially useful for controlling the distribution of impurities and other defects in solids.

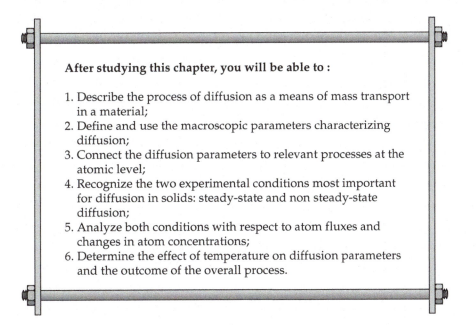

After studying this chapter, you will be able to :

1. Describe the process of diffusion as a means of mass transport in a material;
2. Define and use the macroscopic parameters characterizing diffusion;
3. Connect the diffusion parameters to relevant processes at the atomic level;
4. Recognize the two experimental conditions most important for diffusion in solids: steady-state and non steady-state diffusion;
5. Analyze both conditions with respect to atom fluxes and changes in atom concentrations;
6. Determine the effect of temperature on diffusion parameters and the outcome of the overall process.

Everyday objects made from case-hardened steel

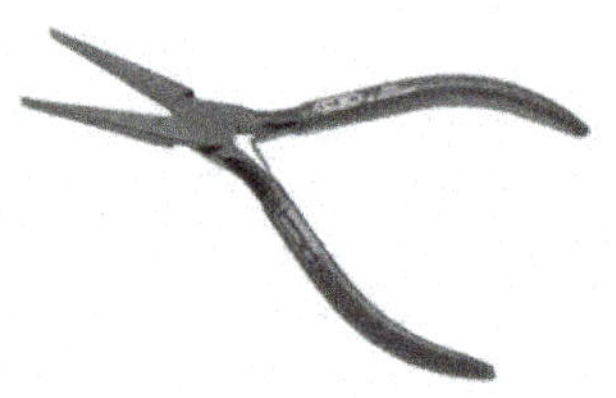

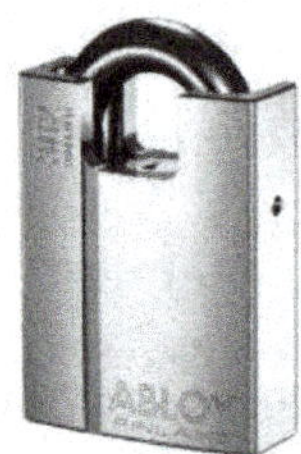

The picture shows a pair of pliers, a padlock, and a motorcycle sprocket, all of which make use of steel parts that have been enhanced by a diffusion process known as case hardening. The process creates a hardened outer layer on the steel by increasing the carbon concentration in the surface region. This is realized by holding the workpiece at an elevated temperature in a carbon-rich ambient for a certain period of time. Under these conditions, carbon atoms are able to migrate into the surface of the steel, making it harder, more abrasion-resistant and more durable, and thus extending the lifetime of the components.

6 Diffusion

By now you have seen several instances where impurities turned out to have a crucial influence on the properties of materials, yet we have not addressed yet the question of how to put impurities into a material in the first place. One way, of course, is to mix them in at the outset, when preparing the material. However, in some cases it is desirable to add them after the fact, by a process known as **diffusion**.

Diffusion is the first of a number of temperature-dependent phenomena in solids that we will examine in the next several chapters. It is a process which occurs whenever the temperature of a solid is high enough, and thus atoms vibrate about their normal positions strongly enough, so that occasionally they can move to another position nearby. An example of where this process will manifest itself is in a solid containing impurity atoms that are distributed in a non-uniform fashion. In that case, diffusion will cause the impurity atoms to become distributed more evenly with time.

In this chapter you will become familiar with the macroscopic characteristics and relevant parameters of the diffusion process, the underlying atomic mechanisms, as well as some practically important experimental applications.

6.1 Description of the Process and Mechanism

The process is illustrated qualitatively with the figure below. Imagine that you have brought two materials, say two metals A and B such as Cu and Ni, together into intimate contact at the atomic level. This could be achieved, for example, by taking a piece of A and depositing B directly on top of it.

Now, raise the temperature of this joined sample to a point where atoms A and B vibrate so vigorously about their equilibrium positions that occasionally they can jump to an adjacent site. This means that occasionally an atom B will cross the boundary to the A side, and vice versa. If you wait a while, more and more atoms B will move to the left, and more and more atoms A will move to the right of the original boundary. There is a steady flux of atoms of either type going from the side of higher concentration to the side of lower concentration.

If atoms A were black and atoms B were white, and if you could observe the color of the sample as atoms move across the center line, you would see the originally sharp **interface** between A and B become blurred and gray, and wider with time (see panels (b) and (c) on the left of Fig. 6.1). If you waited a very long time, the sample would approach a state where A and B are mixed uniformly.

In terms of atom concentrations C_A and C_B you could describe the situation as indicated in the panels on the right of Fig. 6.1. Initially, the atom concen-

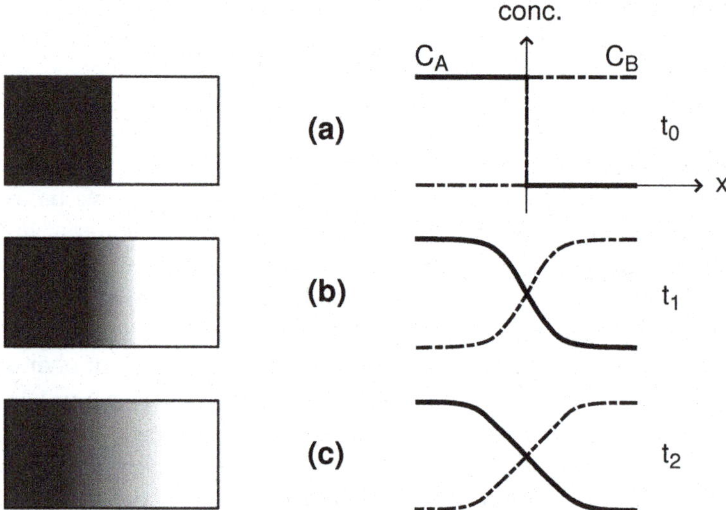

Figure 6.1: Schematic of diffusion profiles as a function of time. (a) Materials A and B joined initially, time = t_0; (b) Atom concentrations at a later time t_1; (c): Atom concentrations after a longer time t_2.

trations drop off with an abrupt step at the interface. As time increases, this drop-off becomes more and more gradual, and the interface becomes wider and wider. If you waited a very long time, both C_A and C_B would end up being uniform in the entire sample.

The truly interesting aspect of this process of diffusion is that individual atoms jump about in an entirely random fashion, yet their combined action results in a directed mass flux. The direction of this flux is determined by the non-uniform distribution of the atoms, from high to low concentration.

Note also that at least in the early stages of this process, you could say that atoms B are impurities on side A, and vice versa.

For atoms A and B to be able to jump to different positions, **two conditions** must be fulfilled:

1. *There must be sites available for the atoms to jump to.*
   Since we are talking about impurity atoms moving about, keep in mind that if B is an interstitial impurity in A, there will always be many other interstitial sites available (See Fig. 4.1). On the other hand, if A and B were Cu and Ni, then there needs to be a vacancy nearby for the impurity to jump into (See Figs. 4.1 and 4.5).

2. *Extra energy is required for an atom to make a jump.*
   As an atom moves to a neighbor site, it must push other atoms out of the

way, and this costs energy. Equivalently, one can say that an atom jumping to a neighbor site causes a temporary local distortion in the lattice.

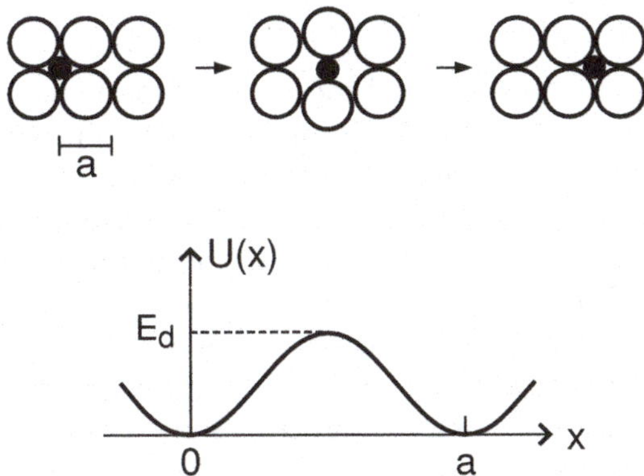

Figure 6.2: Diffusion path and potential energy for diffusion of an interstitial impurity

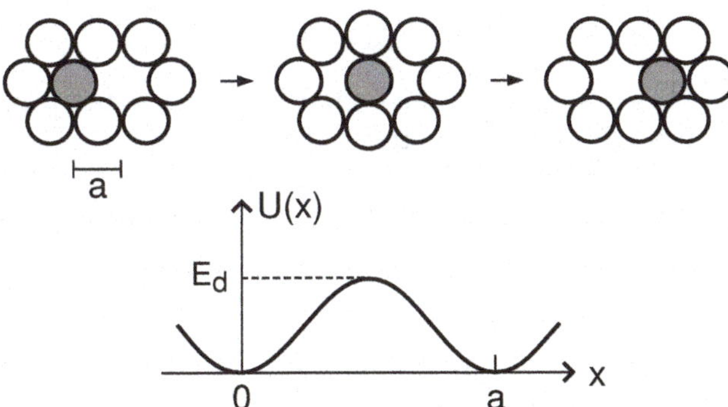

Figure 6.3: Diffusion path and potential energy for diffusion of a substitutional impurity

These two cases are referred to as ***interstitial diffusion*** and ***vacancy diffusion***. They are illustrated schematically in Figs. 6.2 and 6.3. Take note of the similarity in the potential energy $U(x)$ vs. position for jumping atoms and the potential energy for homogeneous shear in Figs. 5.20 and 5.21.

The extra energy required for a jump has the consequence that an atom vibrating about its equilibrium position makes many attempts to jump, but is successful only rarely, i.e. only when it just happens to have attained the energy to overcome the potential energy well E_d indicated in Figs. 6.2 and 6.3.

This situation can be expressed mathematically as follows. Let us define the jump frequency f_j as the the number of times per second that an atom makes an actual jump to a neighbor site. Then we can write

$$f_j = \nu \exp\left(-\frac{E_d}{RT}\right) \tag{6.1}$$

where E_d is the height of the potential energy well mentioned above (in kJ/mole) and ν is the number of attempts for a jump the atom makes per second. So ν is in effect the vibrational frequency of the atom about its equilibrium position.

Note that Eq. 6.1 has a transparent physical interpretation: The number of successful jumps is equal to the number of attempts times the probability that an attempt is successful. The probability is embodied in the exponential factor $\exp(-\frac{E_d}{RT})$. This is a universal expression for the occurrence of a configuration with extra energy, which in this case is E_d. We saw the same type of exponential factor in action when we discussed vacancies as point defects (see e.g. Eq. 4.3).

It turns out that when it comes to numerical values, ν does not vary very much from one material to another. A good number for ν is $10^{-13}s^{-1}$, give or take a factor 3. On the other hand, E_d is very much a function of the material and the impurity. For example, for C in BCC Fe, E_d = 80 kJ/mole, which gives a probability factor $\exp(-\frac{E_d}{RT})$ at 600°C of

$$\exp\left(-\frac{E_d}{RT}\right) = \exp\left(-\frac{80 \times 10^3}{8.314 \times 873}\right) = 1.63 \times 10^{-5} \tag{6.2}$$

Due to the T^{-1} dependence of the exponential factor, the probability for a successful jump is a very strong function of temperature:

Table 6.1: Sample values for jump probabilities

T (°C)	$\exp(-\frac{E_d}{RT})$
20	5.46×10^{-15}
300	5.09×10^{-8}
600	1.63×10^{-5}

Hence, it is safe to conclude that no measurable diffusion will take place at around room temperature.

6.2 Fick's First Law: Steady-State Diffusion

We will now proceed to formulate a quantitative description of the diffusion process. For simplicity, we will only consider one-dimensional diffusion problems. That is, we will assume that impurity concentrations $C(x)$ vary only in one direction, say the x-direction, and are constant in the y- and z-direction.

Let us focus our attention first on a detailed description of the diffusional flux of particles.

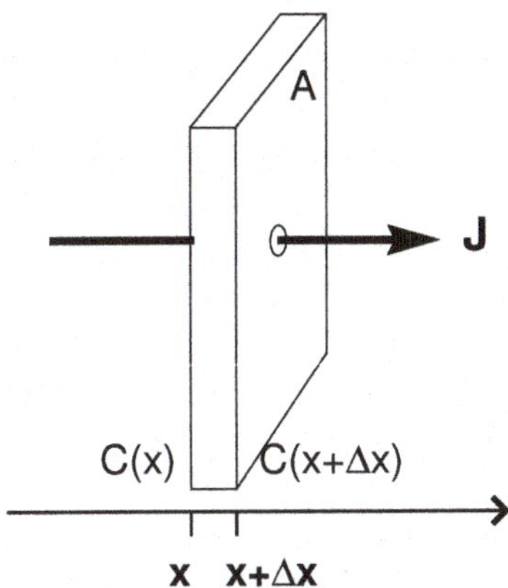

Figure 6.4: Diffusional flux J through a reference plane. The concentration of the diffusing species is constant parallel to the plane and varies only in the x-direction.

Fig. 6.4 illustrates the geometry: We are interested in the particle flux J going across the reference plane. We define J as the amount of mass flowing across this plane per unit area and time:

$$J = \frac{\text{mass flux}}{\text{area time}} = \frac{M}{At}, \quad \text{or} \quad \frac{1}{A}\frac{dM}{dt} \tag{6.3}$$

It is observed experimentally that the magnitude of J is proportional to the gradient of $C(x)$:

$$J \propto \frac{dC}{dx} \tag{6.4}$$

Once again, this proportionality relation can be written as an equivalent equation:

$$J = -D\frac{dC}{dx} \tag{6.5}$$

where the proportionality factor D is known as the **diffusion coefficient**, or sometimes also the **diffusivity**. Eq. 6.5 is known as ***Fick's First Law***.

Note the minus sign in Eq. 6.5: It simply expresses the fact that when $C(x)$ increases in the negative x-direction, there will be a diffusion flux J in the positive x-direction (see Fig. 6.1). Also, beware of units: If we define the concentration $C(x)$ as mass per unit volume, i.e. with units of kg m^{-3}, J will be in units of kg m^{-2} s^{-1}, and therefore D in m^2 s^{-1}.

We could have also taken $C(x)$ to be the number of particles per volume, i.e. in units of particles m^{-3}. In that case, J would be a particle flux and would be in particles m^{-2} s^{-1}. The units of D are the same either way. Of course, one type of concentration can be converted into the other because (mass per unit volume) is equal to (number of particles per unit volume) × (mass per particle).

As regards the relationship of Eq. 6.4, I said that it follows from experiments. However, we can also make a plausible analytical argument for its validity. We mentioned above that diffusional motion of particles is the result of random jumps from site to site performed by a large numbers of particles. If this is so, then from Fig. 6.4 we conclude that the number of particles jumping across the plane from left to right must be proportional to $C(x)$, and the number of particles going right to left must be proportional to $C(x + \Delta x)$. So the *net flux* of particles across the plane must go like $C(x + \Delta x) - C(x)$ or, in the limit of very small Δx, like dC/dx.

It turns out that Fick's First Law, Eq. 6.5 is the general relationship describing diffusional flux in terms of the concentration gradient. However, by itself it is only useful in a **steady state** situation, i.e when the particle concentration $C(x)$ is independent of time. But you may rightfully ask yourself: If particles are in motion from high to low concentration, why is this a time-independent process? How can $C(x)$ not vary with time?

The answer is that a diffusional steady state, with time-independent $C(x)$, can indeed be realized, as illustrated in Fig. 6.5:

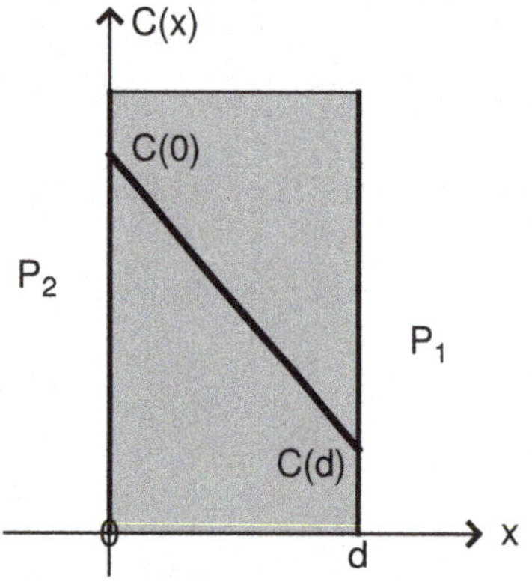

Figure 6.5: Steady state diffusion of a gas through a thin membrane of thickness d that separates a region at high gas pressure P_2 from another region at low pressure P_1.

In the setup, a thin membrane separates the experimental chamber into two sides. On Side 2, the diffusing gas is present at a high pressure P_2, whereas Side 1 is being pumped continuously so as to maintain a very low pressure P_1. If the membrane is raised to a suitably high temperature, gas from Side 2 will enter the membrane, will diffuse through it, and will exit into the vacuum on Side 1. The concentrations $C(0)$ and $C(d)$ at the surfaces inside the membrane are functions of the respective pressures P_2 and P_1 and the temperature of the membrane. $C(0)$ and $C(d)$ will change initially, but will adjust themselves in due time so that a steady state is achieved.

In this steady state, $C(x)$ is a linear function inside the membrane, thus giving rise to a constant diffusional particle flux $J = -D\, dC/dx$. This flux J is exactly equal to the flux of particles entering the membrane at $x = 0$ and, at the same time, leaving the membrane at $x = d$. Under this condition, $C(x)$ is indeed independent of time, yet there is a steady, constant flux of particles from Side 2 through the membrane to Side 1.

This type of situation is not merely hypothetical. It can be set up quite readily with a piece of Pd (palladium) foil and hydrogen gas. H atoms just happen to have the property of dissolving easily into Pd, thus being capable of setting up the necessary $C(x)$ that results in steady state diffusion of H atoms through the Pd membrane (H_2 molecules dissociate into H atoms upon entering the Pd and recombine on exiting the Pd on the other side).

Let us finish this section with a comment meant to shed further light on the meaning of the diffusion constant D in Fick's First Law, Eq. 6.5. Its form makes it clear that the flux J depends in equal fashion on D and on the concentration gradient dC/dx. So, from our earlier discussion of the atomic mechanism underlying diffusion, it should be evident that there must be a connection between D and atoms trying to jump from one site to another.

It can be shown that D is proportional to the jump frequency f_j in Eq. 6.1. Therefore, D can be expressed as

$$D(T) = D_0 \exp\left(-\frac{E_d}{RT}\right) \tag{6.6}$$

This implies that the constant D_0 is itself proportional to ν, the number of attempts for an atom to jump per second (cf. Eq. 6.1). It also means that D is the same strong function of T as is f_j (cf. Eq. 6.2).

In a diffusion process, clearly the diffusion coefficient D is of primary interest, i.e. the parameters D_0 and E_d. The parameter E_d in the exponential factor is often termed the **activation energy** for diffusion, as it represents the extra energy atoms must have in order to be able to jump to a neighboring site. In the analysis of data, Eq. 6.6 is often used in logarithmic form:

$$\ln D = \ln D_0 - \frac{E_d}{RT} \tag{6.7}$$

This form is especially convenient because it says that ln D should be a linear function of T^{-1}, since D_0, E_d, and R are all constants. In other words, if you plotted ln D vs. the inverse of the absolute temperature, you should get a straight line with slope $-E_d/R$.

An example of this type of analysis is shown below in the table with simulated data and the corresponding plot.

Table 6.2: Example of simulated data for $D(T)$

T(K)	$D(m^2/s)$	ln D	1000/T
600	1.01×10^{-11}	-25.3	1.14
700	3.14×10^{-11}	-24.2	1.03
800	7.90×10^{-11}	-23.3	0.932
900	1.01×10^{-10}	-22.5	0.853

The data are plotted as ln D vs. 1000/T rather than vs. 1/T because 1000/T produces more convenient numbers on the x-axis.

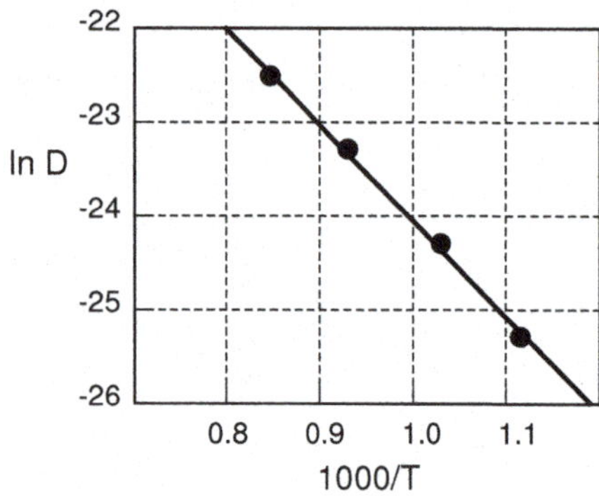

Figure 6.6: Plot of diffusion data from Table 6.2

Using the end points, the slope of the straight line can be deduced as follows:

$$\text{slope} = -\frac{E_d}{R} = \frac{4.0}{(1.19 - 0.80) \times 10^{-3}} \tag{6.8}$$

which yields

$$E_d = \frac{4R}{0.39 \times 10^{-3}} = 85 \text{ kJ/mol} \tag{6.9}$$

With E_d in hand we can obtain D_0 directly from Eq. 6.6 using any one of the data points, e.g. the one at 900 K:

$$D_0 = D \exp(\frac{E_d}{RT}) = 1.7 \times 10^{10} \exp \frac{85000}{8.31 \times 1173} = 1.0 \times 10^{-6}\, m^2/s \tag{6.10}$$

Note that, from a practical point of view, you should use data over as wide a temperature range as possible in order to get good accuracy for E_d and D_0.

6.3 Fick's Second Law: Non-Steady-State Diffusion

Steady-state diffusion is clearly a very special case, and in general you have to deal with the concentration C being a function of both time t and position x: $C = C(x,t)$. Again, we will confine ourselves to one-dimensional problems. In the most general situation, C would be a function of all three coordinates and time: $C = C(x,y,z,t)$.

The equation that describes the one-dimensional non-steady-state case is known as **Fick's Second Law**:

$$\frac{\partial C}{\partial T} = D \frac{\partial^2 C}{\partial x^2} \tag{6.11}$$

You will note that this equation involves so-called partial derivatives of $C(x,t)$ with respect to x and t. If you are not familiar with them, the recipe to take a partial derivative of a function is to just think of it as taking a normal derivative with respect to one variable while treating the other variable as a constant. Also, in Eq. 6.11 the assumption is made that D itself does not depend on $C(x)$, or indirectly on x, but is only a function of T as in Eq. 6.6.

Fick's Second Law is in essence an expression of the continuity of particles (or continuity of mass): It amounts to stating mathematically that the time rate of change in concentration in a small volume element is due to the sum total of particle fluxes into and out of the volume element.

Solving Eq. 6.11 is not simple, even in only one dimension. Most importantly, there is not just one solution, i.e. THE solution. Instead, there are many possible solutions, depending on the particulars of the situation. In general, a solution to Eq. 6.11 depends on the so-called initial condition, i.e. the form of $C(x,t)$ at $t = 0$, when your observation starts, and on the boundary conditions, i.e. on how the solution must behave at the physical boundaries of your observations.

We will not get into technical details of finding solutions to Eq. 6.11. Rather, we will examine the solution and its consequences for a certain important case of diffusion in materials. Based upon this discussion, we will be able to make a few further comments about non-steady-state diffusion in general.

6.3.1 Carburizing: An Approximate Solution

The case of interest involves the process of **carburizing**, which is a form of **case-hardening**. In case-hardening, the outermost layer of a workpiece is hardened selectively by local modification of the material. Carburizing realizes this effect in steel by increasing the carbon concentration in the near-surface region via diffusion. The advantage of performing this type of material modification only in the near-surface region is that the workpiece is strengthened where it is needed, at the surface, while the interior remains at its original strength thus maintaining higher toughness.

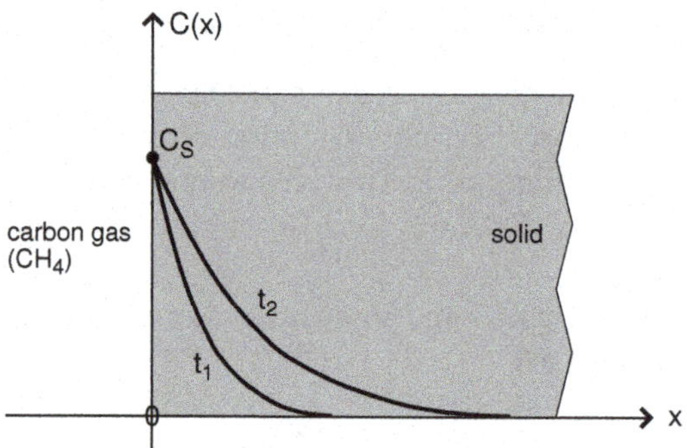

Figure 6.7: Schematic experimental setup for the carburizing process. The surface of the workpiece is at $x = 0$ and extends to very large x.

The experimental setup for the carburizing process is sketched in Fig. 6.7. The surface of the piece of iron exposed to the carbon-containing gas is located at $x = 0$, and the thickness of the piece of steel is large. (Mathematically speaking, the solid is assumed to be semi-infinite.) Most importantly, the pressure of the gas and the temperature of the iron are maintained constant, such that the concentration of carbon just inside the surface has a constant value C_S. For simplicity, we also assume that at the beginning of the process, the interior of the piece of iron does not contain any carbon. (It will be easy to relax this assumption later).

Under these conditions, carbon atoms are dissolved into the surface of the iron sample and are present at a constant concentration C_S at $x = 0$. Thus a gradient in $C(x)$ exist near the surface, so that carbon atoms diffuse gradually into the bulk of the sample. The longer we wait, the more extended the concentration profile $C(x)$ becomes, as the carbon atoms move farther and farther into the sample. Representative shapes of $C(x)$ at times t_1 and t_2 (with $t_1 < t_2$) are displayed in Fig. 6.7, .

An exact analytical solution for $C(x)$ in this problem can be given and will be discussed later, but we will examine first an approximate solution: We will assume that $C(x)$ is a straight line just as in the steady state case, except that now the slope of this line is allowed to vary with time. Fig. 6.8 suggests that this may be a reasonable approximation near the surface.

This amounts to assuming that $C(x)$ can be specified by the surface concentration C_S and the increasing length $l(t)$, which is a measure of the progress of the diffusion process. $l(t)$ can be determined as follows.

Let $C(x)$ here refer to diffusing solute atoms. Thus the total number $N(t)$ of these atoms in the sample per unit area is given by the area of the triangle with base $l(t)$ and height C_S (see Fig. 6.8).

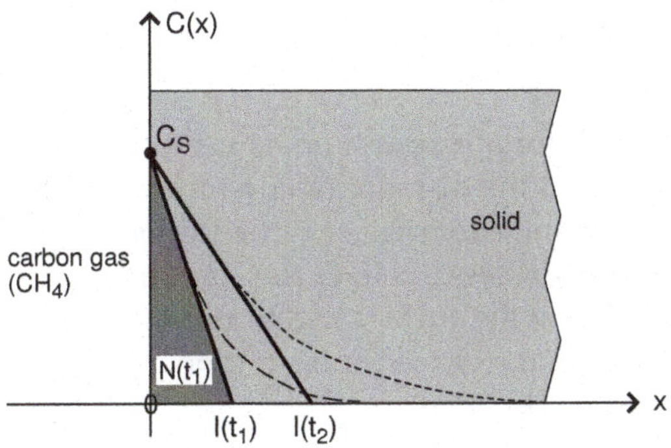

Figure 6.8: Linear approximation for $C(x)$ in the carburizing process. The change of the length l with t indicates the progress of diffusion.

$$N(t) = \frac{1}{2} C_S l(t) \tag{6.12}$$

$N(t)$ changes continually because more atoms are moving into the solid. Specifically, the rate of change of $N(t)$ is equal to the flux J of diffusing atoms:

$$\frac{dN}{dt} = J = -D\frac{dC}{dx} \tag{6.13}$$

where we have used Fick's First Law for the flux J. The gradient of $C(x)$ can be expressed in terms of C_S and l from Fig. 6.8, and the rate of change of $N(t)$ from Eq. 6.12:

$$\frac{1}{2} C_S \frac{dl}{dt} = -D(-\frac{C_S}{l}) \tag{6.14}$$

which can be simplified to

$$\frac{1}{2}\frac{dl}{dt} = \frac{D}{l} \tag{6.15}$$

This last equation is equivalent to

$$\frac{1}{2} l \, dl = D dt \tag{6.16}$$

Eq. 6.16 can be integrated easily to yield

$$\frac{1}{4} l^2 = Dt \tag{6.17}$$

and, solving for $l(t)$:

$$l(t) = 2\sqrt{Dt} \tag{6.18}$$

This is the final result for our approximate treatment of the carbonizing diffusion process: It says that the distance over which the diffusing species move into the solid with time increases as $\sqrt{t}$. In addition, $\sqrt{Dt}$ is a characteristic length for the diffusion process: It indicates the distance at which $C(x)$ has fallen off to half its value at the surface and thus denotes the width of $C(x)$.

If you are interested in the entire concentration profile $C(x)$ as a function of time t, this can be written down as

$$C(x,t) = C_S(1 - \frac{x}{l(t)}) = C_S(1 - \frac{x}{2\sqrt{Dt}}) \tag{6.19}$$

Of course it is understood that this equation is only valid for $0 \leqslant x \leqslant l(t)$. Note that in principle x and t in Eq. 6.19 are independent variables, yet they occur only as the special combination $x/2\sqrt{Dt}$ in the expression for $C(x,t)$.

6.3.2 Carburizing: The Exact Solution

Now you are ready for the exact solution of our problem. I will simply present it and discuss its consequences. The initial condition and the boundary conditions are the same as spelled out above: In essence, $C(0,t)$ remains constant at C_S for all t, and $C(x,t)$ must taper off to 0 as x becomes very large. Again, for the time being we assume that the sample contains no carbon at the start of the diffusion process.

The exact solution cannot be given as an algebraic formula, but only in the form:

$$C(x,t) = C_S(1 - \text{erf}(\frac{x}{2\sqrt{Dt}})) \tag{6.20}$$

where

$$\text{erf}(z) = \frac{2}{\sqrt{\pi}} \int_0^z e^{-u^2} du \tag{6.21}$$

The function $\text{erf}(z)$ is known as the Gaussian error function. As you apply it to Eq. 6.20 and a diffusion problem, you need to set $z = \frac{x}{2\sqrt{Dt}}$.

You can see that x and t enter into the solution $C(x,t)$ as parts of the upper limit of an integral. Furthermore, and most interestingly, again they occur only in the combination of $x/2\sqrt{Dt}$! Also, note the remarkable similarity between Eqs. 6.19 and 6.20.

Let us now compare the approximate and the exact solution for $C(x,t)/C_0$, as per Eqs. 6.19 and 6.20, and using $z = x/2\sqrt{Dt}$ as a generalized independent variable. That is, we will compare the functions 1-z and 1 - erf(z), where the latter needs to be evaluated numerically.

It is apparent from the graph that the linear approximation is remarkably accurate to $z \approx 0.7$, which translates to $x \leqslant 1.4\sqrt{Dt}$. Of course, at large z (i.e. basically large x) the linear approximation fails as it extends only to $z = 1$.

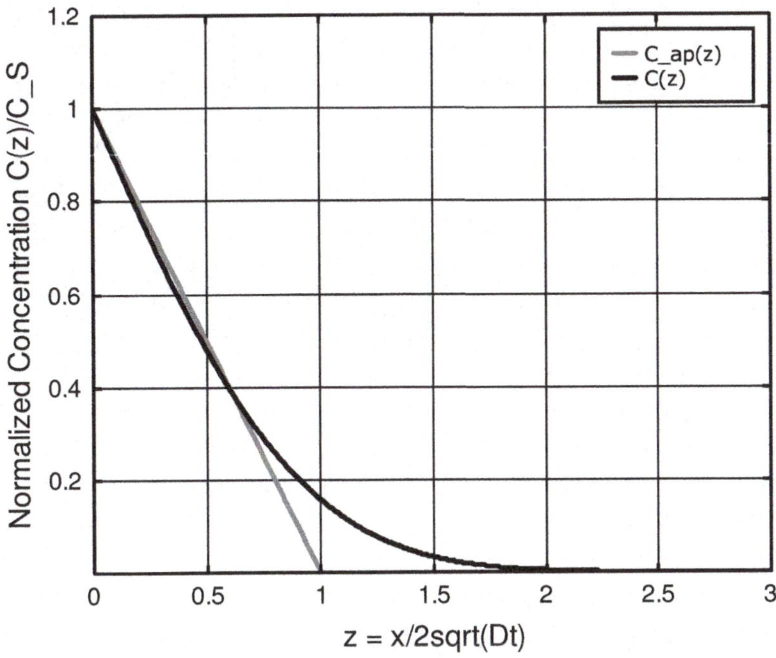

Figure 6.9: Linear plots of $C(x)/C_S$ from Eqs. 6.19 (gray) and 6.20 (black)

If the region $C(x)/C_S \ll 1$ (large z) is of interest, it is better to use a semilog graph (Fig. 6.10). For the carburizing process, the region of large z is generally not important, but it is crucial in certain applications of diffusion in electronic materials.

The most important property of the solution to the carburizing problem is that, although in principle x and t are independent variables and D is an additional experimental parameter, the variable which really counts is their combination in the form $x/2\sqrt{Dt}$. So, it is almost as if there were in fact only one solution, but many different combinations of x, D and t that can produce that solution.

For example, if you need to know the actual diffusion profile $C(x)$, then for any given x, $C(x)$ can be achieved with any number of combinations of D and t as long as their product Dt stays the same. Since D is a function of T, this also means that a particular $C(x)$ can be realized at any number of different

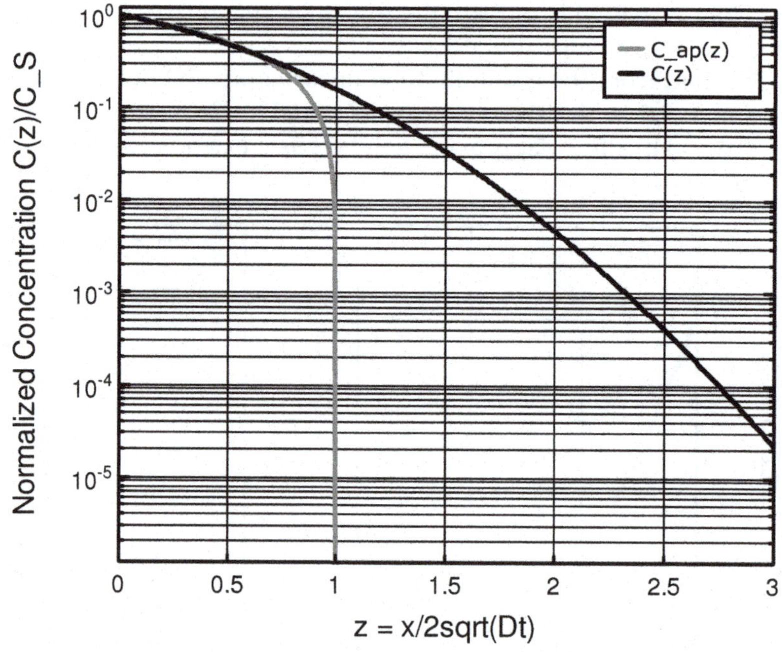

Figure 6.10: Semilog plots of $C(x)/C_S$ from Eqs. 6.19 (gray curve) and 6.20 (black curve)

temperatures, if the diffusion time t is chosen accordingly. We will explore this aspect further in the next section with some numerical examples.

Throughout the discussion up to this point, we have assumed that at the beginning of the diffusion process, the sample does not contain any solute (the diffusing species). This restriction can be relaxed easily, with the result that $C(x)/C_S$ in Eqs. 6.19 and 6.20 needs to be replaced with $(C(x) - C_0)/(C_S - C_0)$, where C_0 is the initial concentration of the solute.

Explicitly, the generalized forms of Eqs. 6.19 and 6.20 would read:

Approximate solution:

$$\frac{C(x) - C_0}{C_S - C_0} = 1 - \frac{x}{2\sqrt{Dt}} \tag{6.22}$$

Exact solution:

$$\frac{C(x) - C_0}{C_S - C_0} = 1 - \text{erf}(\frac{x}{2\sqrt{Dt}}) \tag{6.23}$$

So C_0 is the reference value against which other concentrations are compared. For a carburizing process, C_0 is typically not negligible, because even very-low carbon steel does contain a small amount of carbon.

6.3.3 Carburizing: Numerical Examples

We will do some numerical exercises for the carburizing system, C in BCC Fe. The relevant diffusion parameters are:

$$D_0 = 6.0 \times 10^{-7}\,\mathrm{m^2/s};\quad E_d = 80\,\mathrm{kJ/mol}$$

In formulae we will use SI units throughout: distances in m, times in s, and temperatures in K, even if the question is formulated in terms of other units.

(1) Calculate numerical values for $2\sqrt{Dt}$

As a warm-up, let us develop a sense for the orders of magnitude involved by calculating the characteristic length $l(t) = 2\sqrt{Dt}$ for a time of 1 hour at 600°C. We obtain for $l(t)$:

$$D = 6.0 \times 10^{-7} \exp - \frac{8 \times 10^4}{8.314 \times 873} = 9.8 \times 10^{-12}\,\mathrm{m^2/s}$$

$$l(t = 1\mathrm{hr}) = 2\sqrt{Dt} = 2\sqrt{9.8 \times 10^{-12} \times 3600} = 3.8 \times 10^{-4}\,\mathrm{m}$$

Values for D and $2\sqrt{Dt}$ at some temperatures of interest are listed in the table below. The time t is 1 hour, or 3600 s:

Table 6.3: Parameters for Diffusion of C in BCC Fe

T (K)	D ($\mathrm{m^2/s}$)	$2\sqrt{Dt}$ (m)
293	3.3×10^{-21}	6.9×10^{-9}
773	2.4×10^{-12}	1.8×10^{-4}
873	9.8×10^{-12}	3.8×10^{-4}
973	3.0×10^{-11}	6.6×10^{-4}

It is evident that measurable diffusion does not take place on a macroscopic scale in an hour at room temperature, but it does in the range of 500-700°C.

(2) Calculate the diffusion time to achieve $C(x)$

You need to carburize some steel workpieces so that the carbon concentration at a depth of 0.1 mm is 0.5 wt%. Your supervisor suggests using a temperature of 700°C. The surface concentration will be maintained at 1 wt% and the initial concentration in the steel is 0.2 wt%. You decide that under these conditions, the approximate solution Eq. 6.22 will be good enough:

$$\frac{C(x) - C_0}{C_s - C_0} = 1 - \frac{x}{2\sqrt{Dt}}$$

This works out numerically as follows:

$$\frac{0.5 - 0.2}{1.0 - 0.2} = 0.375 = 1 - \frac{x}{2\sqrt{Dt}}$$

$$\frac{x}{2\sqrt{Dt}} = 0.625$$

$$x = 10^{-4} = 0.625 \times 2\sqrt{Dt}$$

$$t = \frac{x^2}{0.625^2 \times 4D} = \frac{10^{-8}}{1.56 \times 3.0 \times 10^{-11}} = 213\,\text{s}$$

(3) Recalculate the diffusion time for different conditions

You conclude that a time of 850 s is too short, but 1200 s would be about right. So what should the temperature be in that case?

Here we can take advantage of the fact that the critical variable in the process is $x/2\sqrt{Dt}$. We want to keep the concentration profile $C(x)$ fixed. So, for a given x that means the quantity $\sqrt{Dt}$ has to stay the same, and therefore so does Dt. If D_1 and t_1 refer to the temperature $T_1 = 973\,\text{K}$, and D_2 and t_2 to the unknown temperature T_2, the condition on D_2 follows from

$$D_1 t_1 = D_2 t_2$$

Therefore

$$D_2 = D_1 \frac{t_1}{t_2} = 3.0 \times 10^{-11} \frac{213}{1200} = 5.3 \times 10^{-12}\,\text{m}^2/\text{s}$$

What remains now is to determine the T_2 that gives rise to the new D_2. We must solve

$$5.3 \times 10^{-12} = 6 \times 10^{-7} \exp\left(-\frac{8 \times 10^4}{8.314 \times T_2}\right)$$

for T_2. After taking the natural log and some additional algebra, we obtain

$$T_2 = \frac{8 \times 10^4}{11.6 \times 8.314} = 830\,\text{K}$$

Hence, the specified concentration profile $C(x)$ is achieved with a diffusion time of 1200 s at a temperature of 830 K, or 557°C.

(4) Determine x at a point where $C(x)$ is very low.

For the concentration profile achieved in the previous process, at what distance would $C(x)$ be 10% above the background carbon concentration?

What I am asking is: For which x is $C(x) = 0.22$ wt%, since $C_0 = 0.2$ wt%. This low a concentration requires that we use the exact solution, Eq. 6.23. We are looking for the value of z that yields $(C(x) - C_0)/(C_S - C_0) = (0.22 - 0.2)/(1.0 - 0.2) = 0.025$:

$$\frac{C(z) - C_0}{C_S - C_0} = 1 - \text{erf}(z) = 0.025$$

We get this z value from the semilog plot, Fig. 6.10: My reading of the figure is that

$$1 - \text{erf}(z) = 0.025 \text{ for } z = 1.6$$

Go to 0.025 on the y-axis, and read the corresponding x-value from the black curve. So this means that

$$x/2\sqrt{D_2 t_2} = 1.6$$

and therefore

$$x = 3.2\sqrt{D_2 t_2} = 3.2\sqrt{5.3 \times 10^{-12} \times 1200}$$

where I have used the values for D_2 and t_2 from above. The final result is:

$$x = 2.4 \times 10^{-4}\text{m}$$

The concentration $C(x)$ will be 0.22 w% at a depth x of 0.24 mm. By comparison, remember that above we found $C(x)$ to be 0.5 w% at a depth x of 0.1 mm.

(5) Determine t so that a very low $C(x)$ is reached at a certain x.

This is a somewhat hypothetical scenario for carburizing, but it is encountered frequently in diffusion applications for electronic materials, e.g. the diffusion of impurities in Si.

Let us take the diffusion coefficient D_2 from Exercise 3 above: $D_2 = 5.3 \times 10^{-12}\text{m}^2/\text{s}$ and ask: What time would it take for $C(x)$ to be 1 % above its initial value at a depth $x = 0.4$ mm? Or in other words: At what t will $C(x = 0.4\text{mm}) = 0.202$ w% ?

As before, we let $C_S = 1$ w% and $C_0 = 0.2$ w%. Of course, this exercise requires that we use the exact diffusion solution.

These parameters imply that we are looking for the value of z at which

$$1 - \text{erf}(z) = \frac{C(x) - C_0}{C_S - C_0} = \frac{0.202 - 0.2}{1.0 - 0.2} = 0.0025$$

From the black curve on Fig. 6.10 we see that this condition is fulfilled at z = 2.14. Thus we must have

$$\frac{x}{2\sqrt{Dt}} = 2.14$$

We solve this for t to arrive at the final result,

$$t = (\frac{x}{2 \times 2.14})^2 \frac{1}{D_2}$$

which in numerical terms is

$$t = (\frac{4 \times 10^{-4}}{4.28})^2 \frac{1}{5.3 \times 10^{-12}} = 1650 \text{ s}$$

(6) Determine E_d and D_0 from measurements of $D(T)$

A very common outcome of diffusion measurements is a set of data for the diffusion coefficient D as a function of temperature T. Of interest are the two parameters determining $D(T)$, namely E_d and D_0. Speaking mathematically, what we have is a system with two unknowns, and thus we need two independent equations for it.

Assume that we have two data points for $D(T)$ at two temperatures: $D(T_1)$ and $D(T_2)$. From the general form of $D(T)$ in Eq. 6.6 it follows that

$$\frac{D(T_2)}{D(T_1)} = \frac{D_0 \exp(-E_d/RT_2)}{D_0 \exp(-E_d/RT_1)} = \exp(\frac{E_d}{RT_1} - \frac{E_d}{RT_2})$$

By taking the ratio of the two values for D we have eliminated one unknown, D_0, and have arrived at one equation for the other unknown, E_d. Solving for E_d is straightforward by taking the natural log on both sides. This gives

$$E_d = \frac{\ln \frac{D(T_2)}{D(T_1)}}{\frac{1}{RT_1} - \frac{1}{RT_2}}$$

To check our work numerically, let us take the values for D from the table in Exercise 1 above: $D_1 = 2.4 \times 10^{-12}$ at $T_1 = 773$ K, and $D_2 = 3.0 \times 10^{-11}$ at $T = 973$ K:

$$E_d = \frac{\ln \frac{3.0 \times 10^{-11}}{2.4 \times 10^{-12}}}{\frac{1}{8.314 \times 773} - \frac{1}{8.314 \times 973}} = 79 \text{ kJ/mol}$$

This is close enough to the 80 kJ/mol we started out with when calculating the values for D in the first place. The difference is due to rounding errors and shows that you need accurate data for this kind of analysis.

For determining the yet unknown D_0, we take Eq. 6.6, solve it for D_0, and plug in one of the two data points for $D(T)$. For example, with T = 773 K,

$$D_0 = D \exp \frac{E_d}{RT} = 2.4 \times 10^{-12} \exp \frac{79 \times 10^3}{8.314 \times 773} = 5.2 \times 10^{-7} m^2/s$$

The original D_0 was 6.0×10^{-7} m^2/s. Of course, we would have got the same result for D_0 using the other data point at 993 K. You will have noted the similarity of this analysis and what we did earlier, with Fig. 6.6.

6.3.4 More on Non-Steady-State Diffusion: Drive-in

In the two previous sections, we examined in great detail a certain case of non-steady-state diffusion, namely ***carburizing***. Recall that its key feature is the ***constant surface concentration*** of the solute, i.e. the diffusing impurity. This causes the total amount of solute in the sample to increase during the diffusion, as the concentration profile extends further and further into the sample.

Here we will look briefly at another scenario, which is less common in metallurgy, but is very important in the fabrication of integrated electronic circuits. The process is referred to as drive-in in the electronics industry.

The interesting point with respect to diffusion is that both cases are described by the exact same equation, namely Fick's Second Law of Eq. 6.11. Yet, they differ in the experimental particulars, specifically in the boundary and initial conditions, and these different conditions yield mathematically and physically different solutions.

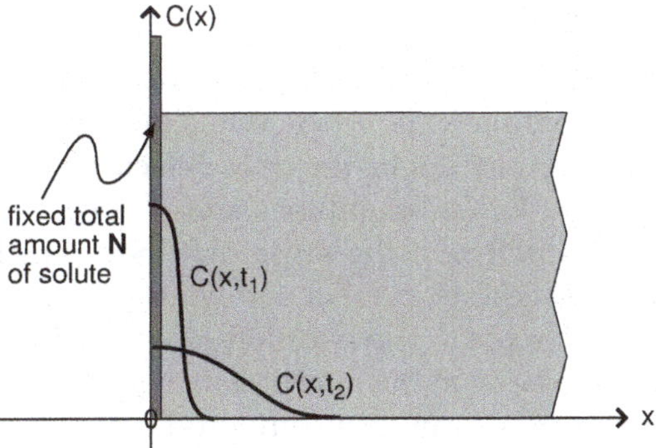

Figure 6.11: Qualitative description of the progress of diffusion in Drive-in. The total amount of diffusant is fixed at a value N.

The main feature of the ***drive-in*** scenario is that the ***total amount of solute to diffuse into the sample is constant*** and fixed at the outset. For example, imagine a finite amount of solute having been deposited onto the surface of the sample, and then diffusion simply redistributes the amount that is there, driving it gradually deeper into the sample. Hence the name drive-in. (We also assume, as before, that no solute is present in the sample at the beginning of the diffusion and that the sample is very thick).

We will not discuss the quantitative solution of the problem here, but we will simply describe how the diffusion process evolves qualitatively.

Think of the initial distribution $C(x)$ as a sharp spike at $x = 0$ with area N. This is displayed schematically in Fig. 6.11 by the dark gray rectangle. The area under the curve $C(x)$ represents the total amount of solute and remains fixed at a value N through the entire process. Therefore, the concentration profile becomes lower and wider with increasing diffusion time.

In order to appreciate the difference between drive-in and the previous case of carburizing, please compare Figs. 6.7 and 6.11. Clearly, the behavior of $C(x,t)$ at $x = 0$ is very different. On the other hand, the profiles appear to taper off in a similar fashion towards large x.

Additionally, note that the two cases of diffusion resemble each other closely in one important respect. It turns out that the exact solution to drive-in again depends on x and t only via the universal variable $x/2\sqrt{Dt}$. The major conclusion we can draw from this result is that the width of the diffusion profile increases as $\sqrt{t}$ in both cases.

If you think of this width as the average distance x_{av} traveled by the diffusing atoms, then $x_{av} \propto \sqrt{t}$ is a unique feature of diffusion. It comes about due to x_{av} being the result of a huge number of *random jumps* by a huge number of atoms. If the atoms all jumped in the same direction, at a constant rate, you would end up with uniform directed motion where $x \propto t$.

6.3.5 Concluding Remarks on Diffusion

Throughout our discussion of diffusion, I emphasized the random nature of the events occurring at the atomic scale when atoms exchange places with vacancies. However, vacancies may not be the only defects enabling atoms to jump from one site to another. Grain boundaries and dislocation lines may in fact provide easier diffusion paths, in the sense that they constitute defects with more extended disorder and thus present pathways with a lower activation energy for the atoms to overcome. The question of which diffusion path is most prevalent is often a main issue in the interpretation of diffusion experiments.

Also, our discussion has focused mostly on the diffusion of impurities in metals. Yet many of our considerations have been generic and thus apply to other materials as well. However, it is useful to add a few specific remarks with respect to diffusion in other classes of materials.

For diffusion in ceramics, the overriding principle is, of course, that whatever happens in diffusion, charge neutrality must be preserved. If you look at the various types of point defects in ionic materials that we discussed in Chapter 4, you will realize readily, that constraints will be placed on what type of ion can move where. Also, as a rule of thumb, it is generally easier for cations to diffuse than for anions, due to the size difference between the two. Anions must induce a greater distortion of the lattice in order to be able to move than do the cations.

Diffusion can also occur in polymers. For those materials, one needs to distinguish between the diffusion of very large molecules, such as those making up the polymer itself, and diffusion of small molecules or atoms. The first situation occurs, for example, when two types of polymers are in contact and are heated to a temperature where they can interdiffuse, or when polymers crystallize or are annealed. The second situation is more similar to what we considered in the main part of our discussion, as it involves motion of small species from one small local region of lower density to another. An example for the second case would be the permeation of gases through a polymer film in food packaging.

In electronic materials, the diffusion of charged species such as electrons is often of great interest. As long as we did not have to worry about such charges, it was appropriate so say that the concentration gradient alone provided the driving force for the diffusion process, as embodied in Fick's First Law. However, a concentration gradient of charged particles also brings with it an electric field. Thus the motion of electrons, especially in a semiconductor, is a consequence of the combined forces from a concentration gradient and an electric field. Because of this dual influence, it is entirely possible that electrons end up moving in opposition to the concentration gradient.

Image sources

Title pictures: Everyday objects made from case-hardened steel
`http://www.engineershandbook.com/MfgMethods/casehardening.htm`
`http://www.m-i.be/EN/Abloy.html`
`http://www.motorcycle-superstore.com/BRTP/nickel-countershaft-sprocket.aspx`

References

Both Guy and Van Vlack provide diffusion parameters for additional systems. Guy goes into greater depth on some of the diffusion topics and also describes other applications, such as sintering in ceramics, soldering and brazing of metals, and joining of metals and ceramics.

A more extensive tutorial on the physical fundamentals of diffusion, at a somewhat higher level, can be found at
`http://www.tf.uni-kiel.de/matwis/amat/def_en/`

The same site also has nice 2D diffusion animations illustrating different diffusion mechanisms
`http://www.tf.uni-kiel.de/matwis/amat/def_en/kap_3/backbone/r3_2_1.html`
and a discussion of methods to determine diffusion profiles, e.g. radioactive tracers, etc.
`http://www.tf.uni-kiel.de/matwis/amat/def_en/kap_3/backbone/r3_3_1.html`

Commercial companies offering diffusion processing services sometimes provide useful practical information. See for example:
`http://www.edwardsheattreating.com/services.htm`
`http://www.imoa.info/moly_uses/moly_grade_alloy_steels_irons/case_hardening_steel.php`

Exercises

1. For the diffusion of Zn in Cu, E_d = 189 kJ/mole. Calculate the exponential probability factor for Zn in Cu at 500°C and compare it to the example given in the text for C in Fe. For both systems also calculate the exponential probability factor at 900°C. Comment on the relative changes.

2. What might account for the difference in the temperature dependence between the diffusion of C in BCC Fe and Zn in Cu? (This question has a simple superficial answer, but a more subtle explanation of it.)

3. You need to carburize a piece of steel so that the carbon concentration at a depth of 0.5 mm is 0.4 wt%. The surface concentration will be maintained at 1.2 wt% and the initial carbon concentration in the steel is reported to be 0.2 wt%. You decide to perform the procedure at 560°C. How long will it take? You can use the approximate treatment.

4. You found out after you finished your procedure of Exercise 3 that the piece of steel had had an initial carbon concentration of 0.3 wt%. What difference did that make to the outcome of your experiment?

5. Using the same conditions as in Exercise 3 and the concentration profile they produce, how could you achieve the same result in half the time?

6. In a certain diffusion experiment it is found that the diffusion coefficient increases to 3 times its original value when the temperature is raised from 800°C to 900°C. What is the activation energy E_d in this case?

7. For the diffusion of Zn into pure Cu, the parameters are:

   $D_0 = 2.4 \times 10^{-5}$ m^2/s and $E_d = 189$ kJ/mol.

Use the approximate treatment for the following:
a) Determine the diffusion profile $C(x)$ after diffusion at 850°C for 10 minutes if the surface concentration of Zn is 10%.
b) You conclude that this layer is much too thin, and you need to produce a diffused layer with a width of 0.1 mm (where $C(x) = 0$). What would the temperature have to be to create this new layer in 15 minutes?

8. The diffusion of B into Si is performed with the same type of process as carburizing steel.
The initial concentration of B in Si is negligible. At the process temperature of 1100°C, the surface concentration of the B is 2×10^{20} atoms/cm^3. The diffusion coefficient of B in Si has the form

$$D_{B,Si} = 7.6 \times 10^{-5} \exp -(333 \times 10^3/RT)$$

where D_0 is in m^2/s and E_d in J/mol.
How long will it take for the concentration of boron to reach a value of 2×10^{16} atoms/cm^3 at a depth of 1 μm below the Si surface?

9. For the diffusion of P in Si it has been found that

$$D = 1 \times 10^{-18} \text{m}^2/\text{s} \quad \text{at} \quad 1000° \text{C}$$

$$D = 8 \times 10^{-16} \text{m}^2/\text{s} \quad \text{at} \quad 1300° \text{C}$$

What is the activation energy E_d for this process?

Problems

1. A process related to carburizing is known as carbonitriding. Find out what this process is about, how it is implemented, and how its results compare with carburizing.

2. Research the processes of diffusion bonding and brazing. Describe their differences and give some applications of each.

3. a) Use results from the approximate treatment of the carburizing process to derive an expression for the total amount of solute $N(t)$ diffusing into the solvent as a function of time.
b) Estimate how accurate the approximate treatment is for $N(t)$ by comparing it to the exact treatment. See Fig. 6.9.
c) What kind of a function of t would the exact total amount of solute be?

4. In the main text of this chapter, I mentioned that in the drive-in process the width of the concentration profile increases as $\sqrt{t}$ because $C(x,t)$ is a function of the universal variable $x/2\sqrt{Dt}$. What can you conclude about how $C(x=0,t)$ depends on time in drive-in?

5. From a fundamental physics point of view, a process called self-diffusion is of interest. It refers to the motion of atoms in their own host material by the usual diffusion mechanisms, such as hopping into a vacancy.

   Of course, in a pure metal sample there cannot be a concentration gradient driving diffusion of the host atoms themselves. However, if the sample contains a localized concentration of a radioactive isotope of the host atoms, the spreading out by diffusional motion may be observable.

   For the self-diffusion of Ag, the following parameters have been determined for the mechanisms of bulk crystal and grain boundary diffusion (data from Van Vlack):

   Parameters for the self-diffusion of silver

bulk crystal diff.	$D_0 = 0.4 \times 10^{-4}\ m^2/s$	$E_d = 185\ kJ/mol$
grain bound. diff.	$D_0 = 0.14 \times 10^{-4}\ m^2/s$	$E_d = 90\ kJ/mol$

   a) At what temperature are the diffusion coefficients for the two mechanisms the same?
   b) Which diffusion mechanism dominates at one half the melting temperature?
   c) How far, in an average sense, will Ag atoms diffuse in 1 minute at one half of the melting temperature of Ag?

6. You know by now that the diffusion coefficient D is a strong function of temperature. At a typical temperature where substantial diffusion occurs, for which system would you expect diffusion to be faster: Zn in Cu or Cu in Zn? State your reasons.

7. Why are the activation energies so high for the diffusion of B and P in Si, compared to the examples we gave for diffusion of impurities in metals? (See Exercises 8 and 9).
   Note that these impurities are substitutional.

8. In what process, or processes, mentioned in Chapter 5, does self-diffusion play an important role?

Chapter 7
Phase Changes

Overview

A material of a given composition can exist in several different forms called phases. The simplest example would be an element, where these phases comprise the material as a solid, as a liquid, or as a gas. Certain materials may exist in several solid phases which differ only in their crystal structure. The nature of these phases depends on the number of components making up the material, i.e. on the number of its basic building blocks. The simplest case is a one-component system, such as an element, H_2O, or CO_2. Temperature and pressure determine which of the possible phases is thermodynamically stable. Under special conditions, in a one-component system two or even three phases can coexist, as one phase is transformed into another. The transformation of primary interest is the one involving melting/solidification, especially as it applies to crystalline and amorphous materials.

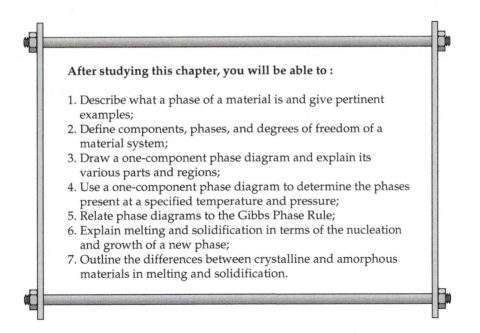

After studying this chapter, you will be able to :

1. Describe what a phase of a material is and give pertinent examples;
2. Define components, phases, and degrees of freedom of a material system;
3. Draw a one-component phase diagram and explain its various parts and regions;
4. Use a one-component phase diagram to determine the phases present at a specified temperature and pressure;
5. Relate phase diagrams to the Gibbs Phase Rule;
6. Explain melting and solidification in terms of the nucleation and growth of a new phase;
7. Outline the differences between crystalline and amorphous materials in melting and solidification.

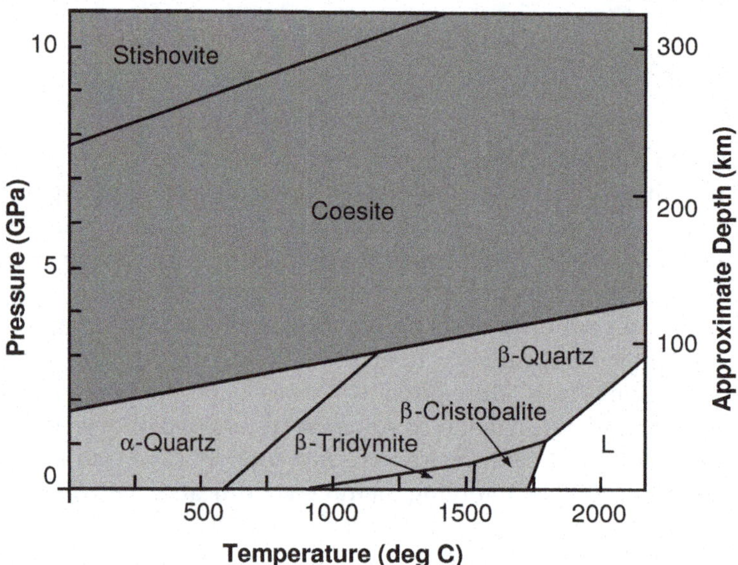

The picture shows a so-called phase diagram for SiO_2. This is a graph which delineates regions of stability for various solid forms of Quartz as well as molten SiO_2 as a function of temperature and pressure in thermal equilibrium. The solid forms are all crystalline, with composition SiO_2, but have different crystal structures, i.e. different unit cells. Note the large range of pressure on the left y-axis, and the corresponding depth below the Earth's surface on the the right y-axis.

7 Phase Changes

In this second of three chapters dealing with temperature-dependent phenomena in solids, we will study the state in which a material exists at different temperatures and pressures. We will be especially interested in situations where two or three phases of a material, meaning the solid, liquid, or gaseous form, are in thermal equilibrium.

We will focus first on so-called one-component systems, i.e. in essence pure materials, for example elemental metals, elemental carbon, simple ceramics like SiO_2, or compounds such as H_2O or CO_2. This is in order to lay the groundwork for the discussion of two-component systems such as metallic alloys in the next chapter.

The practical motivation is to understand the process of melting and solidification, as this is the most important fundamental pathway by which engineering materials are prepared. The details of the solidification process in particular turn out to be crucial in determining the microstructure, and thus the properties, of most engineering materials.

7.1 Phases of Materials

As mentioned above, our ultimate goal is to examine engineering materials such as metallic or ceramic alloys, which are systems containing more than one component, and therefore may consist of more than one solid phase. In order to do so, we need first to introduce some new terminology, including several definitions. Some of these definitions may not seem to be very precise to you, but their meaning will become clear in their application.

A **phase** is a part of a material that is homogeneous but distinct from other parts in its structure and/or composition. A **component** is a basic atomic or molecular unit making up the material. Also, in Chapter 4 we had introduced the notion of a solid solution, consisting of a solid solvent with solute in it. Here we will discuss a **solution** in general, liquid or solid, as a material consisting of a single phase with more than one component, and a **mixture** as a material consisting of more than one phase.

Some examples should clarify the meaning of these terms. First, consider a glass of pure tap water with ice cubes in it. For all practical purposes, that would be a mixture with one component, namely H_2O, and two phases, namely liquid H_2O (water) and solid H_2O (ice). Note that the component here is H_2O, that is the molecule which is the natural building block for the material, and not the individual atoms H and O.

As a second example consider sea water. For all practical purposes, this is a solution of NaCl in H_2O, i.e. a system with a single phase but two components.

The solution is the liquid phase, and note again that even though Na^+ and Cl^- ions are separated in the solution, it is NaCl that is the second component, besides water. Na^+ and Cl^- are added as a unit to the water. A good indicator for something to be truly a component is that its amount in the system is an independent variable by itself. In that respect, Na^+ and Cl^- are not independent, and thus are not separate components.

From the sea water example it is evident that once we deal with more than one component, an important variable for the system is its composition. The other external variables of interest are usually the temperature and the pressure.

For the most part, we will be interested in a system in **thermal equilibrium**. By this we mean a state in which no further changes (e.g. in temperature or pressure) or further reactions take place. Sometimes, we may just use the short term **equilibrium**, and from the context it will be apparent that this refers to the thermal rather than the mechanical state of the system.

A subtle point about phases is that even for a pure solid, consisting of a single component, several solid phases may be possible. A good example is SiO_2 (see the title figure). Depending on the temperature and pressure, it can exist in several different crystalline phases, that is, as a solid with the exact same composition given by the formula SiO_2, but with different crystal structures, i.e. different unit cells. Moreover, depending on the method of preparation, SiO_2 can also be an amorphous solid.

Another example of an even simpler system with different solid phases is the element iron. At room temperature and normal ambient pressure of 1 atm, the stable phase of pure Fe has a BCC structure. If you heat it up to 912°C and above, it transforms into an FCC structure. This change is entirely reversible, if you go down again in temperature. But if you heat it up further, at 1394°C it transforms into yet another BCC structure, before it melts at 1538°C. We will discuss the phases of Fe and in particular Fe-C alloys, also known as steels, in more detail in the next chapter.

A third example is elemental carbon. It has several solid phases, the best-known ones being diamond and graphite. It turns out that in the thermodynamic sense, the most stable phase is in fact graphite, and not diamond. However, from a practical point of view, diamonds are forever because their transformation to graphite takes a very long time. In this type of situation we would call diamond a **metastable phase**. We will encounter a number of other examples of metastable phases when we deal with metal alloys (two-component systems). The common characteristic of such systems is that the rate at which the metastable phase is transformed into the true equilibrium state is extremely slow under normal ambient conditions, so that for all practical purposes the metastable phase can persist indefinitely.

Finally, it should be pointed out that thermodynamic equilibrium, as defined above, is defined by a special kind of thermal energy called **free energy**. This free energy contains contributions from the internal interactions between the atoms or molecules of the system, and also from the amount of order or disorder in the system. We will not deal with these subtleties here but simply think of equilibrium as the state in which the system has attained the lowest possible energy relevant for the particular circumstances.

7.2 The Generic One-component Phase Diagram

A phase diagram is a graph that delineates regions of stability for various phases of a system. If we deal with a one-component system, the graph is usually a temperature-pressure diagram. A generic example is shown in Fig. 7.1:

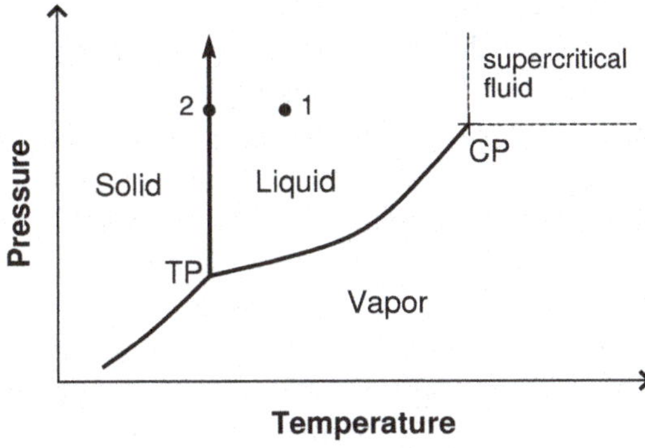

Figure 7.1: Generic phase diagram for a one-component system

The main feature of the diagram is the three regions labeled Solid, Liquid, and Vapor. These regions are separated by three lines which come together in one point called **triple point** and labeled TP. For the time being, we can assume that one line goes up from the triple point essentially in the vertical direction for as far as the graph extends, whereas the second line pointing towards the upper right ends in a well-defined point called critical point and labeled CP.

The location of any point specified by a combination of pressure p and temperature T indicates what state the system is in. The way to read this diagram is as follows:

1. If a point (T,p) is *inside one of the three regions*, it designates a state where only one phase is present, as per the label of the region. For example, the point 1 in Fig. 7.1 refers to a system which is entirely liquid.

2. If a point (T,p) is *exactly on a line*, it designates a state where the two phases on either side of the line coexist and are in equilibrium. Thus, at point 2 in Fig. 7.1 liquid and solid coexist, and therefore point 2 is the melting point of the system at that pressure.

3. The triple point TP is the one combination of T and p at which all three phases can coexist.

The nearly vertical line going up from the triple point is sometimes called the melting line, and the line extending from TP to the critical point CP is sometimes called the boiling line. These names are self-evident. On the melting line solid and liquid coexist, and hence the line defines the melting/freezing point. On the boiling line liquid and vapor coexist, and when that line is crossed, evaporation or condensation take place. The line extending downward from TP designates points where solid and vapor coexist. Along that line sublimation or deposition occur, depending on whether you go towards the high P or low P side.

The critical point CP deserves further comment. As I mentioned, the line from TP to CP describes the conditions under which boiling takes place, i.e. liquid and vapor are in equilibrium. But what is really the difference between these two phases? It is their density, since they both are in a disordered state. When you follow the boiling line towards higher T and P, the density of the liquid decreases while the density of the vapor increases, until at the critical point the two densities are the same. Thus, at and above the critical point, the liquid and vapor phase become one and are indistinguishable.

In the region where both T and p are above their critical values, the system is sometimes said to be in the state of a supercritical fluid (Fig. 7.1). This type of fluid has properties that are very different from both the normal liquid and the normal vapor. It is of practical importance for example for CO_2: Supercritical CO_2 is an excellent solvent for many compounds. One of its uses is for extracting caffeine from coffee.

7.3 Examples of One-component Phase Diagrams

Here we will examine three examples of one-component phase diagrams of actual materials. The first one is the one displayed for SiO_2 in the title figure above. It is of interest because of the many possible solid phases, all different crystalline forms of Quartz.

As an aside, you should be aware that sometimes variations can be found between different published versions of nominally the same phase diagram. Establishing crystal structures and phase boundaries under extreme conditions may be difficult and is an area of current research. For example, there seems to be some uncertainty in the literature as to where exactly the melting line lies between β-Quartz and molten SiO_2.

The second diagram we want to highlight is the one for water (Fig. 7.2). The diagram as drawn is only partial, but at least 11 different crystalline forms of ice have been identified! The ice known to us that exists just below 0°C at around 1 atm pressure is known as Ice Ih. The diagram is being extended to still larger pressures. Some of the phase boundaries between the higher-numbered forms of ice appear to be still uncertain.

The feature most noteworthy for us about the water diagram is that, especially with a logarithmic pressure scale, the melting line looks vertical. Yet,

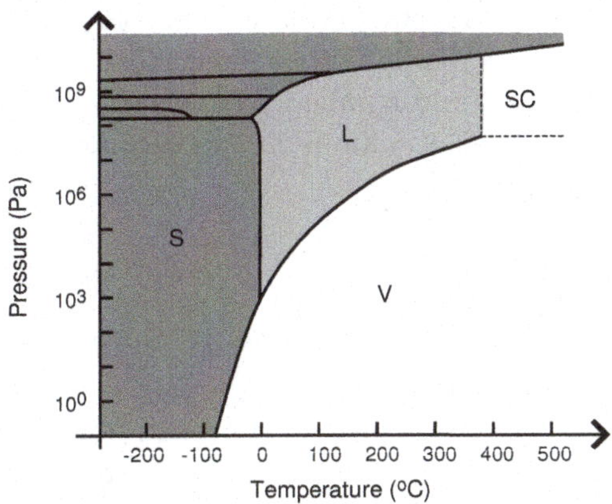

Figure 7.2: Phase diagram for water. The scale for the pressure is logarithmic. The many solid phase regions are in dark gray, and the liquid region is in light gray.

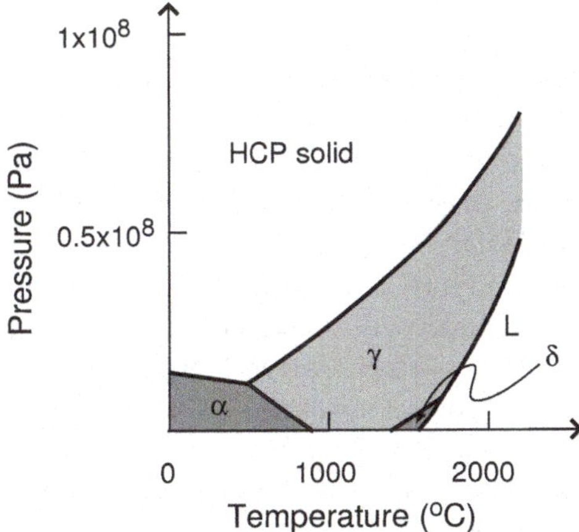

Figure 7.3: Phase diagram for iron. Note the α (BCC) and δ (BCC) solid phases in dark gray, the γ (FCC) solid phase in light gray, and the liquid region in white. There is also a high-pressure HCP solid phase.

upon inspection on a finer scale, it turns out that for water the melting line is actually tilted very slightly to the left, whereas for most other materials it is tilted very slightly to the right. This becomes apparent from accurate measurements of the triple point and the normal melting point: The triple point of water is at 0.01°C and 0.006 atm (611 Pa), and the normal melting point is at 0.00°C and 1 atm (101325 Pa).

The third diagram to consider is the one for iron. It has a linear pressure scale (y-axis). As mentioned above, at 1 atm pressure (essentially on the x-axis) it exhibits three solid phases. The stable phase from room temperature on up to 912°C is called α-Fe and has the BCC crystal structure. Between 912°C and 1394°C, the stable phase is FCC and is called γ-Fe. From 1394°C to the melting point at 1538°C, the stable phase is again BCC and is called δ-Fe. There is also an HCP phase at very high pressure.

7.4 The Gibbs Phase Rule

There is a formula known as the Gibbs Phase Rule, which describes a general relationship between equilibrium phases and components in a materials system. Let P be the number of phases present (you will not confuse it with the pressure p) and C the number of components in the system. Also, let F be the number of degrees of freedom, meaning the number of independent external variables that can be varied without changing the number of phases present. The degrees of freedom typically include temperature, pressure, and composition. Then the following relationship holds:

$$P + F = C + N \tag{7.1}$$

where N is the number of non-compositional variables. That is, $N = 2$ if both temperature and pressure are varied, or $N = 1$ if the pressure is fixed and only the temperature is varied. (The case of $N = 1$ typically applies for system in which we only deal with solids and their associated liquids).

The Gibbs Phase Rule can be derived by considering the conditions necessary for all components to be in thermodynamic equilibrium with each other, together with constraints imposed by stoichiometric conditions between the different components. We will not derive it here, but it can be applied and verified easily for one-component systems.

As an example, look at Fig. 7.1. For a one-component system, $C = 1$. First, at Point 1 it is clear that we can vary both T and p independently, and thus $F = 2$. Therefore, at Point 1 the Gibbs Phase Rule reads:

$$P + 2 = 1 + 2$$

We conclude that $P = 1$ at Point 1, which is at it should, since Point 1 is inside a single-phase region.

At Point 2, on the melting line, we have $F = 1$ because only one of the two variables T and p is independent, and the other one is fixed by the condition that Point 2 be on the line. In this case the Gibbs Phase Rule reads:

$$P + 1 = 1 + 2$$

We conclude that $P = 2$ at Point 1, i.e. two phases are present. Again that is at it should be, since on the melting line liquid and solid coexist.

By the same argument, the Gibbs Phase Rule says that for our one-component system, at the triple point $F = 0$ and $P = 3$.

You may think at this point that the Gibbs Phase Rule adds nothing new but expresses only the obvious. However, it will prove useful later, and will provide additional insights, for two-component systems.

7.5 Phase Changes: Melting and Solidification

In discussing one-component phase diagrams, so far we have excluded explicitly any issues of time or rates of change, since phase diagrams, by definition, describe a system in thermal equilibrium, and equilibrium refers to a state in which no further changes take place.

From a practical point of view, in Materials Engineering melting and **solidification** are of primary interest. In terms of a phase diagram, this means the melting line as the locus of points at which the solid and its liquid are in equilibrium. Now we want to explore further what exactly happens as the change from solid to liquid takes place, or vice versa.

At the same time, we will be cognizant of the fact that it may take a certain amount of time for a system to reach equilibrium, perhaps even a rather long time. In particular, we will see that the relevant time scales are very different for crystalline and for amorphous materials, and this makes a big difference in their melting/solidification behavior.

Also, for our future arguments we will consider the pressure to be fixed at 1 atm. This is not too restrictive an assumption, since the phenomena we will be interested in turn out not to depend strongly on pressure.

7.5.1 Melting of Crystalline Solids

We shall begin our discussion by focusing on crystalline solids, i.e. metals and ceramics. We will include inorganic glasses and polymers in the next section on amorphous materials, keeping in mind that polymers may be partially crystalline.

If you were to describe the main features of a solid and how it melts eventually, as the temperature is increased, the first thing that comes to your mind may well be that the solid has a sharp, well-defined melting point (or melting temperature). And what about the main difference between the solid and the liquid state? Perhaps it is this: The liquid flows easily and can take on any shape.

The fact that a liquid flows easily has a fancier explanation in terms of mechanical properties, which you can appreciate readily now. Flowing easily means that the liquid does not present resistance to shear deformation: Its shear modulus is zero. (This is a useful first approximation and will be qualified somewhat shortly). Please note that the same is not true at all for another type of deformation, namely compression: Liquids are generally almost incompressible.

The other major difference between the solid and the liquid state is, of course, that the crystalline solid has long-range order extending over thousands of atomic distances, whereas the liquid does not. As we mentioned in the chapter on structure, the liquid is disordered over the long range, but has some short-range order relative to its close neighbors.

An additional difference between the solid and the liquid state can be gleaned from data on the density of a material as a function of temperature:

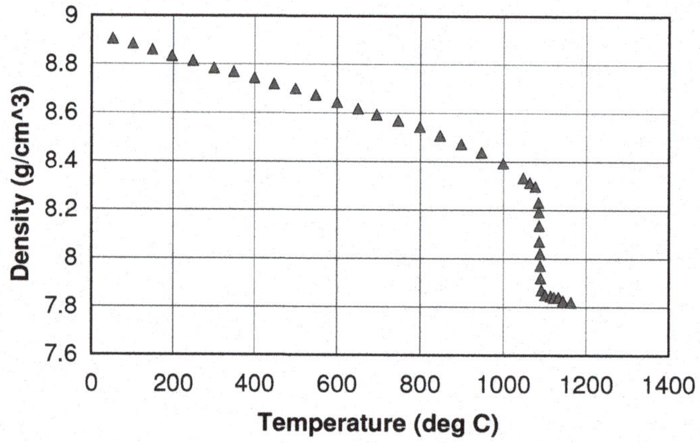

Figure 7.4: Experimental data for the density of copper as a function of temperature.

The raw data were for volumetric expansion vs. T and yielded a sudden increase in volume by 5.3% at the melting temperature T_m of copper at 1088°C. This converts into an abrupt drop in density from about 8.24 g/cm^3 to about 7.88 g/cm^3, assuming a room temperature density of 8.91 g/cm^3. This change in density at the melting point of the crystalline material mirrors the transition to the less ordered liquid state and reflects the smaller average number of nearest neighbors in the liquid.

Note that the loss of long-range order at T_m comes at a price. If one were to plot the temperature of the sample as a function of heating time, one would observe what is displayed schematically in Fig. 7.5 below.

Going from the solid to the liquid state means going from a state of lower free energy to a state of higher free energy. This energy difference is called the **heat of fusion**, ΔH_f, and is supplied by the external heating source. Thus, if the sample is being heated at a slow constant rate, its temperature increases gradually up to T_m, then remains constant from the onset of melting to its completion, before continuing to rise again.

We have yet to answer the question of what happens on the atom scale when

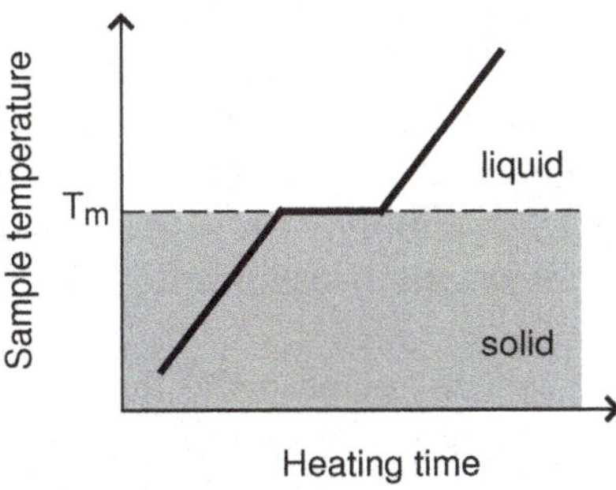

Figure 7.5: Schematic plot of sample temperature vs. heating time. T remains constant at T_m while the sample is melting.

a solid melts. This is a complicated matter, even for a relatively simple system such as a pure metal, and is still poorly understood. Several types of models have been advanced. In one type, melting is attributed to an instability arising from the thermal motion of the atoms about their equilibrium lattice positions. If this motion grows too large, it is thought that the crystal structure becomes unstable and turns into a liquid. Another type of model focuses on the shear modulus and identifies the onset of melting with the point at which the shear modulus becomes zero. A third class of models suggests that melting initiates at defects: vacancies, dislocations, and most of all at interfaces or surfaces.

It is clear from materials data that the stronger the interatomic bonds in the solid, the higher is the melting point. This was illustrated in Fig. 5.17 by the strong correlation between the melting temperature T_m and the heat of atomization ΔH_a for elemental metals. A similar correlation exists for the Group IV covalent crystals. Furthermore, heats of formation for vacancies are related directly to heats of atomization. Therefore, it is not surprising that the concentration of vacancies in the solid just below T_m is typically between about 0.2×10^{-3} and 1×10^{-3}.

However, you also know that the type of atom-atom bond in a solid varies greatly not only in terms of strength, but also in terms of directionality. Just compare metals (metallic bonding), ceramics (ionic bonding), covalent solids, and water/ice. So perhaps there cannot be a single mechanism describing melting in general for all kinds of crystalline solids.

7.5.2 Melting of Amorphous Solids

Having pondered some of the features of the melting of a crystalline solid, what would you say is the main difference in the melting of an amorphous solid, i.e. a glass? If you have ever done any glassblowing in a chemistry lab, or if you have watched a glassblower work his or her magic, I imagine you would say that the glass does not really have a distinct melting point. Rather, it gradually becomes softer and softer as its temperature is increased.

The reason for this behavior is not hard to understand, given what you know about the structure of a glass. We have described it before as amorphous, i.e. without long-range order. But an equivalent way of expressing this fact would be to say that a glass is basically a frozen-in, rigid liquid. That is, when it comes to structure, there is hardly a difference between the glass in its nominally solid state at low temperature and its state at a higher temperature where it behaves more like a liquid. The difference lies mainly with its deformation properties, i.e. with how easily it flows, or how little resistance it presents to shear stress.

What happens to this type of structure as you increase the temperature from, say, room temperature on, is that a rising temperature will impart continually increasing mobility to the basic building blocks of the material, be that SiO_4 tetrahedra in a silicate glass or polymer chain molecules in a polymeric glass, so that deformation of the material becomes easier and easier. The nature of this deformation is, of course, rather different in silicate, or inorganic, glasses and in polymers. When a soft silicate glass is deformed, Si-O bonds are broken and reformed, whereas in a soft polymer the geometric constraints for sliding and uncoiling of the chains are being reduced.

One way in which this behavior can be illustrated and quantified is to look again at the volume expansion of the material as a function of temperature, as with the raw data for Cu in Fig. 7.4 above.

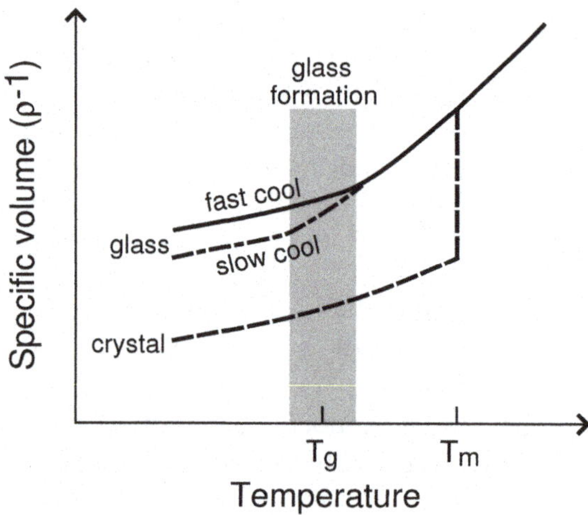

Figure 7.6: Specific volume vs. temperature for a material which can either form a crystal or a glass.

The figure displays the so-called **specific volume** as a function of temperature for a hypothetical material that can be a crystalline or a glassy solid. The specific volume is the inverse of the mass density ρ, i.e. volume per mass. As such it is a thermodynamically **intensive variable**, meaning a variable whose value does not depend on the amount of material present. Other such variables would be e.g. the temperature T or the pressure p. The opposite of an intensive variable is an **extensive variable**, which depends on the amount of material present, e.g. the volume. The point of using the specific volume is that it represents a property which is inherent to the material.

Please note first that the curve for the crystal is the inverse of the data curve in Fig. 7.4: It shows the abrupt change of the specific volume of the crystal at T_m. For the case of the glass there are two curves. (This shows up when one cools a glass-forming material from the melt at different rates. More about that below, in the section on solidification). You can see that the glass curves start out higher than the crystal at low T. This is because the density of the glass, being like a liquid, is lower than the density of the crystal. The glass curves then increase continuously with increasing temperature. Above T_m the crystalline and the glass curves become one, as at that point the structural differences have vanished.

The primary feature of interest with the glass curves is that in a region around the temperature marked T_g, the slope of the curves changes. T_g is known as the **glass transition temperature**. You can think of it as the temperature above which the glass looses its rigidity, but from the earlier discussion it is evident that this is at best a qualitative definition. Similarly, and with some simplification, you can say that below T_g the material will be brittle, and above T_g it will be ductile and tough.

Representative values for the glass transition temperature and the melting temperature, where applicable, are listed in the table below:

Table 7.1: Typical glass transition and melting temperatures

Material	T_g (°C)	T_m (°C)
Polyethylene	-90	130
Polypropylene	-10	175
Nylon	50	250
Polyvinyl chloride	80	220
Polystyrene	100	240
Polycarbonate	150	265
Soda-lime glass	525	
Borosilicate glass	575	
96% silica	925	
pure silica	1125	

You should be aware that the numbers for the polymeric glasses in Table 7.1 are approximate because in some cases they represent average values. T_g and T_m for a certain material depend on the chain length of the polymer molecules and possibly on other structural parameters. In addition, T_g signifies more properly a certain temperature interval rather than an exact value.

Since a glassy material increases its ease of flow continuously as a function of increasing temperature, we really need a measure for ease of flow whose value itself varies in a continuous fashion. The relevant property is the viscosity. It can be visualized with Fig. 7.7 below.

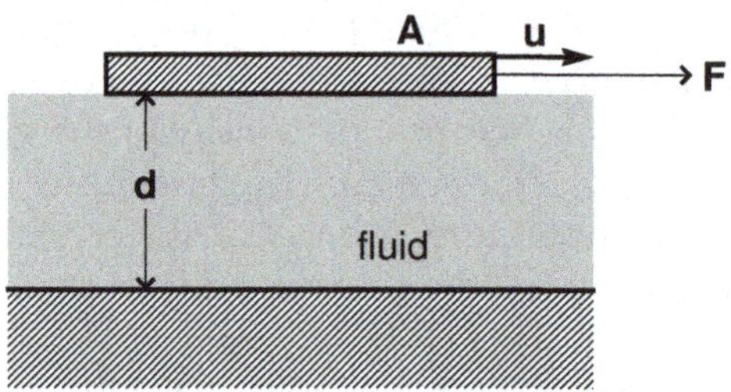

Figure 7.7: Setup to explain the definition of the viscosity η

Assume that the top plate with area A is being pulled by a force F across the fluid with a constant velocity u_x. The fluid has a thickness d and rests on a fixed bottom plate. We make the common assumption that the very bottom layer of the fluid sticks to the lower plate and thus has velocity 0, whereas the topmost layer of the fluid sticks to the top plate and thus has the same velocity u_x. We also assume that the velocity in the fluid changes linearly from top to bottom.

Under these conditions it turns out that the force F required to move the top plate is proportional to the area A of the top plate. Furthermore, F is proportional to u_x and inversely proportional to the distance d between the plates. Thus we can write

$$F \propto A \frac{u_x}{d} \tag{7.2}$$

As we have done in the past, this type of proportionality can be converted to an equation:

$$F = \eta A \frac{u_x}{d} \tag{7.3}$$

where the proportionality constant η is known as the **viscosity** of the fluid. The connection to materials becomes apparent when you write $u_x = \Delta x/\Delta t$, with x being the direction parallel to the plates. Consequently,

$$\frac{u_x}{d} = \frac{1}{d}\frac{\Delta x}{\Delta t} = \frac{1}{\Delta t}\frac{\Delta x}{d} = \frac{\Delta \gamma}{\Delta t} \tag{7.4}$$

where $\gamma = x/d$ is the shear strain exerted by the shear stress $\tau = F/A$ on the fluid, and $\frac{\Delta\gamma}{\Delta t}$ is the ***shear strain rate*** (see Fig. 5.1 and Eq. 5.10).

If we plug this into Eq. 7.3 and solve for η, we obtain:

$$\eta = \frac{F/A}{\frac{\Delta\gamma}{\Delta t}} = \frac{\tau}{\frac{\Delta\gamma}{\Delta t}} \tag{7.5}$$

This means that η, our measure of how easily a fluid can flow, is the ratio of (shear stress)/(shear strain rate). Eq. 7.5 says that for a given shear stress τ, the lower the viscosity η, the larger is the strain rate of the induced deformation. Please take note that the units of η are Pa ṡ (or N s/m^2).

From Fig. 7.6 you can expect that η will be a strong function of the temperature. This is illustrated on Fig. 7.8:

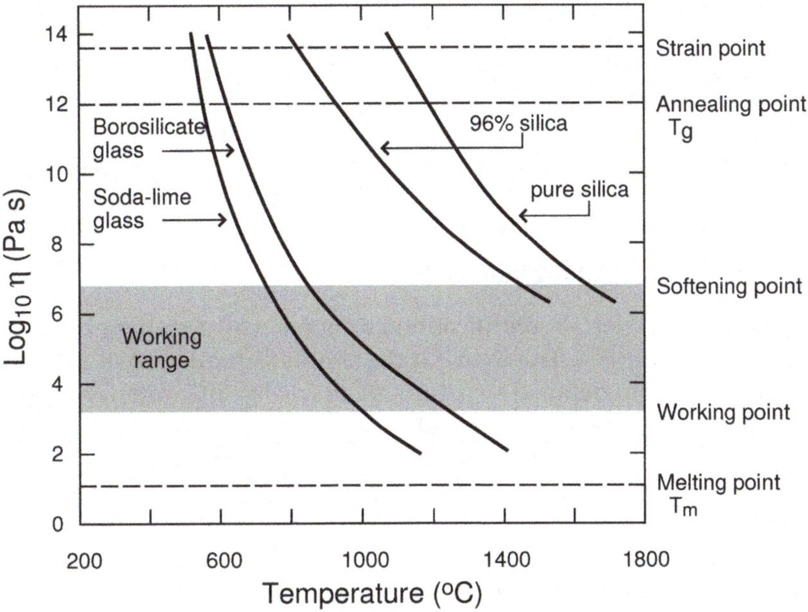

Figure 7.8: Viscosity η of some silicate glasses as a function of temperature T. Note the logarithmic scale for η.

In fact, $\eta(T)^{-1}$ has the usual shape of a thermally activated process, so that for $\eta(T)$ itself one can write

$$\log_{10} \eta = C + \frac{B}{T} \tag{7.6}$$

where B and C are suitable constants.

Note that the various points labeled on Fig. 7.8 are approximate. T_g is usually identified with the temperature at which $\eta = 10^{12}$ Pa s, and $\eta = 10$ Pa s is a representative value for the "melting point" of a glass. The point $\eta = 10^{12}$

Pa s is sometimes also referred to as the annealing point, because at that point internal stresses in the material are relieved on a time scale of minutes. The working range denotes the range of conditions typically used in glass-forming processes. At temperatures above the strain point, a glass is rigid enough so that it can be handled easily without introducing new stresses.

Now we are in a position to give a more complete description of the deformation of a material that can undergo both elastic and viscous deformation (see Section 5.4.4). To a first approximation, we might choose to simply combine the two effects (Eqs 5.20 and 7.5). I.e. for the shear deformation of a viscoelastic material one could set

$$\tau = G\gamma + \eta \frac{\Delta\gamma}{\Delta t} \tag{7.7}$$

The first term in Eq. 7.7 is of course the one from the linear, time-independent Hooke's Law, and the second term represents the time-dependent part as described earlier in Section 5.4.4 (see Fig.5.14). Eq. 7.7 is known as the Kelvin-Voight model for viscoelastic deformation. Other similar models have also been formulated.

7.5.3 Solidification: Amorphous Solids

We begin our discussion of solidification as the reverse of melting by considering amorphous materials. This is by far the simpler situation. As I emphasized above, with respect to detailed structure, there is very little difference between a molten glass at high temperature and its solid, rigid form at much lower temperature. However, one minor structural difference is evident from the plot of specific volume vs. T in Fig. 7.6: The solid glass has a higher density than the molten glass.

Keep in mind that the curve for specific volume vs. T in Fig. 7.6 is in principle reversible: It applies for going up as well as going down in temperature. But in the process of cooling a new effect comes into play. Even though what happens during cooling is mainly that the building blocks of the glass loose mobility, how closely packed they end up depends on how rapidly you cool. The SiO_4 tetrahedra, or the polymer chains, try to achieve the lowest energy configuration possible, but this requires time! Therefore, as the glass becomes rigid, the final configuration is the more densely packed the slower the cooling rate. This is why the $\rho(T)^{-1}$ curve in Fig. 7.6 shows two branches below T_g, an upper one for a higher cooling rate and a lower one for a slower cooling rate. In this light, it becomes clear why T_g is not an exact number: T_g depends on the cooling rate.

7.5.4 Solidification: Crystalline Solids

Forming a crystalline solid from the melt upon cooling is a much more complex process than simply freezing in the disordered structure of a glass. In a crystal, the structure with the lowest energy has long-range order. Achieving this highly ordered state requires coordinated motion of a large number of building blocks, atoms or molecules, to align themselves precisely with each other.

When you think of metals, you may conclude that it is almost impossible not to end up with a crystalline, or at least polycrystalline, material. When you think of SiO_2 or ionic materials, you realize that it could go either way. For the case of polymers, forming crystals may be the exception, or it may not be possible at all! So, what does it take for a material to solidify as a crystal, rather than as an amorphous solid?

The answer is that time, i.e. the cooling time, is again a factor in many systems, in that the atoms or molecules need enough time to be able to find their preferred crystal positions. But, most importantly, the formation of a crystalline phase, i.e. a large crystal, is greatly facilitated if one starts out with a small crystal **nucleus**! This requires some explaining.

First, with regard to the relevant time scale, this depends on the type of material in question. As you recall, metals are characterized by a strong non-directional interaction which is the same between any atoms. It turns out that near the melting point, the mobility of metal atoms is large enough so that the time scale on which atoms are able to find their preferred crystal positions is generally much smaller than the time scale on which cooling takes place. This is certainly true for processes relevant in engineering practice, although it has been possible to create small amounts of amorphous metals in the laboratory, under conditions of extremely rapid cooling (rapid meaning a cooling rate of something like 10^6 Ks^{-1}, depending on the composition of the metal).

For an ionic solid, the time scale to allow for crystallization is longer, because of the additional constraints due to the different ionic sizes and the conditions of electrostatic repulsion/attraction between like/unlike ions. For polymers to form crystals, the time scale is longer yet due to the constraints imposed by the geometry of the chain molecules.

Second, the formation of a crystal is really envisioned to involve two linked processes:

1) ***formation of a crystalline nucleus,*** followed by
2) ***further growth*** of the initial nucleus.

In order for atoms to know where they are supposed to go, namely into a regular array of lattice sites, one should start out with a small template which already has the required order and symmetry. Then, adding more atoms in the right places is simply a matter of letting them find the lowest energy positions proscribed by the template.

So, forming a crystalline solid amounts to maintaining conditions which allow for crystal templates, so-called nuclei, to form in the disordered liquid and, at the same time, for nuclei to keep growing to sizes of many thousands of atom-atom distances. In short, forming a crystalline solid involves **nucleation** followed by **growth** of crystallites.

As we will explain shortly, nucleation and growth are in a sense competing processes. Conditions may be such that nucleation occurs readily, but growth is impeded greatly. On the other hand, conditions may be suitable for growth yet not enough nuclei are being formed. In either case, the formation of a crystalline phase would not be favored. If the conditions are such that they do lead to formation of a crystalline phase, then it is clear that how the competition between nucleation and growth plays out in practice will have a great deal of influence on the final microstructure of the solid phase.

Nucleation

Let us pause briefly to review the process we are trying to describe. We are interested in how a crystalline solid forms as we cool down a piece of molten material. For the sake of simplicity let us assume we are dealing with pure metal.

If during the cool-down we stopped exactly at the melting point T_m, nothing would happen because right at T_m liquid and solid would be in equilibrium if there were any solid, but no solid has formed yet. In order to provide a driving force for solidification, we have to create a condition called **undercooling**, i.e. we have to take T a little below T_m,. Yet even under that condition, crystalline nuclei do not form automatically, and here is why:

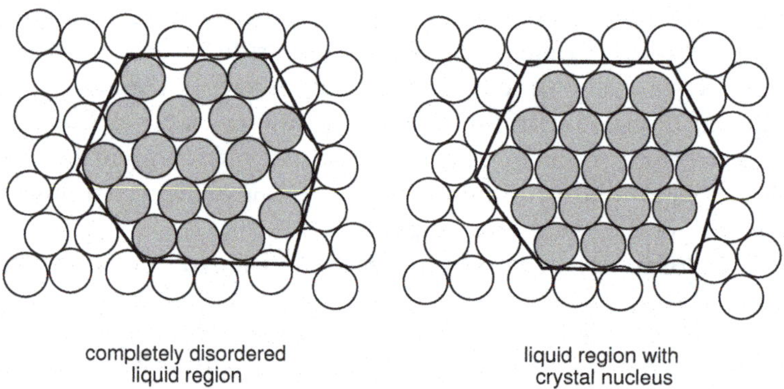

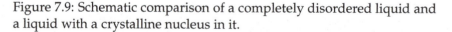

Figure 7.9: Schematic comparison of a completely disordered liquid and a liquid with a crystalline nucleus in it.

The panel on the left in Fig. 7.9 shows an entire region of disordered liquid. The gray atoms are singled out because they are the ones that will form a crystalline nucleus in the right panel. The white disordered atoms are in their original disordered configuration in both panels. The polygon is drawn to circumscribe an approximate boundary around the disordered gray atoms, and it is the same as well on both sides.

When you examine the ordered gray atoms in the panel on the right, you will realize that two things have happened to them compared to the liquid state. First, by forming a crystal they have lowered their energy: All else being the same, hexagonal close-packing is the most favored configuration. However, these atoms have also had to pull away a little from their still disordered neighbors around the periphery. This is perhaps most evident in the upper left-hand and lower right-hand corners of the polygon. This second effect basically amounts to creating a new surface for the crystal nucleus, thereby increasing the energy of the gray atoms.

Therefore, the fate of the transient crystal nucleus will depend on the balance between the two effects of lowering the "bulk" energy and raising the "surface" energy of the gray atoms. Most interestingly, as we will show next, there exists a critical size for a nucleus, in the sense that a nucleus smaller than the critical size will revert to a disordered state, but a nucleus larger than the critical size will continue to grow as a crystal.

Consider now a spherical crystal nucleus with radius r and its change in free energy ΔG relative to the liquid state. That is, ΔG is the difference between the free energy of the gray atoms of Fig. 7.9 aggregated into a nucleus and the disordered gray atoms. We argued above that ΔG has two parts, a bulk part due to the lowered energy of the ordered state, and a surface part due to the extra energy involved in creating a new surface for the nucleus.

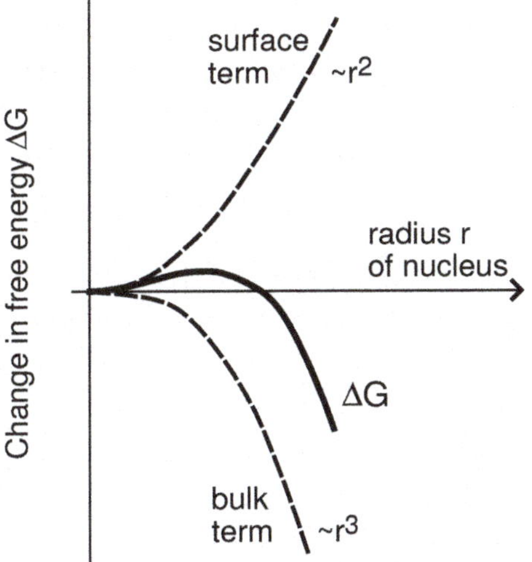

Figure 7.10: Contributions to the change in free energy ΔG of a spherical nucleus. Note the bulk and surface parts, and their characteristic dependencies on the radius r of the nucleus.

What we can say immediately is that the bulk part must be proportional to the volume of the nucleus, and the surface part proportional to the surface area of the nucleus. Explicitly, assuming spherical nuclei with radius r:

$$\Delta G = \frac{4}{3}\pi r^3 \Delta G_v + 4\pi r^2 \gamma \tag{7.8}$$

where ΔG_v is the bulk free energy difference per unit volume and γ is the surface free energy difference per unit area. Note that for $T < T_m$, as per our earlier argument, ΔG_v must be < 0.

Clearly, $\Delta G = 0$ at $r = 0$. Furthermore, because of these dependencies on r, it follows that for r very close to 0, the surface terms dominates, i.e. ΔG must increase initially. On the other hand, for large r the bulk term dominates, and thus ΔG must turn negative at large enough r. Therefore, ΔG *must have a maximum* as a function of r !

The condition for that maximum can be obtained easily by taking the derivative of Eq. 7.8 with respect to r, and setting $\frac{d}{dr}(\Delta G) = 0$:

$$\frac{d}{dr}(\Delta G) = 4\pi r^2 \Delta G_v + 8\pi r \gamma = 0 \tag{7.9}$$

This equation can be solved for the critical radius r_c at which it is fulfilled:

$$r_c = -\frac{2\gamma}{\Delta G_v} \tag{7.10}$$

In turn, r_c can be substituted into Eq. 7.9 in order to yield the critical value ΔG_m at the maximum of $\Delta G(r)$:

$$\Delta G_m = \Delta G(r = r_c) = \frac{16\pi\gamma^3}{3(\Delta G_v)^2} \tag{7.11}$$

In thermodynamics it can be shown that ΔG_v is itself a function of temperature, namely:

$$\Delta G_v = \Delta H_f \left(1 - \frac{T}{T_m}\right) \tag{7.12}$$

where ΔH_f is the heat of fusion we introduced earlier and T_m is the melting temperature as usual. We may conclude, combining Eqs. 7.10, 7.11, and 7.12, that r_c and ΔG_m are both functions of temperature:

$$r_c = \left(-\frac{2\gamma}{\Delta H_f}\right)\frac{1}{1 - \frac{T}{T_m}} = \left(-\frac{2\gamma}{\Delta H_f}\right)\frac{T_m}{\Delta T} \tag{7.13}$$

$$\Delta G_m = \frac{16\pi\gamma^3}{3\Delta H_f^2}\frac{1}{1 - \left(\frac{T}{T_m}\right)^2} = \frac{16\pi\gamma^3}{3\Delta H_f^2}\left(\frac{T_m}{\Delta T}\right)^2 \tag{7.14}$$

In these two equations, we have used the shorthand $\Delta T = T_m - T$, and you can consider either T or ΔT as the independent variable. Note that $\Delta H_f < 0$.

We can conclude that the farther T is below T_m, the smaller are the critical radius r_c and the critical maximum ΔG_m of the free energy of a nucleus. In other words, the greater the amount of undercooling $\Delta T/T_m = 1 - T/T_m$, the easier it is for nucleation to take place.

Note also that ΔG_m represents a type of activation energy, i.e. an energy barrier the would-be nuclei have to overcome in order that they are able to continue growing. Fig. 7.10 makes it clear that at the earliest stage of nucleation, the energy of a very small nucleus is in fact slightly larger than the energy of the same atoms in their disordered configuration.

Although Eqs. 7.13 and 7.14 provide the essential picture of nucleation when T is below but close to T_m, they are not quite the whole story of nucleation yet. Just imagine that instead of lowering the temperature slowly from above T_m to below T_m, you lower it very rapidly to a temperature way below T_m, where the disordered positions of the atoms are frozen in instantly. At this temperature extreme, when all atomic motion is blocked, the rate of nucleation must be zero. The atoms cannot make even the slight changes in position necessary to rearrange into a small crystal.

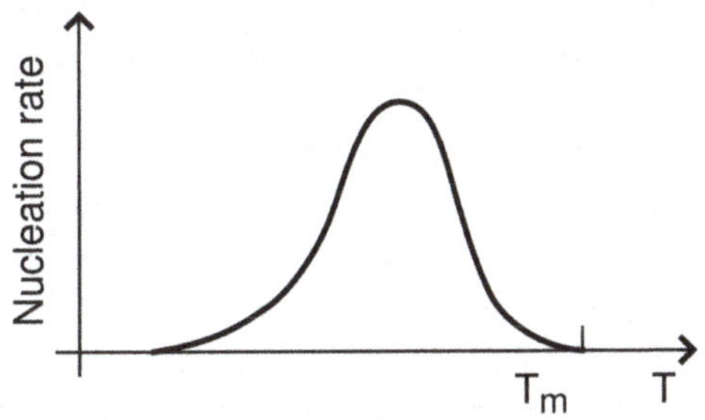

Figure 7.11: Schematic illustration of the nucleation rate as a function of temperature.

If we combine the behavior of the nucleation rate near T_m and much below T_m, it follows that the nucleation rate must have a maximum at some temperature in-between. This is illustrated schematically in Fig. 7.11.

To conclude our discussion of nucleation, let me put these arguments in perspective by pointing out that our main assumption in deriving Eqs. 7.10 and 7.11 was that nuclei would be spherical. The underlying assumption is that the solid material of a nucleus can be considered continuous and isotropic.

However, nuclei supposedly serve as little crystal templates for the further growth of the solid. So you should really think of nuclei as mini-crystals, bounded by crystal planes. It should also be clear that different crystal planes will have different surface energies. Thus we should not take the spherical shape of a nucleus for granted. Still, our derivation rests basically on the competing effects of changes in surface vs. bulk free energy, and that argument should still hold even if the shapes of nuclei are more complex.

You should also be aware that the process of nucleation we have examined in detail in this section involved spontaneous formation of solid nuclei inside the melt of the same pure material. This would be referred to, more accurately, as **homogeneous nucleation**.

In practice, homogeneous nucleation is probably the exception, rather than the rule, since in most applications the melt will come in contact with other solids during solidification. In that case one deals with what is referred to as **heterogeneous nucleation**. Nuclei are being formed around or on a pre-existing "template" such as impurity atoms in the melt, impurity particles in the melt, a solid surface exposed to the melt, etc. Examples of heterogeneous nucleation would be what happens when liquid metal is poured into a solid form for a casting, or when the molten bead of a piece of solder solidifies on a piece of Cu circuit board.

The general considerations from above will apply to either situation. The main difference in heterogeneous nucleation is that nuclei would form more easily, since ΔG_m would be lower. In turn, this would cause r_c to be smaller.

Growth

As you will discover in doing the Homework Problems, the number of atoms in a nucleus of critical size is of the order of a few hundred to a few thousand, depending on the material and the amount of undercooling. This means that a typical nucleus is quite large on an atomic scale, but still very small on an engineering scale, i.e on the scale of "engineering crystals" also known as grains. (You recall that grains have sizes of the order of a few μm.)

These different length scales indicate that a nucleus still has to grow a great deal until it reaches the size of a typical grain. This is why it is useful to think of solidification as occurring in two stages: nucleation followed by growth.

The most important characteristic of the step of growth is that it, too, is a strong function of temperature. However, its temperature dependence is simpler than that of the nucleation step, because growth relies primarily on the continued transport of atoms onto the surface of the expanding nucleus. That is, the rate of growth is determined by the rate of diffusion in the melt. Hence we can expect the temperature dependence of the growth rate to be essentially that of the diffusion coefficient in the melt.

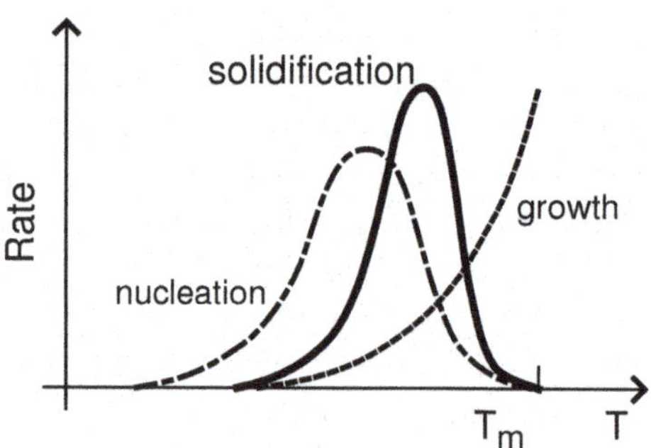

Figure 7.12: Schematic comparison of nucleation rate, growth rate, and overall rate of solidification

Moreover, the transformation rate overall can be thought of as the product of the nucleation rate and the diffusion rate. This is displayed schematically in Fig. 7.12. Note that the shapes of the nucleation and total solidification curves are similar, but one is skewed relative to the other because of the diffusion rate.

From the shape of these curves we can glean one important consequence:

- If, roughly speaking, the transformation occurs above the peak of the overall solidification curve, where nucleation is slow but growth is rapid, the number of nuclei formed will be small but they will grow to a large size. This means that the solid material will have fewer but larger grains.
- If, on the other hand, the transformation occurs below the peak of the overall solidification curve, where nucleation is rapid but growth is slow, the number of nuclei formed will be large but they will grow more slowly and thus remain small. This means that the solid material will have more but smaller grains.

This trade-off between the rates of nucleation and growth is one of the essential tools in controlling the grain size, and the microstructure in general, of a solid material. We will explore this issue in greater detail in the next chapter, where we will apply these same principle to two-component materials.

The final outcome of a growth process depends sensitively on a number of experimental parameters, primarily because growth occurs over a large length scale. This implies that gradients in concentration and temperature, in the melt and in the growing solid, will play an important role. Here I will just mention a few effects, and we will not pursue them in any detail.

One effect has to do with the fact that solidification is an exothermic process: For every atom moving from the liquid into a crystal lattice site, the heat of fusion ΔH_f per atom is released and taken up by the solid and liquid. This heat can be substantial, to the point where it changes the temperature locally at the liquid/solid interface. If this leads to a non-uniform temperature, the growth rate may become non-uniform accordingly.

Also, given that the growing solid is crystalline, we should expect that different crystal planes may grow at different rates. So it is not obvious what the final shape of the solid will be. The evolving solid surface may not be a single plane, but may develop a rough texture exposing different crystal planes.

If the process is heterogeneous and nucleation/growth takes place on a separate substrate, the growth process, thus the overall structure, may well be controlled by constraints imposed by the symmetry of the substrate.

The growth process may lead to an amazing variety of structures. A few examples are given below.

Growth Structures 1: Spherulites in Polymers

You may recall from Fig. 3.34 that polymers can form very thin but long crystalline regions called lamellae, where chain molecules are aligned parallel to each other to form a repeating structure.

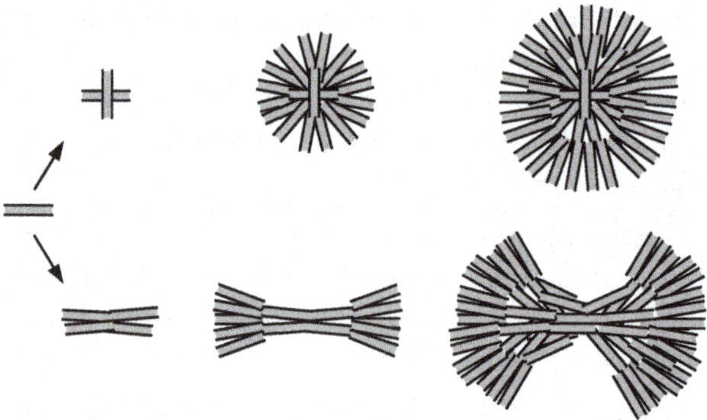

Figure 7.13: Two growth modes for spherulites in polymers. A set of two black lines with gray fill represents a crystalline polymer lamella.

It is possible for such lamellae to grow into fairly large, more or less spherical aggregates called **spherulites**, such that a large portion of the volume occupied by the material is in fact crystalline. This is how certain polymers can achieve a high degree of crystallinity, up to around 90% . A schematic representation of this process is shown in Fig. 7.13, where two possible growth modes are illustrated.

The top panel of Fig. 7.13 depicts what I will call growth mode 1, i.e. isotropic growth, whereas the bottom panel shows growth mode 2, i.e. directional growth, e.g. due to a thermal gradient or the presence of a planar substrate. In either case, you should imagine a very small crystalline lamella, or perhaps a few lamellae, at the center. As these platelets grow farther outward, they may develop rather complex branched structures

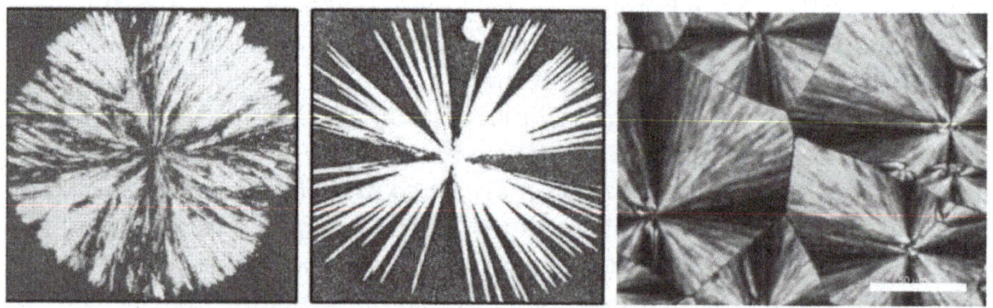

Figure 7.14: Experimental data for mode 1 spherulites. Left: Individual spherulites. Right: Spherulites grown to fill space. Bar marks 50 μm.

Experimental examples are shown in Figs. 7.14 and 7.15. The first two pictures in Fig. 7.14 are optical micrographs of individual growth mode 1 polymer spherulites. The first one displays lamellae with branches. In the second one the lamellae are very long and straight and show no branches.

The third picture in Fig. 7.14 shows the end result: Spherulites have nucleated individually, at different times, and have kept growing until their edges touch and they fill the entire region. Each domain with a cross-shaped pattern is a spherulite, and the center of the cross indicates the location of the nucleus. The scale bar in the lower right hand corner on the third picture marks 50 μm. We can conclude that spherulites can grow to macroscopic dimensions.

You will note that these spherulites resemble grains in metals. Their boundaries are clearly delineated and indicate distinct regions of the material. However, spherulites are only partially crystalline: The lamellae are separated from each other by amorphous material.

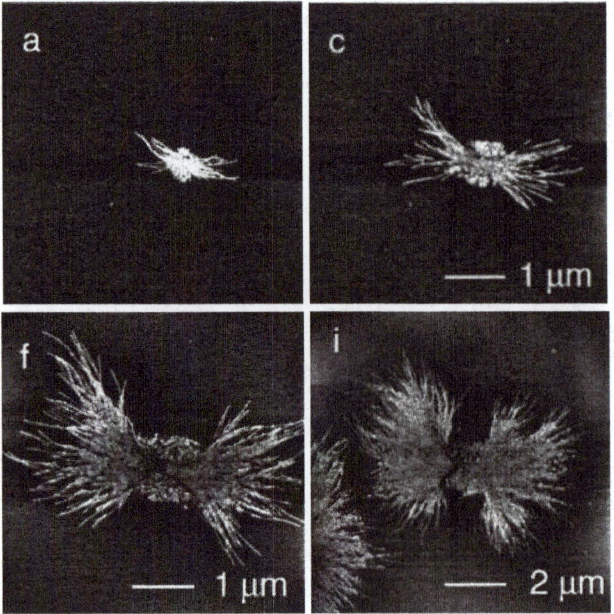

Figure 7.15: Atomic Force Microscope pictures of a growth mode 2 polymer spherulite.

The pictures in Fig. 7.15 show an excerpt of the growth sequence of a mode 2 spherulite. The growth directionality in this case was provided by the silicon substrate on which the polymer grew as a thin film. These pictures were taken with an Atomic Force Microscope. This is a device whose sensing part consists of an atomically sharp movable tip which can follow the outline of a surface or, in this case, the growing spherulite as an object on a surface, with resolution approaching the atomic scale. (We will discuss the Atomic Force Microscope, with further applications, in the last chapter of this book).

Note the much smaller scale in these pictures than in Fig. 7.14. Also, as this mode 2 spherulite grows larger yet, its ultimate shape overall again becomes more nearly spherical, consisting of crystalline as well as amorphous parts.

Growth Structures 2: Dendrites

In our prior discussion of nucleation and growth we have assumed, for the most part, that the processes occurred in a uniform medium, and thus isotropically in the three dimensions. In many practical processes of solidification, this is not the case. A solidifying melt is usually in contact with another surface, for example when a cast metal part is produced, or when liquid solder cools and solidifies on the copper conductor of a printed circuit board.

The process thus becomes directional: Nucleation is heterogeneous on a separate surface, and growth occurs presumably in a direction normal to this surface. Furthermore, such a geometry almost automatically will cause gradients to occur, certainly a gradient in temperature, and possibly also a gradient in concentration. So how will the solidifying sample grow under these conditions? The brief answer is that things may get very complicated, but you are now in a position to get a grasp on the essential parameters and phenomena. In particular, we will see that growth may not simply proceed perpendicular the original surface.

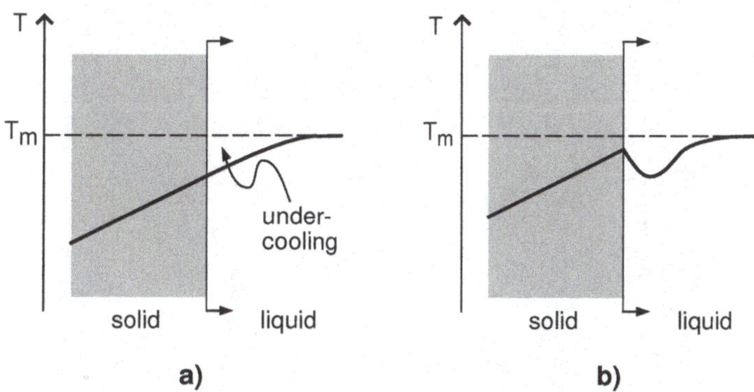

Figure 7.16: Schematic temperature profiles for two situations of solidification of a pure liquid on a surface. a) Normal temperature gradient. Note the melting temperature T_m and the undercooling at the moving interface. b) Inverted temperature gradient in liquid at interface.

Consider first the experimental setup in Fig. 7.16 a). This corresponds to a normal temperature gradient, such that heat flows from the liquid to the solid, and is dissipated in the growing solid. This situation also leads to a stable planar solid-liquid interface that moves to the right, into the liquid. Imagine a perturbation in the planar interface that sticks out into the melt. Now, according to Fig. 7.12, growth occurs most readily where the amount of undercooling is highest. In the present case, this means right at the interface. Thus, a protruding perturbation will not persist, but will dissolve, as it reaches into a region where the solid phase is less stable.

Occasionally, a situation as in Fig. 7.16b) may arise, where heat is not dissipated efficiently into the solid, but also flows back from the interface into the liquid. This will cause an inverted temperature gradient into the liquid. In that case, a local disturbance in the solid sticking out into the melt will in fact continue growing faster than rest of the solid, and thus the the planar interface will become unstable.

The previous argument is essentially one-dimensional, relying on the shape of the temperature gradient perpendicular to the growing solid surface. Other effects may play an important role in three dimensions.

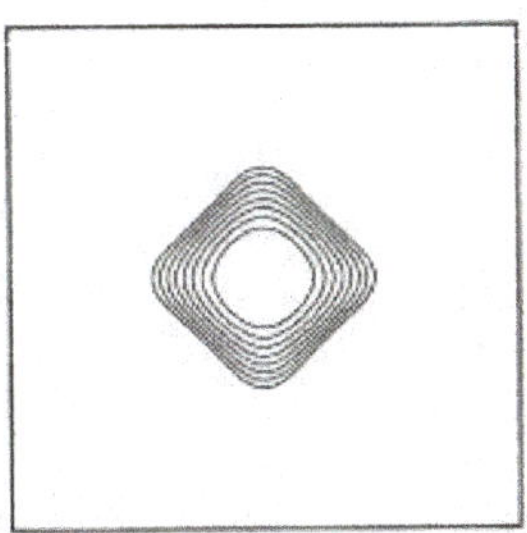 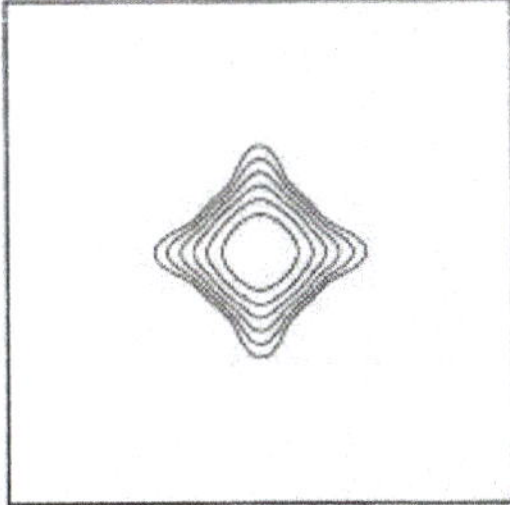 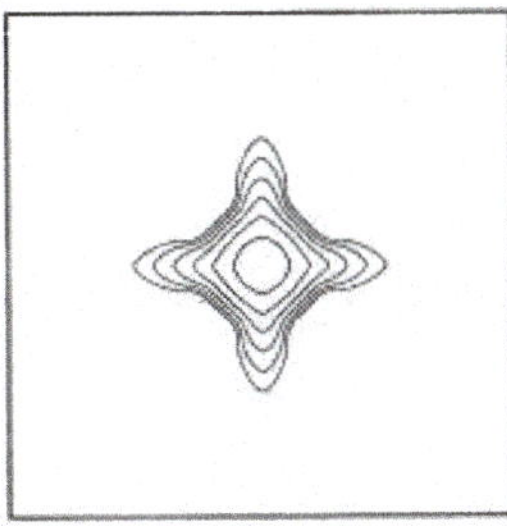

Figure 7.17: Effect of anisotropy of growth rates on crystal shape. The degree of anisotropy increases from left to right.

Imagine that you start out with a small particle of nearly spherical shape, as in Fig. 7.17, but the growth rates in the ±x- and ±y-directions are higher than in the diagonal directions. In crystallographic terms, we can say that the growth rates of the 2D {1 0} family of planes is larger than for the {1 1} family of planes. When this anisotropy is small, as in the first panel, with continued growth the crystal planes growing the fastest gradually disappear! However, for sufficiently large anisotropy in the growth rates, the faster-growing planes outgrow the other ones and give rise to a new form (panels 2 and 3)! Crystal shapes such as the one in panel 3 of Fig. 7.17 are called **dendrites**.

Examples of dendrites, or dendritic growth, abound in nature. If you have ever found the windshield of your car covered with ice crystals on a cold winter morning, you have observed dendrites.

Dendrites are fairly common in the solidification of metals, especially alloys. However, with alloy systems, yet another factor comes into play, namely concentration gradients in the melt. When you think of an alloy as a solid solution, it should be quite evident that the composition of the melt and the solid may be different. For example, assume as is often the case that the solute is more soluble in the melt than in the solid. This means that in the course of solidification, all of the solute in the melt cannot be incorporated into the solid. Therefore, the melt must become enriched gradually in solute compared to the growing solid. This, in turn, may lead to local variations in the concentration of solute in the melt, and thus to local variations in the growth rate.

In general, one can say that anything causing a deviation from isotropy in the experimental conditions may lead to rather complicated growth structures such as the dendrites mentioned above. The factors contributing to an anisotropic growth rate include a temperature gradient, crystallographic variations as in Fig. 7.17, differences in solute solubility, and mass transport in the melt due to convection. In addition, the amount of undercooling or, in other words, the growth rate overall is very important. All of these effects are coupled to a moving interface which may not be planar, and whose various parts may move in different directions at different speeds.

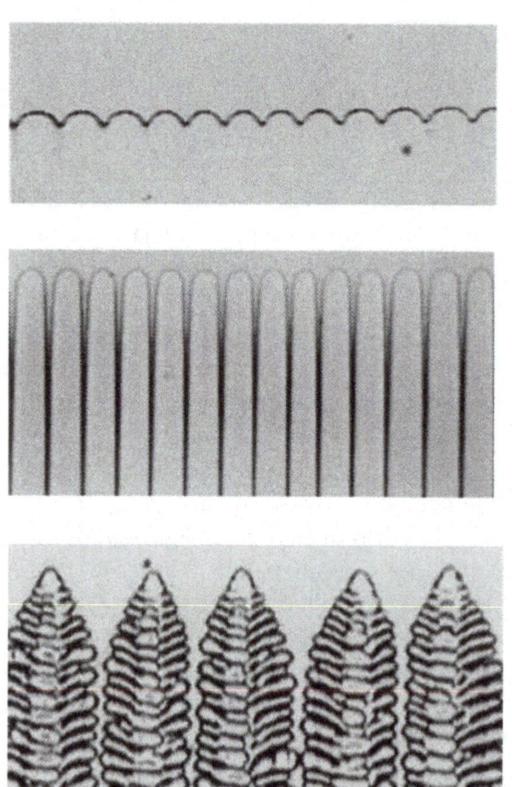

Figure 7.18: Solidification of a succinonitrile- coumarin 152 organic alloy. The pictures show the liquid-solid interface, with the liquid above. The growth rate increases from top to bottom. At very low growth rate, the solid-liquid interface is planar. It then becomes rough, develops deep pillars, and eventually yields a complex pattern of dendrites.

A dramatic example of these effects is shown in Fig. 7.18. The pictures show solidification experiments with a succinonitrile-coumarin152 organic alloy. Succinonitrile ($NC-(CH_2)_2-CN$), abbreviated as SCN, has been used as a model system for metal alloys because it is transparent, and thus easy to study by optical means, and because it melts at a convenient temperature of around

55°C. In the top panel, at a very low growth rate the solid-liquid interface is planar. As the growth rate is increased, the interface becomes rougher. First it develops a periodic system of cells, then deep pillars, and eventually a complex pattern of dendrites.

The experiments of Fig. 7.18 were done in a terrestrial lab, under conditions where variations in concentration of the melt and convection in the melt probably played a role. These complications were effectively eliminated in the so-called Isothermal Dendritic Growth Experiment (IDGE), a solidification experiment run with pure SCN on the space shuttle Columbia in 1994.

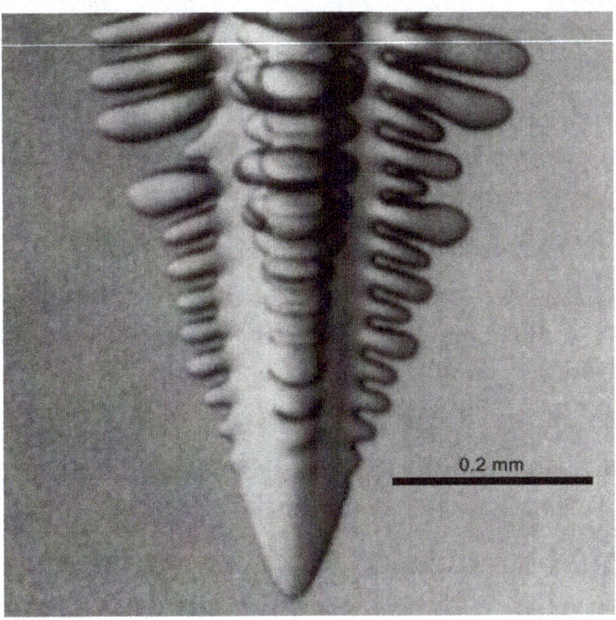

Figure 7.19: Dendrite grown with pure SCN in the IDGE on the space shuttle Columbia in 1994 (Figure courtesy of M. Glicksman).

Fig. 7.19 shows an SCN dendrite grown in the microgravity environment of the space shuttle. The growth velocity of the tip of the dendrite was relayed to earth and was observed there in real time. It showed substantial differences to terrestrial measurements. Photographic pictures taken on the space shuttles were analyzed later with greater precision, in order to account for all the details of the growth rate and structural features of the dendrites, including the shape of the tip, the dense core, and the numerous side lobes.

Growth Structures 3: Single-crystal silicon

Single-crystal silicon in a way represents the opposite in growth compared to dendrites. Silicon is grown as large, perfect single crystals on an industrial scale. It is the base material used for the fabrication of large-scale integrated circuits that power our personal computers, laptops, cell phones, etc.

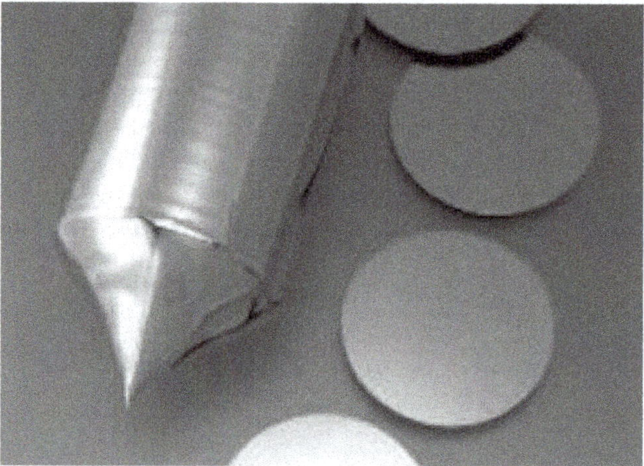

Figure 7.20: Single crystal silicon ingot, with Si wafers cut from it.

A raw Si crystal, also referred to as a Si ingot, is a cylinder either 20 cm or 30 cm in diameter and about 150 cm long as grown . The entire Si ingot is one perfect crystal: It has a definite crystallographic orientation, e.g. [1 0 0] along the axis. There are no grains and grain boundaries, no dislocations; just a few unavoidable vacancies here and there.

The Si wafers, upon which electronic circuits are formed, are cut from the original ingot as thin slices, and then are polished and chemically etched, so as to end up with a perfectly smooth, defect-free surface.

Si single crystals are grown either by the so-called Czochralski or the Floating Zone method. The Czochralski method is the one used primarily for growing large-diameter crystals. It involves using a small single crystal seed with the proper orientation. The seed is lowered into the molten Si, then pulled out very, very slowly, so that Si can crystallize onto the seed.

The diameter of the growing crystal is controlled by the pulling rate. When the seed makes initial contact with melt, it experiences a thermal shock, which causes dislocations to form. With an initial pulling rate of several mm per minute, a small Si neck is formed in which the dislocations are concentrated. The pulling rate is then reduced gradually, leading to the conical shape at the end of the ingot. In the steady state, the pulling rate is of the order of 1mm per minute. As it is being pulled, the crystal is rotated slowly. At the same time, the quartz crucible holding the molten Si is rotated slowly in the opposite direction, in order to eliminate any inhomogeneities.

A single crystal Si ingot is also perhaps the purest material produced on an industrial scale. Impurities, such as O and C, are held typically at a level of one part in 10^6, whereas certain metal impurities must be more strictly controlled because they interfere with the electrical properties of the Si. The level of such metal impurities is of the order of one part in 10^{10}. This low impurity level is necessary because other impurities are introduced into Si on purpose, for modifying its electrical properties, at a level of one part in 10^8 and up.

Image sources

Title picture: adapted with permission by A. Akhavan from:
http://www.quartzpage.de/gen_mod.html

Fig. 7.2: adapted from:
http://ergodic.ugr.es/termo/lecciones/water1.html

Fig. 7.3: adapted from:
http://www.spring8.or.jp/en/news_publications/press_release/2010/101015

Fig. 7.4: Data by Netzsch, used with permission
http://www.netzsch-thermal-analysis.com/download/dil-liquidmetals-e_190.pdf

Fig. 7.8: adapted from:
Essentials of Materials Science, by A. G. Guy

Fig. 7.14: reproduced with permission by the American Institute of Physics
H.D. Keith, F. J. Padden, J. Appl. Phys. 34, (1963) 2409 and 35 (1964) 1270.
http://www.szfki.hu/~grana/pre05.pdf
http://www.pages.drexel.edu/~hjs25/photos_4.html

Fig. 7.15: reproduced with permission by the American Chemical Society
Lin Li et al., Macromolecules 34 (2001) 316-325
http://ihome.ust.hk/~kyeung/pdf/macro3401314.pdf

Fig. 7.17: reproduced with permission by Elsevier
C. Beckerman, Q. Li, and X. Tong, Sci. Technol. Adv. Mat. 2 (2001) 117-126
http://www.engineering.uiowa.edu/~becker/documents.dir/equiaxedoverview.pdf

Fig. 7.18: reproduced with permission by the Natl. Academy of Sciences
W. Losert, B. Q. Shi, and H. Z. Cummins, Proc. Natl. Acad. Sci. U.S.A. 95 (1998) 431-438
http://www.ncbi.nlm.nih.gov/pmc/articles/PMC18437/

Fig. 7.19: Courtesy of Prof. M. Glicksman (Florida Institute of Technology).

Fig. 7.20: reproduced with permission by Shinetsu
http://www.shinetsu.co.jp/en/products/pdf/c401.pdf

References

Guy is an excellent reference, with considerably more detail than is presented here.

Most recent textbooks deal with one-component phase diagrams more briefly, with fewer examples. They also treat matters related to melting and solidification separately from one-component phase diagrams.

For much more on all things Quartz, visit:
`http://www.quartzpage.de/gen_mod.html`

For movies on dendrite growth and solidification:
`http://math.nist.gov/mcsd/savg/vis/dendrite/`
`http://web.mechse.illinois.edu/research/dantzig/`
`solidification.org/Movies/movies.html`

For a summary of NASA microgravity work, see:
`www.nasa.gov/pdf/315963main_Microgravity_Science_Space_Flights.pdf`

This is a high school teachers' guide, but with an excellent overview.

For more information on the growth of single-crystal silicon, see:
`http://www.tf.uni-kiel.de/matwis/amat/elmat_en/kap_6/backbone/r6_1_2.html`
`http://www.tf.uni-kiel.de/matwis/amat/elmat_en/kap_6/illustr/i6_1_2.html`

For a picture of a 6 foot tall, 20 cm diameter Si ingot, go to:
`http://www.tf.uni-kiel.de/matwis/amat/elmat_en/kap_6/illustr/i6_1_2.html`

For examples of the growth of crystals and solids of a totally different, namely biological kind, visit:
`http://gower.mse.ufl.edu/research.html`

Exercises

1. In the early days, the pioneer women were able to dry their washed clothes by hanging them outside even in the winter time, when the temperature was below freezing. Why was that possible?

2. How many triple points are shown on the phase diagram of SiO_2 in the title figure?

3. What are the approximate conditions for the critical point of water (see the relevant figure in the text) ?

4. Name at least five components in the Coca Cola drink in a sealed can.

5. How many components are there in each of the following materials systems:
   a) the metal alloy of brass,
   b) the ceramic Al_2O_3,
   c) a solid ceramic alloy of NaCl and KCl,
   d) an aqueous solution of NaCl and KCl,
   e) semicrystalline polyethylene ?

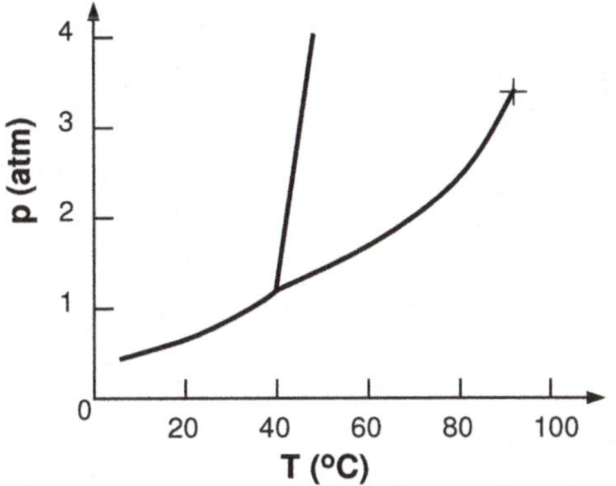

Figure 7.21:
Please answer the questions of Exercise 6 below based on the phase diagram to the left:

6. a) What is the normal boiling point of this material (normal = at 1 atm) ?
   b) At what pressure and temperature is the critical point?
   c) What phases coexist, if any, at 2.6 atm and 44°C (on the line) ?

7. Consider the phase diagram for SiO_2 below.

   a) Is this diagram consistent with the report in Ref. 2 below that the boiling point of SiO_2 is at 2230°C?
   b) Is the phase diagram from the title page consistent with a boiling point of 2230°C for SiO_2 ?
   c) What are the main differences between the two SiO_2 phase diagrams?

   Ref. 1:
   `http://serc.carleton.edu/research_education/equilibria/simplephasediagrams.html`

   Ref. 2:
   `http://www.chemicalbook.com/ProductMSDSDetailCB8138262_EN.htm`

8. Based on the data of Fig. 7.4, if you assumed that the drop in density at T_m were due to a sudden increase in the number of vacancies, what would be the concentration of vacancies just above T_m? Comment on this number.

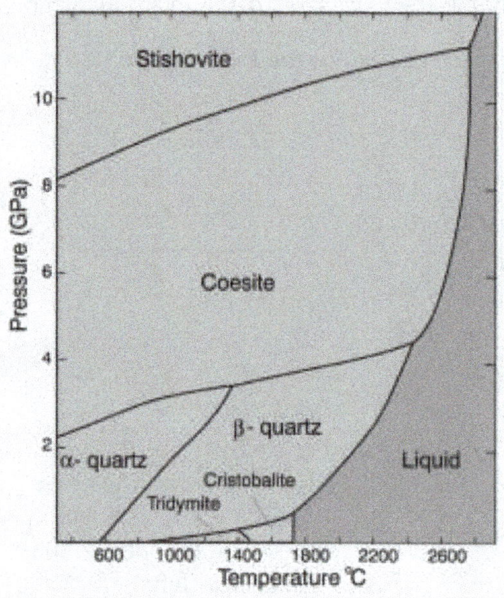

Figure 7.22:
The phase diagram for SiO_2 shown here is from Ref. 1 below.

Compare it to the one on the title page at the beginning of this chapter.

9. Pick three points on the curve for soda-lime glass in Fig. 7.8 and show that $\eta(T)$ obeys Eq. 7.6. What is the activation energy?

10. Consider the particle shown below.

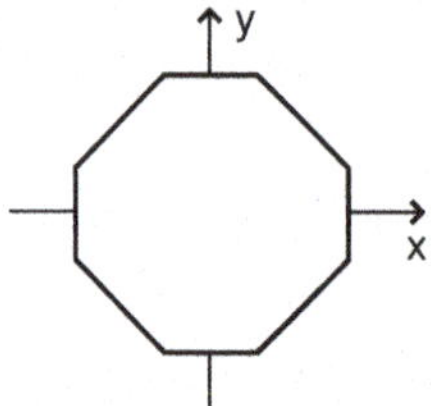

Figure 7.23: Crystalline particle with planar facets formed by crystal planes.

Assume that the growth rates in the $\pm$x- and $\pm$y-directions are slightly higher than in the diagonal directions. Sketch a sequence of shapes as the particle continues growing.

Problems

1. Formulate in words what experimental observation one would make following the melting line for H_2O to higher pressures. What is anomalous about this, and why?

2. How would the phase diagram of SiO_2 in the title figure have to be completed if it were extended to larger temperatures?

3. How do you think the densities of the various polymorphs of quartz compare ?

4. A tank filled with propane fuel contains both liquid and vapor.
   a) What should the pressure be in the tank for normal use?
   b) Why should the conditions in the tank be such that liquid and vapor coexist? Why not simply use compressed gas ?

5. Based on the data of Fig. 7.4, come up with an estimate for the average number of nearest neighbors for Cu in the liquid state.

6. Calculate the critical radius r_c of a Cu nucleus for three values of undercooling:

   $\Delta T = 50$ K, 10 K, and 2 K,

   given that the surface energy $\gamma = 0.2\ J/m^2$ and the energy of fusion $\Delta H_f = 13$ kJ/mol. Pay particular attention to using consistent units so that r_c comes out the way it should. Discuss your numerical results for r_c.

7. An estimate used occasionally for the surface energy γ is the following:

   $$\gamma = 0.2\,\Delta H_a/A$$

   where ΔH_a is the heat of atomization and A the surface area per atom. Check how good this estimate is by applying it to Cu (see the Problem above). Again, be careful with units.
   For further discussion of the surface energy, see:
   `http:\\www.virginia.edu/ep/SurfaceScience/Thermodynamics.html`

8. How large, or how small, should a nucleus be that it could usefully serve as a *crystalline* template for further growth of the solid? For example, one unit cell would appear to be too small.

   Make an educated engineering guess as to how many unit cells such a nucleus should contain, at a minimum, and determine what undercooling would be required for Cu so that r_c corresponds to this nucleus.

9. Discuss in general terms how our treatment of critical size nuclei would have to be modified if allowance were made for the fact that nuclei will be small crystals, with crystal planes, so that the shape of the nuclei could not be assumed as spherical.

   What would be the shape of real crystalline nuclei?

Chapter 8
Phases and Microstructure

Overview

When a material consists of two components, for example a metallic alloy, it typically melts to form a single solution at high enough temperature. However, in the solid state, the two components usually do not mix very well because the miscibility of one component in the other is limited. This leads to rather complex behavior upon solidification, and in particular to the formation of different solid phases, which are characterized by different compositions and possibly different crystal structures. Moreover, the final microstructure is a function of certain kinetic variables of the solidification process, and it can also be modified by additional heat treatment. The microstructure of such materials, and thus their mechanical properties, depend sensitively on the composition and the thermal history. These phenomena are illustrated with help of a detailed discussion of the Cu-Ni, Cu-Ag, and Fe-C material systems, among others.

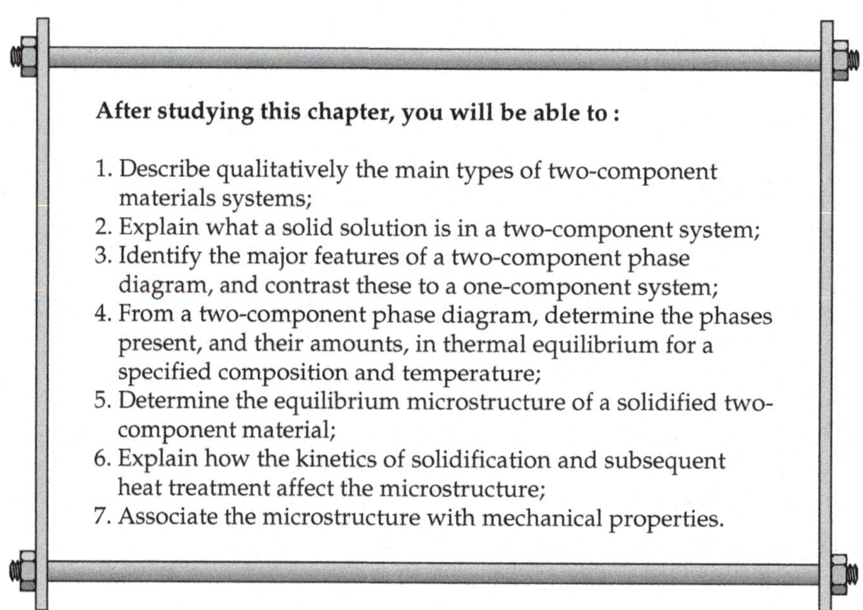

After studying this chapter, you will be able to :

1. Describe qualitatively the main types of two-component materials systems;
2. Explain what a solid solution is in a two-component system;
3. Identify the major features of a two-component phase diagram, and contrast these to a one-component system;
4. From a two-component phase diagram, determine the phases present, and their amounts, in thermal equilibrium for a specified composition and temperature;
5. Determine the equilibrium microstructure of a solidified two-component material;
6. Explain how the kinetics of solidification and subsequent heat treatment affect the microstructure;
7. Associate the microstructure with mechanical properties.

Microstructures of Pb-Sn alloys

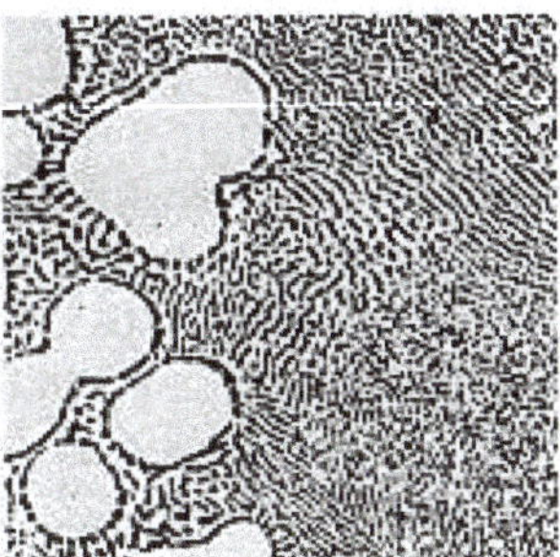

The picture shows optical micrographs at 500x magnification of two Pb-Sn alloys used as solder. The sample on the left has a composition of 37% Pb and 63% Sn, and the sample on the right has a composition of 30% Pb and 70% Sn. The seemingly small difference in composition leads to rather different microstructures. Both samples have similar alternating bright and dark layers. The bright areas depict Sn-rich material and the dark areas Pb-rich material. However, only the sample on the right shows additional large rounded inclusions of the bright material.

8 Phases and Microstructure

In this third of three chapters dealing with temperature-dependent phenomena in solids, we will revisit all the issues we addressed in the previous chapters, but this time as applied to ***two component alloys*** instead of to pure substances.

You will see shortly that once two components are involved, even such a seemingly simple process as melting is considerably more complicated. And the complexities go up even more when it comes to solidification. This is a consequence of the compatibility, or lack thereof, between the two components. Depending on how similar they are, they may mix readily in the molten state, but not in the solid state. If they don't, that means the solid will consist of different phases, i.e. parts that differ in composition and possibly crystal structure. What these phases are, and how they combine to form a microstructure, depends on the overall material composition and on the thermal history. All of this, in turn, determines the properties of the final material in general, and the mechanical properties in particular.

8.1 Two-Component Phase Diagrams: Generic Issues

You recall that with one-component systems, we were able to represent everything we needed to know about the thermal equilibrium state of the material by a phase diagram, i.e. 2D-plot in which the phases present were listed as a function of the pressure p and the temperature T. If we deal with a two-component system, there is clearly another equally important variable, namely the composition, that we need to keep track of. Also, the question is whether we can use phase diagrams again, and if so how?

First, with regard to composition, if we call the two components A and B, then for the composition we will simply state what percentage of the entire material is A, and what percentage is B. The only minor complication is that there are in fact two ways in which to specify a percentage of composition. We will use mostly what is referred to as **weight percent**, that is percentage of the total in units of weight (or mass), abbreviated by wt %.

Sometimes it may be advantageous to give concentrations as **atom percent**, or at % in short, that is as percentages expressed in terms of number of atoms (or moles) involved. Moreover, in certain calculations it may be simpler yet to use fractions rather than %, and once again fractions can be formulated in terms of weight (mass) or number of atoms.

These different concentration scales are readily interchangeable. As an example, consider the compound Fe_3C, a very important marker in the iron-carbon system, which is the foundation of all steel materials.

The composition in at % or atom fractions follows in a straightforward fashion from the atomic formula:

$$\text{at \% Fe} = \frac{3}{3+1} = \frac{3}{4} = 75\,\text{at \%} \tag{8.1}$$

$$\text{at \% C} = \frac{3}{3+1} = \frac{1}{4} = 25\,\text{at \%} \tag{8.2}$$

That is, 3 out of every 4 atoms are Fe, and 1 out of every 4 atoms is C. The weight % or weight fractions can be expressed as follows:

$$\text{wt \% Fe} = \frac{3 \times 55.8}{3 \times 55.8 + 12.0} = 0.933 = 93.3\,\text{wt \%} \tag{8.3}$$

$$\text{wt \% C} = \frac{12.01}{3 \times 55.8 + 12.0} = 0.067 = 6.7\,\text{wt \%} \tag{8.4}$$

where 55.8 and 12.0 are the atomic weights of Fe and C. Weight % are more common in engineering practice because this is how commercial alloys are prepared: by mixing certain weights of the various components. The advantage of at % is that they are directly related to chemical formulae.

Comparing these numbers you will realize that the wt % of the component with the lower atomic weight are much smaller than the at % of the same component.

Second, with regard to phase diagrams, the key question is how to represent information about the presence, and coexistence, of phases in a 2D graph when there are three independent variables. This difficulty is resolved in principle by noting that for melting and solidification, the effect of pressure is negligible compared to temperature and composition. So for practical purposes the pressure can be considered constant at 1 atm.

Therefore, a two-component phase diagram is a (temperature, composition) plot delineating the presence of liquid and solid phases, and the boundaries between them. This type of plot can be obtained as outlined below.

Let us assume that we are interested in an alloy $A_{1-x}B_x$. The parameter x is the atom fraction of component B defining the composition of the alloy. Clearly we can record normal (p, T) phase diagrams for the pure components A and B, as displayed in Fig. 8.1 (cf. Fig. 7.1). Let us also assume that we have recorded similar data for a series of alloys with varying compositions x. We can combine all these plots as follows:

The (p, T) plot farthest in back of Fig. 8.2 is the phase diagram for pure A, with its normal melting point $T_{m,A}$, and the plot in front is the corresponding diagram for pure B, with melting point $T_{m,B}$. In addition, you should picture a set of similar plots, in-between the ones shown explicitly, for alloys $A_{1-x}B_x$ with various compositions x.

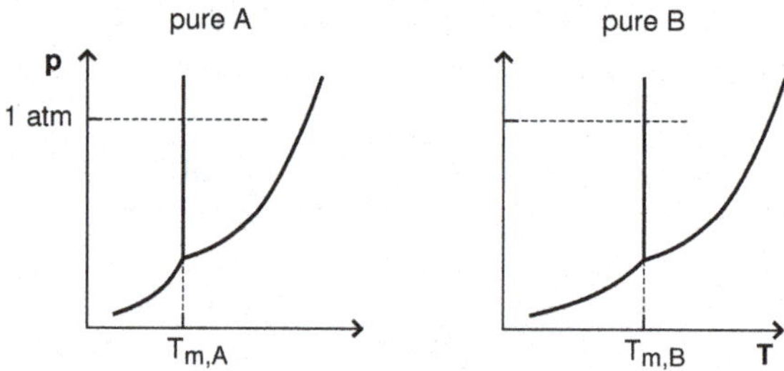

Figure 8.1: One-component phase diagrams for the pure components A and B

Now, the ***two-component phase diagram*** is the horizontal slice at 1 atm pressure through the entire set of curves in Fig. 8.2, with composition as the x-axis and temperature as the y-axis. If we display this diagram separately, it would look like Fig. 8.3 below.

Please note that in Figs. 8.2 and 8.3 we have introduced a general composition variable X. We will keep in mind that it refers to component B, and for the moment we will take its numerical value as wt %. However, in most instances it could also be taken as weight fraction. The important thing is to know which component it points to, and to use consistent units for concentration.

Two additional comments are in order with respect to Fig. 8.3. First, be aware that the line connecting $T_{m,A}$ and $T_{m,B}$ represents the melting points of alloys $A_{1-x}B_x$, but it is not obvious that this should be a *straight* line. However, it turns out that there are practical systems for which, to a good approximation, this line is indeed straight. Second, as we will argue below, the diagram in Fig. 8.3 is not the whole story yet, even for the simplest two-component systems.

8.2 Binary Isomorphous System: Complete Solid Solubility

8.2.1 Cu-Ni Phase Diagram

The first material example we will consider is the two-component, or binary, ***Cu-Ni system***. Let Cu be component A, and Ni component B. As you learned in Chapter 4, Cu and Ni atoms are so similar that they can form a solid solution for any mixture of the two components (see the Hume-Rothery rules). That is, Cu and Ni are completely soluble in each other. It is for such a so-called **isomorphous system** that the line connecting the melting points of the pure components in the two-component phase diagram turns out to be a straight line (Fig. 8.3).

What this implies is that Cu atoms have no preference between Cu and Ni with respect to their nearest neighbors, nor do Ni atoms. Cu and Ni atoms can mix freely, although of course the difference in the melting points indicates

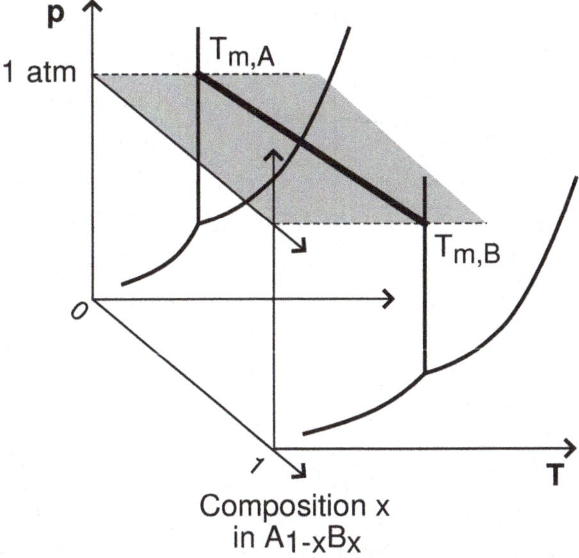

Figure 8.2: Combined (p, T) plots for various alloys $A_{1-x}B_x$, including pure components A and B. The two-component phase diagram is the horizontal slice shaded in gray.

that their interaction is not exactly the same. Recall also that the atom-atom interaction is related directly to the melting temperature (cf. Fig. 5.17). Hence, the straight line between $T_{m,Cu}$ and $T_{m,Ni}$ simply says that the average atom-atom bond strength in the mixture is a linear combination of the bond strengths of the pure components.

However, the diagram in Fig. 8.3 is incomplete, and you can appreciate readily why that is so with help of the Gibbs Phase Rule (cf. Eq. 7.1):

$$P + F = C + N \tag{8.5}$$

As a reminder, note what this equation says for a pure substance: For pure Cu at the melting point $T_{m,Cu}$ where liquid and solid are in equilibrium, $C = 1$ and $P = 2$. Furthermore, $N = 1$, since the pressure p is fixed. Therefore

$$F = C + N - P = 1 + 1 - 2 = 0 \tag{8.6}$$

That is, at the melting point of a pure substance there is no degree of freedom; everything is set.

Now, consider a Cu-Ni solid solution with a given composition and assume you are exactly on the straight line connecting $T_{m,Cu}$ and $T_{m,Ni}$, i.e. exactly at the presumed melting point of the solid solution. But in this case, $C = 2$ and $N = 1$, and therefore

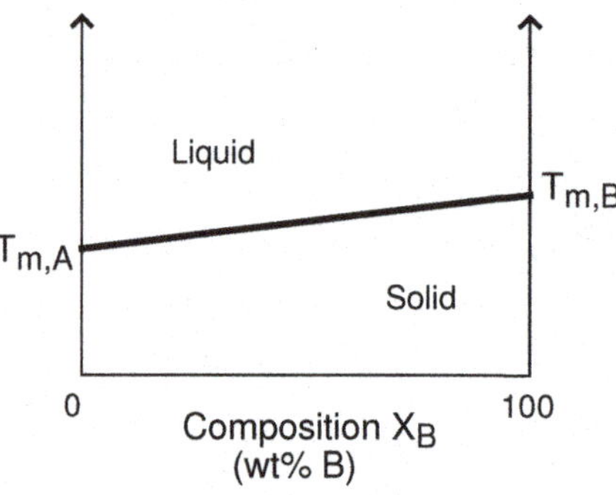

Figure 8.3:
Generic two-component phase diagram.

$$F = C + N - P = 2 + 1 - 2 = 1 \tag{8.7}$$

That is, it appears as if at the melting point of a solid solution there is one degree of freedom left! What this means is that for a solid solution there is no exact ***melting point*** but rather a ***melting range***, i.e. a range of temperature over which the material melts or solidifies gradually. Thus, the complete Cu-Ni phase diagram looks as follows:

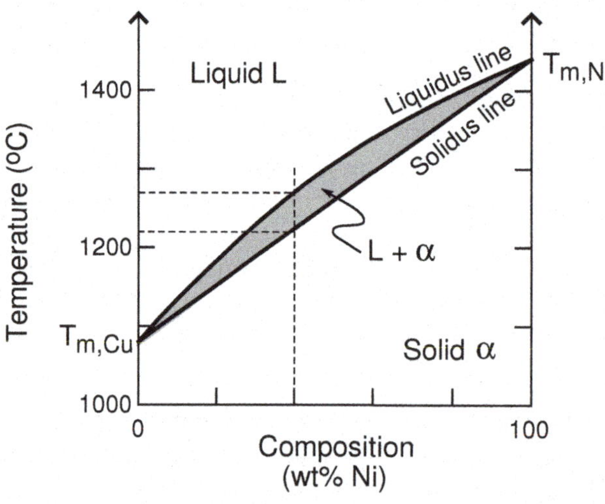

Figure 8.4:
Phase diagram for the Cu-Ni isomorphous system.

The diagram is to be interpreted as follows: Let us take as an example a material containing 40 wt % Ni at a temperature of 1000°C. The diagram says that at that temperature the sample is entirely solid. If we now raise the temperature slowly, when we reach the lower line called **solidus line** at about 1220°C, the sample starts to melt. By this I mean that a small amount of liquid starts to appear. If we continue raising the temperature, the higher we go, the more of the solid melts. When we reach the upper line called **liquidus line** at about

1270°C, all the solid has melted. Above that temperature the sample is entirely liquid.

So for this two-component system, two phases, i.e. solid and liquid, can coexist over a range of temperature, not just at a precise temperature point. The gray region in Fig. 8.4 outlines this **two-phase region**, and the liquidus and solidus lines mark the boundaries between two-phase and **one-phase regions**. This implies also that for a two-component system in a two phase-region at a given temperature and composition, a given fraction of the sample is liquid and the rest solid. By contrast, in a one-component system at the melting point, an arbitrary fraction of liquid and solid can be in equilibrium.

The liquidus and solidus lines are clearly important parts of such a phase diagram. Another feature of the diagram concerns notation. It is common to use L to designate a liquid region and lower case Greek letters α, β, etc., for solid phase regions. In Fig. 8.4 there is one solid region labeled α.

As regards solidification, the opposite sequence of events will occur. If we start at a temperature where the system is liquid and then lower the temperature slowly, the first solid will appear when we reach the liquidus line. The lower we go, the more liquid will turn solid. Once we reach the solidus line, the entire sample will have been converted to solid.

8.2.2 Tie Line and Lever Rule

The two-phase region contains more information than simply that solid and liquid are present. One can also determine how much of the entire sample is liquid and how much is solid.

Let us assume that the composition X_0 of our sample is 50 wt % Ni and that the sample is at 1270°C. This is marked by the black dot in Fig. 8.5. We know that at this condition both liquid and solid are present. We now draw a horizontal line through his point out to the liquidus and solidus lines. This line is called **tie line**. Where the tie line intersects the liquidus line, on the left, is found the composition X_L of the liquid phase, and where the tie line intersects the liquidus line, on the right, is found the composition X_α of the solid phase. Thus $X_L = 39$ wt % Ni and $X_\alpha = 54$ wt % Ni.

The tie line also enables us to determine how much of the sample is liquid and how much is solid. Let us say that at the point chosen, $X_0 = 50$ wt % Ni and 1270°C, m_L is the amount (i.e. the mass) of liquid, m_α is the amount of solid, and let m_t designate the total amount of material. Then the following relation clearly holds:

$$m_L + m_\alpha = m_t \tag{8.8}$$

In addition, the total amount of Ni in the sample can be expressed as follows:

$$m_L X_L + m_\alpha X_\alpha = m_t X_0 \tag{8.9}$$

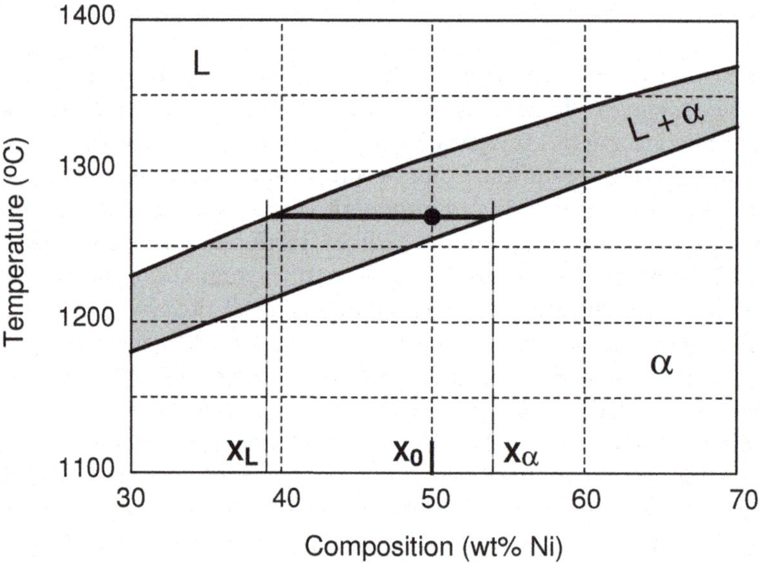

Figure 8.5: Central portion of the Cu-Ni phase diagram.

Now we take the total amount m_t as prescribed and solve the two equations above for the two unknowns m_L and m_α. Remember: We know X_L and X_α from the phase diagram. After some simple algebra the result is

$$\frac{m_L}{m_t} = \frac{X_\alpha - X_0}{X_\alpha - X_L} \tag{8.10}$$

$$\frac{m_\alpha}{m_t} = \frac{X_0 - X_L}{X_\alpha - X_L} \tag{8.11}$$

Numerically our example in Fig. 8.5 works out as follows:

$$\frac{m_L}{m_t} = \frac{54 - 50}{54 - 39} = 0.27 \tag{8.12}$$

$$\frac{m_\alpha}{m_t} = \frac{50 - 39}{54 - 39} = 0.73 \tag{8.13}$$

Equations 8.10 and 8.11 express the fractional amounts of liquid and solid phase in terms of the respective concentrations. These are not formulae to remember literally, but you should note their structure. The denominator is the length of the tie line for both phases, and the numerator is the piece of tie line opposite to the phase in question. For example, in order to get the amount of solid phase, you divide the part of the tie line extending to the liquid (to the opposite phase) by the entire tie line.

These equations are also referred to sometimes as the **lever rule**, although I do not find this name particularly instructive. The name comes from visualizing the tie line as a lever with the fulcrum at the system point. Then the lever is in mechanical equilibrium if it is loaded with weights according to the phase amounts given in the two equations.

8.2.3 Microstructure Development

Let us now go a step further and ask what the material would look like as a function of temperature. Imagine that we can view the material in an optical microscope as we raise the temperature very slowly from 1200°C. Again let the average composition X_0 of the material be 50 wt % Ni (Fig. 8.6).

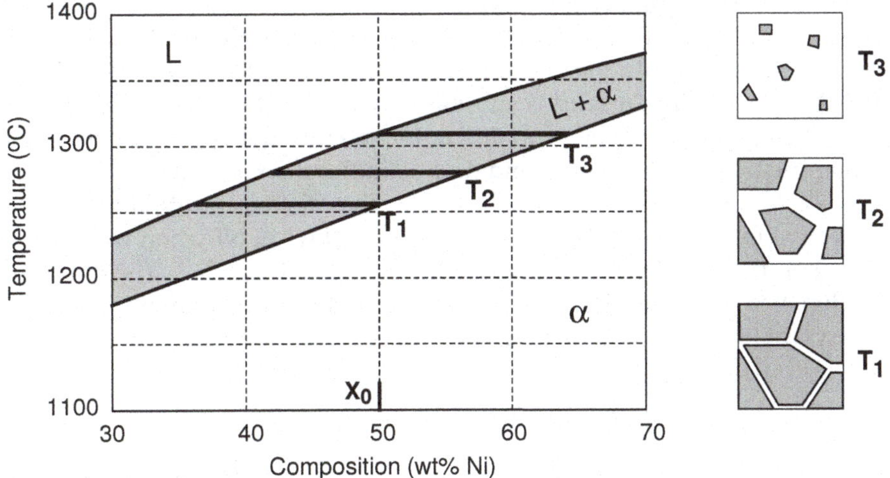

Figure 8.6: Microstructure development in a Cu-Ni sample. Panels on the right show schematic microstructure of Cu-Ni alloy.

Let us say that with X_0 = 50 wt % Ni, the temperature T_1 of about 1255°C is just a little *above* the solidus line. The phase diagram then tells us that there is a small amount of liquid present, but most of the sample is still solid. This is displayed schematically in the lowest panel on the right of Fig. 8.6, with the initial solid grains almost intact but a bit of liquid in-between them. Incidentally, the phase diagram also tells us that the composition of this very first liquid is 36 wt % Ni, at the left end of the tie line.

The next snapshot we take is at an intermediate temperature T_2 of 1280°C. Convince yourself from the lever rule that at T_2 about half of the sample is now liquid and half solid. The solid grains have shrunk considerably at this point.

The top panel on the right of Fig. 8.6 displays the microstructure at the temperature T_3 of 1310°C, just *below* the liquidus line. Almost the entire sample is liquid now. The solid grains still left are tiny and, incidentally, their composition is 64 wt % Ni, at the right end of the tie line. Once the liquidus line is crossed, of course the entire sample is liquid.

The above description illustrates the gradual melting of the alloy. If you performed the process in reverse, solidification would also be gradual. At T_3 the first solid nuclei would appear. They would continue growing as the temperature is lowered, until at T_1 almost no liquid is left. Below the solidus line, the entire sample would be solid.

At the beginning of this discussion I said that we would perform this thought experiment very slowly, meaning so slowly that all parts of the system are in thermal equilibrium at all times. In practice this may not be realized, especially in solidification, as it may require an impractically slow solidification rate. A careful examination of the phase diagram reveals the difficulty.

During solidification, at T_3 the Ni concentration in the very first solid grains is X_α = 64 wt % Ni. At the same time, the concentration in the remaining liquid, which is practically the entire material, is essentially 50 wt % Ni, or just a little less. So the initial solid is substantially enriched in Ni compared to the remaining liquid or the average material. Indeed this makes sense, since Ni has a higher melting point than Cu, and thus solidifies more readily.

As the temperature is lowered below T_3, the Ni concentration in the solid goes down, from 64 wt % Ni to an average of 50 wt % Ni when we reach T_1. But look at the liquid concentration: X_L = 36 wt % Ni at T_1 ! The liquid has been depleted in Ni continually because Ni solidifies preferentially, and hence the very last bit of material turning solid has a composition of 36 wt % Ni.

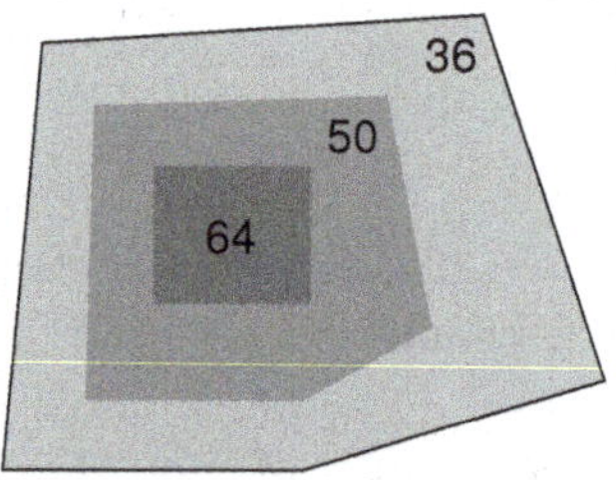

Figure 8.7:
Schematic grain with non-uniform Ni distribution due to non-equilibrium solidification.

The consequence of all this is that the solid grains may have a non-uniform composition: At the core it is 64 wt % Ni, on the outside 36 wt % Ni, and in-between it changes smoothly from one to the other (Fig. 8.7 shows a simplified view with three discrete layers). If we wanted true equilibrium throughout at the microscale, the solidification process would have to be slow enough that diffusion can equilibrate the Ni distribution within the growing solid grains and within the liquid at all times. And, as mentioned above, slow enough for maintaining equilibrium may be impractically slow.

8.3 Binary Eutectic System: Partial Solid Solubility

As you can well imagine, complete solid solubility in binary alloys will be the exception, since it requires the two components to be very similar (see the Hume-Rothery rules in Chapter 4). A much more common two component phase diagram is the one found for so-called **eutectic systems** where the solid solubility is limited. As an example, the diagram for the ***Cu-Ag system*** is shown below.

8.3.1 Cu-Ag Phase Diagram

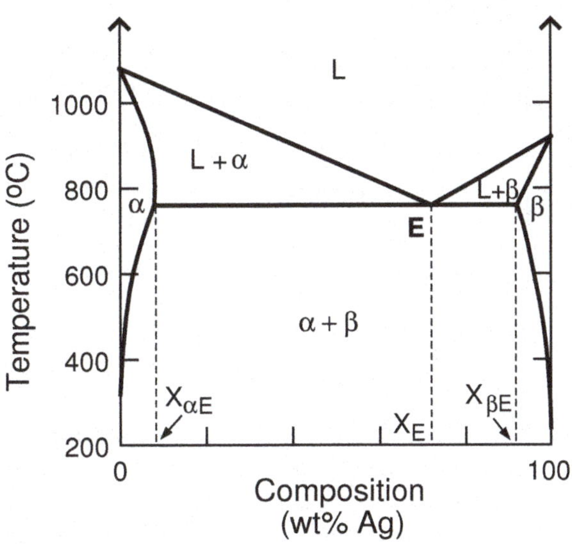

Figure 8.8: Phase diagram for the eutectic Cu-Ag system.

To compare, and put in perspective, this diagram and the previous one for Cu-Ni, you should note that Cu, Ni, and Ag are all fairly similar. In particular, all three are FCC. The only difference is in their atomic radii: 0.36 nm for Cu, 0.35 nm for Ni, and 0.41 nm for Ag. Yet this seemingly minor difference between Ni and Cu gives rise to very different phase diagrams (cf. Fig. 8.4 and (Fig. 8.8).

This is not to say that the two diagrams have nothing in common. In fact, one way of comparing them is to note that the top parts, above the horizontal line, look similar: They show the melting points of the pure components, as well as the nearby liquid, liquid-solid, and single-solid phase regions. What is new here, is the horizontal line and the rest of the diagram below it.

This horizontal line in Fig. 8.8 is called **eutectic line**, and the special point E on it is called **eutectic point**. It occurs at the eutectic composition X_E = 72 wt %

Ag and the eutectic temperature T_E of 779°C. The eutectic line is reminiscent of a tie line, and its ends are at $X_{\alpha E}$ = 8.0 wt % Ag and $X_{\beta E}$ = 91 wt % Ag. From the ends of the eutectic line, two additional lines called **solvus lines** extend to lower compositions. The region delineated by the eutectic line and the two solvus lines is a two-phase region, where the *two solid phases* α and β *coexist.*

Let us investigate now what new phenomena occur because of the eutectic line. First, imagine that your system has a composition X_0 of 5 wt % and is at a temperature of 1100°C (Fig. 8.9) when we begin to lower the temperature:

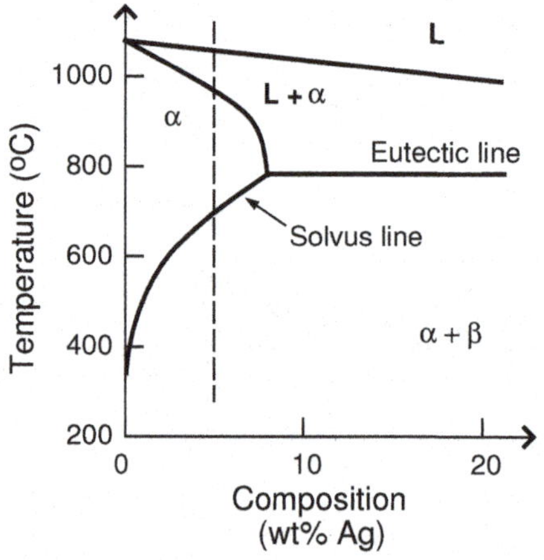

Figure 8.9: Excerpt of Cu-Ag phase diagram in the region of low Ag concentration. The system concentration is at X_0 = 5 wt % .

At 1100°C the system is all liquid. As we lower the temperature slowly, the system begins to solidify at around 1070°C, and solidification is complete at around 960°C. At that point the solid is made up of a single phase α containing 5 wt % Ag. Up to this point everything has been the same as before with Cu-Ni.

However, as we continue lowering the temperature, at about 700°C we reach the solvus line. The phase diagram now tells us that as soon as we are *below the solvus line,* the single solid phase α cannot exist by itself anymore. Rather, the system separates into two phases, α and β. In other words, below the solvus line the system is a mixture of the two solid phases α and β.

The meaning of the two solid phases α and β, with the emphasis on *solid,* is sometime difficult to grasp. What the two phases are in the case of a solid and a liquid is straightforward by comparison. Think of α as mostly component A, i.e. Cu, with just a small amount of component B, i.e. Ag, as impurities in it. For β, it is simply the other way around: β is mostly Ag, with a few Cu impurities in it. If you did X-ray diffraction on this material, you would find that it is a mixture of two parts (i.e. α and β): Both have FCC crystal structure, but their lattice constants are slightly different.

The phase diagram also provides more information regarding the solubilities. As you can see from Fig. 8.9, the solubility of Ag in Cu is at its highest, namely 8 wt %, at the eutectic temperature (at the left end of the eutectic line).

The solubility decreases from there as the temperature is lowered. At 300°C and below, it is so low that it shows as essentially 0 on the diagram. Of course, there is always some finite, although very small solubility, but for practical purposes you can consider the α phase to be almost the same as Cu at low temperature. Again, on the other side of the phase diagram, the situation would be the same with regard to β and Ag.

The most striking feature of the Cu-Ag phase diagram is the eutectic line at T_E = 779°C, and in particular the eutectic point on it at X_E = 72 wt % Ag. Below the eutectic line, no liquid can exist. In particular, at the eutectic point, *all the liquid* of composition X_E *transforms at once* into solids α and β, as if at the normal melting point of a pure material. This transformation may be represented symbolically as a chemical reaction:

$$L(X_E) \rightleftharpoons \alpha(X_{\alpha E}) + \beta(X_{\beta E}) \tag{8.14}$$

Another way of interpreting what happens at the eutectic point is to say that when considering solidification of Cu-Ag alloys, the eutectic composition represents the alloy for which solidification, i.e. the appearance of solid, occurs last. At all other compositions, the first solid appears when crossing the liquidus line, which is at a higher temperature than T_E.

The eutectic line is yet another instance where it is instructive to apply the Gibbs Phase Rule. On the eutectic line we have L, α and β in equilibrium. Hence C = 2, P = 3, N = 1, and

$$F = C + N - P = 2 + 1 - 3 = 0 \tag{8.15}$$

assuming the pressure p is fixed and equal to 1 atm.

You may ask why the number of degrees of freedom is 0 on this line. In the one-component phase diagrams, being on a line meant that there was one degree of freedom. The answer is that here the temperature and all the compositions, specifically $X_{\alpha E}$ and $X_{\beta E}$, are fixed. The only thing that is variable is the amounts of the different phases present, and the Gibbs Phase Rule says nothing about amounts. That is where the lever rule comes in.

8.3.2 Amounts of Phases

When you are asked to determine the phase structure of a material, basically you proceed in three steps. You ask yourself:

1. *Which phases are present?*
   The answer follows from the phase diagram, depending on whether the system point lies in a one-phase or a two-phase region, or possibly on a line between two different regions.

2. *What are the compositions of the phases present?*
   If you are in a one-phase region, the composition of the phase present is the system composition. If you are in a two-phase region, you draw a tie line through the system point and note what the phases and their compositions are at the two ends of the tie line.

3. *What are the amounts of the phases present?*
   If you are in a one-phase region, the answer is 100 % of that phase. If you are in a two-phase region, you apply the lever rule to your tie line.

With respect to point 3 above, it is especially important to note that:

The lever rule applies to any two-phase region,

whether it is a liquid-solid or solid-solid region. As you know already how to deal with a liquid-solid two-phase region, we will work through ***two examples*** for the α - β region in the Cu-Ag diagram.

Example 1: $X_0 = X_E, T_0 =$ **778°C**

Our system point is at the eutectic composition X_0 = 72 wt % and *just below* the eutectic temperature of 779°C (Fig. 8.8).

Therefore, X_α = 8 wt % and X_β = 91 wt %. Now we apply the lever rule just as in Eqs. 8.10 and 8.11, except that the β phase substitutes for the liquid phase:

$$\frac{m_\alpha}{m_t} = \frac{X_\beta - X_0}{X_\beta - X_\alpha} = \frac{91 - 72}{91 - 8} = 0.23 \tag{8.16}$$

$$\frac{m_\beta}{m_t} = \frac{X_0 - X_\alpha}{X_\beta - X_\alpha} = \frac{72 - 8}{91 - 8} = 0.77 \tag{8.17}$$

A solid sample at the average eutectic composition and at 778°C consists of about $\frac{1}{4}$ of α phase and about $\frac{3}{4}$ of β phase.

Example 2: X_0 **= 5 wt %,** T_E **= 20°C**

Our system point is at the composition X_0 = 5 wt % as in Fig. 8.9 and at room temperature .

In this case we can set X_α = 0 wt % and X_β = 100 wt %. Again we apply the lever rule:

$$\frac{m_\alpha}{m_t} = \frac{X_\beta - X_0}{X_\beta - X_\alpha} = \frac{100 - 5}{100 - 0} = 0.95 \tag{8.18}$$

$$\frac{m_\beta}{m_t} = \frac{X_0 - X_\alpha}{X_\beta - X_\alpha} = \frac{5 - 0}{100 - 0} = 0.05 \tag{8.19}$$

A solid sample at the composition of 5 wt % and at room temperature consists of α phase; only a small fraction of 0.05 is β phase.

All of these arguments, of course, refer to thermal equilibrium conditions, as phase diagrams do in general. No matter what the final state is, it must have been reached very slowly, so that the sample had time to equilibrate fully.

8.3.3 Microstructure Development

The final question, and for materials properties the crucial question, is how the features of the phase diagram determine the microstructure of a material with a certain composition, especially in the context of solidification.

We will begin by examining a material with the eutectic composition, X_0 = 72 wt % Ag (cooling path 1 in Fig. 8.10).

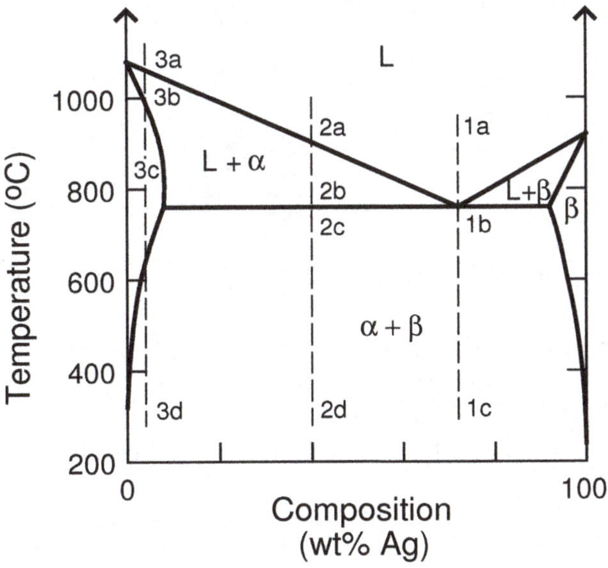

Figure 8.10: Cu-Ag phase diagram with three cooling paths at three different compositions.

We will start our thought experiment, as usual, with the material in the molten state: point 1a in Fig. 8.10. We will cool the sample very slowly, maintaining thermal equilibrium, and we are interested the microstructure that arises at point 1b just below the eutectic temperature T_E and also at point 1c at a much lower temperature. As we said above, when the temperature crosses the eutectic line, the phase diagram requires that the liquid turn into all solid at once. So how does this happen?

I will use a simple color scheme to aid in the explanation. For the sake of argument let us assume that Cu is black and Ag is white. Hence the liquid, being a uniform single phase would look medium gray. But what will the eutectic solid look like just below T_E, where it must separate into two phases α and β while maintaining its average system composition of 72 wt % Ag?

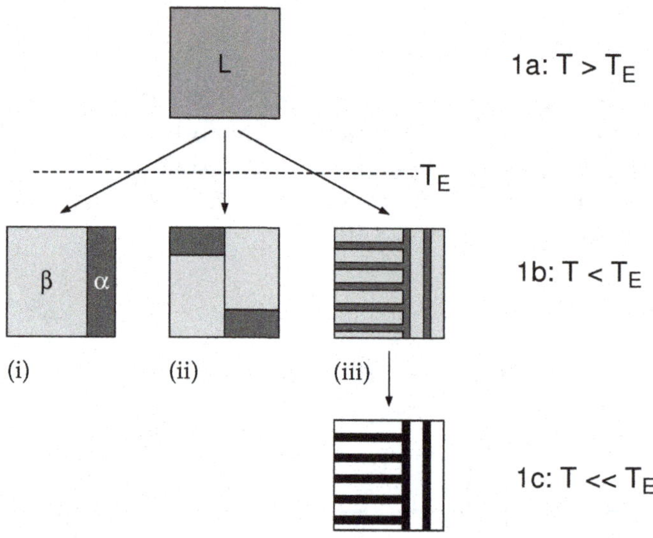

Figure 8.11: Possible microstructures of eutectic solid just below T_E.

If we want to know the compositions of α and β just below T_E, we draw a tie line. This tells us that just below T_E, $X_\alpha = 8$ wt % Ag and $X_\beta = 92$ wt % Ag (see Fig. 8.8). In our color scheme, this makes α dark grey and β light grey. Furthermore, the lever rule tells us that a fraction 0.24 of the sample must be α and 0.76 must be β.

But how are the α and β arranged in the eutectic solid? Fig. 8.11 displays some hypothetical microstructures. Panels (i) and (ii) show the phases as large chunks, with the proper compositions and about the correct fractional amounts. However, these are not correct microstructures. Experience shows that the eutectic solid looks something like Panel (iii): It consists of grains, and each grain is composed of fine alternating layers of α and β phase. (Panel (iii) shows two grains).

The reason why the microstructure of Panel (iii) is favored over Panels (i) and (ii) is that below the eutectic line, Cu and Ag atoms have to separate, and they can only due this by diffusion. If they were to form chunks of largely Cu or Ag atoms, the atoms would have to diffuse over very large distances. By contrast, they can form thin layers by diffusing only over a much smaller distance, in opposite directions, perpendicular to the layers. A layered structure like the one in Panel (iii) is known as a **eutectic microstructure**. It is characteristic of how a liquid of eutectic composition solidifies.

As the eutectic sample cools further, to a final temperature of 200°C or lower, further changes occur but these are fairly subtle. As I mentioned earlier, at that low a temperature α is essentially Cu, with very few Ag impurities, and β is essentially Ag, with very few Cu impurities. So the compositions of α and β continue to change during further cooling: α continues to loose Ag and β continues to loose Cu. You can also convince yourself that at temperatures of 200°C and below, the fractional amounts of α and β phase will be 0.22 and 0.78, respectively, which is very close to what they were just below T_E. The final microstructure is displayed schematically in the bottom panel 1c in Fig. 8.11.

Scanning electron micrographs of two real experimental eutectic Cu-Ag microstructures are shown in Fig. 8.12:

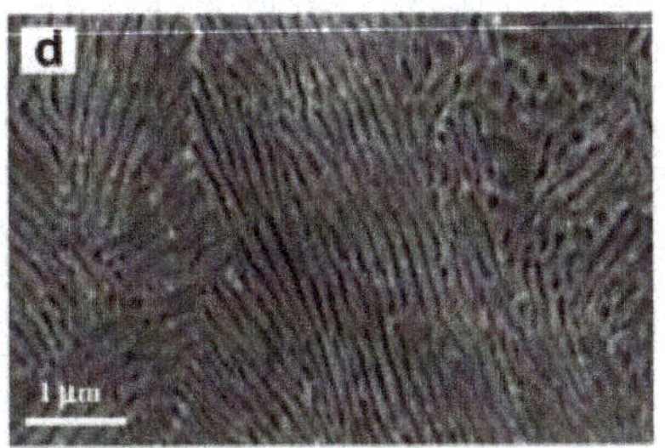

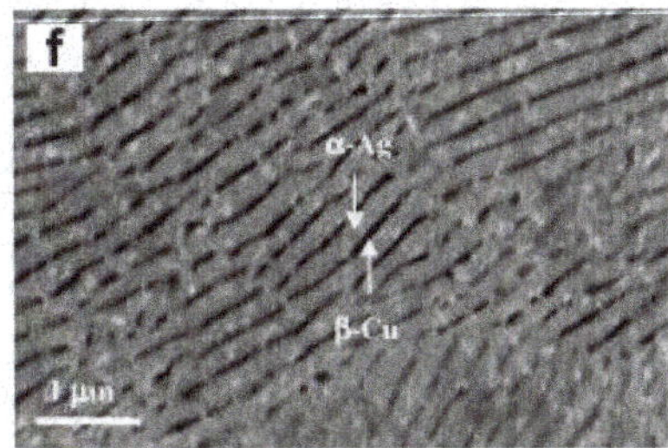

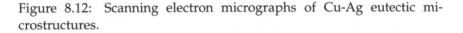

Figure 8.12: Scanning electron micrographs of Cu-Ag eutectic microstructures.

The difference between panel d) and panel f) is that the material for f) was cooled more slowly than for d). In both panels, the α-β layered structure is evident. In addition, you can discern individual grains, i.e. regions where the α and β layers are parallel but oriented differently than in other grains nearby. Note that the notation is reversed in Fig. 8.12: The α phase is Ag-rich, and the β phase is Cu-rich. However, in terms of composition, the dark and light phases have the same meaning in Fig. 8.11 and Fig. 8.12. Also, given the 1 μm scale bar in the lower left corner, note how thin these layers are.

With a good grasp of the eutectic microstructure, we are in a position to deduce the microstructures along the other cooling paths of Fig. 8.10. These are summarized graphically on Fig. 8.13 below.

Focusing now on path 2 you realize that it starts out as in path 1. However, when the system on path 2 approaches the eutectic line, there is a substantial amount of solid α phase present already! This kind of α is sometimes called **proeutectic** (for: prior to eutectic). It is displayed as the two grains in panel 2b of Fig. 8.13. In fact, a glance at the lever rule indicates that, just above T_E, about half the system is solid α and half still liquid. As the temperature now moves just below T_E, the remaining liquid transforms into eutectic solid, i.e. the layered α-β structure (panel 2c). As the temperature is lowered further, the α phase becomes more enriched in Cu, and the β phase more enriched in Ag (panel 2d), as before with path 1.

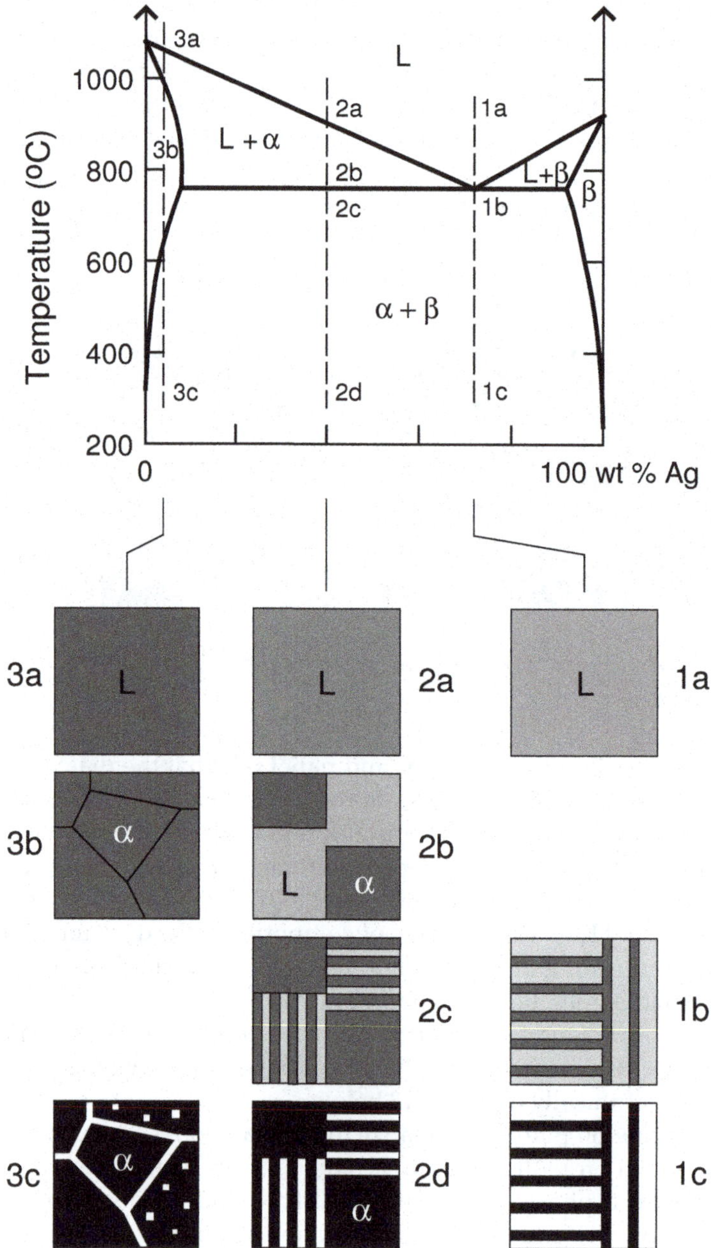

Figure 8.13: Schematic graphical summary of microstructure development in the eutectic Cu-Ag system. Note that, of course, the real grains in conditions 2b, 2c, etc., are not square.

Let us work out again the phase amounts for path 2, using the lever rule at points 2b and 2c. Just *above the eutectic line* we have for the *proeutectic* α:

$$\frac{m_\alpha}{m_t} = \frac{72-40}{72-8} = 0.50 \tag{8.20}$$

Make sure you know why the compositions are what they are. On the other hand, just *below the eutectic line* we have for the *total* α:

$$\frac{m_\alpha}{m_t} = \frac{92-40}{92-8} = 0.62 \tag{8.21}$$

The total α is made up of proeutectic and eutectic parts. Therefore, the fractional amounts are 0.50 for the proeutectic α, 0.12 for the eutectic α (hence a total amount of α of 0.62), and 0.38 for the eutectic β.

Path 3 is, at the beginning, very similar to the case of complete solubility. From panel 3a to 3b, the liquid forms a single α phase without any change in concentration. However, this all changes as the temperature goes below the solvus line. At that point, the complete solubility is lost, and the system begins to separate into two phases, α and β. In this case, the new β phase nucleates and grows in the α solid, at a rather low temperature. This process is initiated at defects, either along the grain boundaries of the α solid, or inside the grains. Both alternatives are indicated schematically in panel 3c. In practical systems, the growth of β usually occurs in one or the other type of location.

Whereas we have concentrated so far on cooling paths for concentrations of $X_0 \leqslant X_E$, similar results are obtained for concentrations $X_0 > X_E$. For example, you can easily convince yourself that for $X_0 \approx 0.85$ the outcome would be very similar to path 2 in Fig. 8.10, but with the roles of α and β reversed.

8.4 Other Binary Systems

In this section, we will examine briefly the phase diagrams of a few other binary systems of interest. This is to give you an idea of the complexities that are possible. For the most part, solubilities are limited in these binary systems. We will not go into any details of analysis, but the methods outlined above will apply to these systems as well.

8.4.1 The Pb-Sn System

The phase diagram of the Pb-Sn system is shown below. You will recognize it as another eutectic system. In its overall shape, it looks much like the phase diagram for Cu-Ag, but the critical parameters are different. For Pb-Sn the eutectic point is at T_E = 183°C and X_E = 62 wt % Sn. The ends of the eutectic line are at $X_{\alpha E}$ = 18 wt % Sn and $X_{\beta E}$ = 18 wt % Sn.

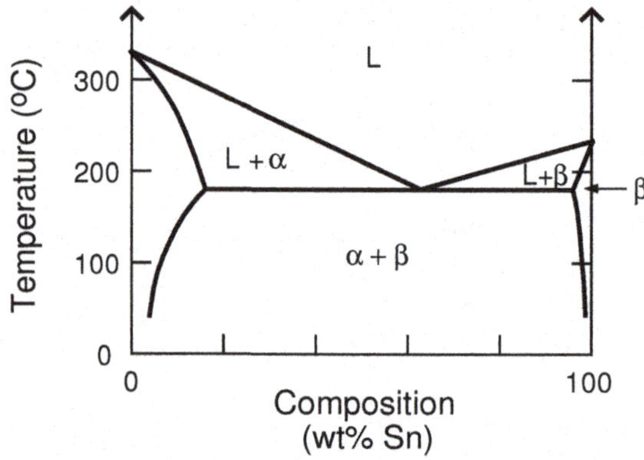

Figure 8.14: Phase diagram of the eutectic Pb-Sn system.

This system is of interest because Pb-Sn alloys with composition around X_E = 62 wt % Sn have been widely used as solders in electronic equipment and for Cu tube home water lines. The main advantages of the material are its low melting point and the strong, electrically conductive bond it forms when used for joining Cu wires, foils or pipes.

Alternative materials suitable as solders which do not contain Pb are available commercially. These types of materials and their phase diagrams are a subject of current interest in electronic materials research. Some examples are Sn-Ag, Sn-Sb, Sn-Bi, and various ternary (three-component) alloys.

8.4.2 The Cu-Zn System

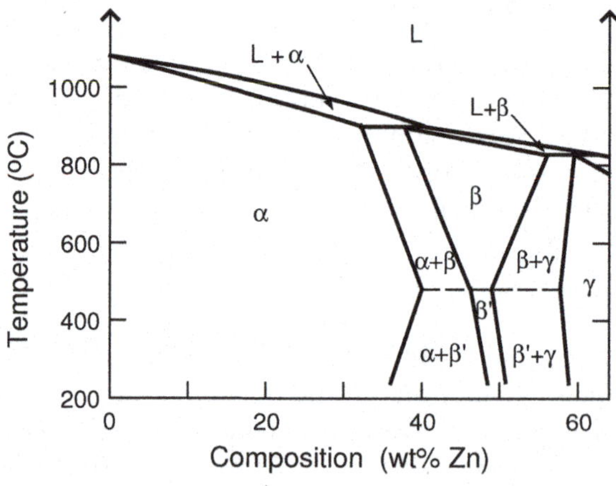

Figure 8.15: Partial phase diagram of the Cu-Zn system.

The phase diagram of the Cu-Zn system is shown in Fig. 8.15. The main purpose is to illustrated how complex phase diagrams can be. Of course, the Cu-Zn system is in wide use as brass, and we have discussed its mechanical properties in some detail. You will note in this system the many different solid phases at higher Zn content. Fortunately, most commercial brasses contain 30 wt % of Zn or less. Thus brass is typically a single-phase material.

8.4.3 The NiO-MgO System

The phase diagram of the NiO-MgO System is shown below.

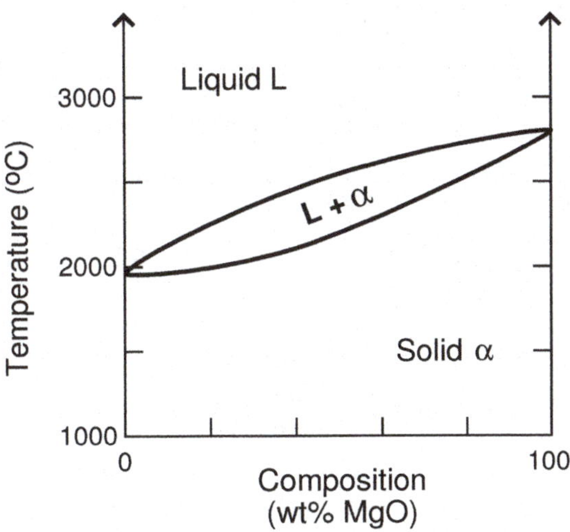

Figure 8.16: Phase diagram of the NiO-MgO system.

This figure is meant to illustrate the fact phase diagrams also apply to ceramic systems, and not just to metallic ones. It represents a case of complete solubility between two ceramics, NiO and MgO. Both these two components have a cubic lattice and the NaCl crystal structure (coordination number of 6, cf. Chapter 3). The ionic radii are as follows:

$$r_{Ni^{2+}} = 0.083\,\text{nm}; \quad r_{Mg^{2+}} = 0.086\,\text{nm}; \quad r_{O^{2-}} = 0.128\,\text{nm}$$

Therefore, Ni^{2+} and Mg^{2+} are very similar ions, and it is not surprising that they form a solid solution at any composition. As regards the crystal structure of the solid solution, it has Ni^{2+} and Mg^{2+} distributed randomly on the cation sites of the NaCl structure.

8.4.4 The SiO_2-Al_2O_3 System

The SiO_2-Al_2O_3 system is an example of a commercially important **refractory** material, i.e. a ceramics-based material that is physically and chemically stable at high temperatures. These types of materials are used for anything from linings of fireplaces and industrial furnaces to the tiles for the space shuttle heat shield.

The SiO_2-Al_2O_3 (alumina-silica) phase diagram is shown below:

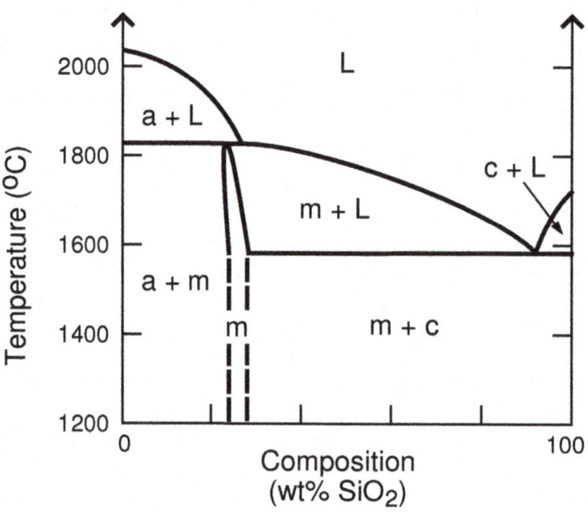

Figure 8.17: Phase diagram of the SiO_2-Al_2O_3 system

This diagram displays some known aspects as well as a new feature. Its components are Al_2O_3, or **alumina**, abbreviated by a in the figure, and the crystalline form of SiO_2 known as cristobalite, abbreviated by c in the figure (see also the title figure of Chapter 7). The diagram also exhibits a third, intermediate solid phase, abbreviated as m for mullite. This phase has a fairly well-defined composition of about 28 wt % SiO_2 and represents an intermediate compound of alumina and silica with the molecular composition 3 Al_2O_3 - 2 SiO_2.

Based on those solid phases, we can say that the lower right side of the phase diagram looks like a eutectic system with components m and c (i.e. the intermediate mullite phase acts as if it were a component), and the components m and c are insoluble in each other. At the lower left in the diagram, i.e. at a low silica content, mullite and alumina coexist.

An interesting side note is that there has been a long controversy as to how exactly the two liquidus curves between a+L and L, and between m+L and L, join in the region around 30 wt % SiO_2 and 1830°C. Ceramic phase diagrams are not easy to determine at high temperature.

8.5 The Fe-C System: Kinetics of Phase Transformation

8.5.1 The Fe-C Phase Diagram

The iron-carbon system has the most complex phase diagram we will encounter in this book. It is also arguably the most important material system, as all **ferrous** (iron-based) alloys derive from it. However, you will find that, based on what you have learned previously, you will easily be able to recognize and interpret its essential features.

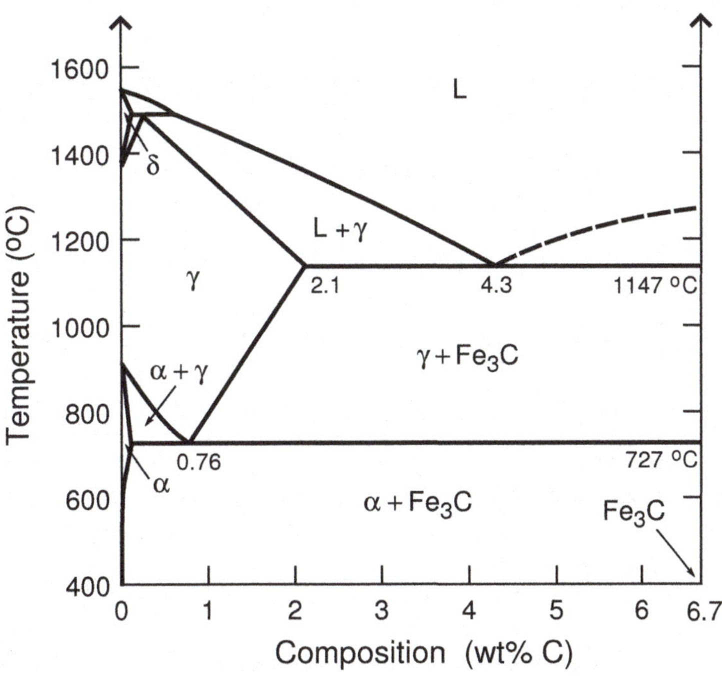

Figure 8.18: The iron-iron carbide phase diagram. Note the composition scale in wt % C even though the components shown are Fe and Fe_3C. The three numbers inside the graph indicate the compositions of important points. The temperatures of the two horizontal lines are marked near the right y-axis.

The first thing to be pointed out is that in terms of the Fe-C system, the diagram of Fig. 8.18 is only partial, extending to $X_C = 6.7$ wt % C. This is because the components of practical importance for ferrous materials are really Fe and the intermediate compound Fe_3C, iron carbide, rather than Fe and C, and the compositional range of interest is only up to 6.7 wt % carbon. The compound Fe_3C also has its own name: It is called **cementite**.

Of course, in principal the phase diagram does extend all the way to 100 wt % carbon, and for that case carbon, i.e. graphite, would be the appropriate second component.

You will also notice that there are three solid solution phases in the diagram in addition to Fe_3C: α, γ, and δ, which may be present at various temperatures. α is the low temperature phase at very low carbon concentration. γ is a high temperature phase that includes carbon concentrations up to 2.1 wt %, depending on the temperature.

These three solid solutions contain C atoms in interstitial sites. They differ in their crystal structure as indicated in Table 8.1 below. Fe_3C also has its own crystal structure.

Table 8.1: Solid solution phases in the Fe-Fe_3C system

Phase	Name	Structure
α	Ferrite	BCC
γ	Austenite	FCC
δ		BCC
Fe_3C	Cementite	orthorhombic

The α and γ phases have their own names: α is called **ferrite**, and γ is called **austenite**. The corresponding two single phase regions α and γ and their adjacent two-phase regions are most important for practical materials, and thus are what we will focus on.

Another important feature of the phase diagram is the two horizontal lines. The line at 1147°C is a eutectic line, similar to eutectic systems you have seen before. If a system goes through the eutectic point at 4.3 wt % carbon and 1147°C, the eutectic reaction

$$L(4.3\text{wt}\,\%\,C) \rightleftharpoons \gamma(2.1\text{wt}\,\%\,C) + Fe_3C(6.7\text{wt}\,\%\,C) \tag{8.22}$$

takes place. The line at 727°C is a so-called **eutectoid line** and the point at 0.76 wt % carbon and 727°C is a **eutectoid point**. Upon cooling a system through the eutectoid point, the reaction

$$\gamma(0.76\text{wt}\,\%\,C) \rightleftharpoons \alpha(0.022\text{wt}\,\%\,C) + Fe_3C(6.7\text{wt}\,\%\,C) \tag{8.23}$$

takes place. This reaction differs from a eutectic reaction in that it involves ***a solid phase*** being transformed at once into ***two other solid phases***. The eutectic reaction, by contrast, involves liquid being transformed at once into two solid phases.

As regards material properties, ferrous alloys can be assigned to three categories based upon their carbon content. ***Pure commercial iron*** contains less than 0.008 wt % carbon, and thus is composed mostly of the α phase (ferrite) at room temperature. (At room temperature, the solubility of carbon in iron is of the order of only 0.005wt % carbon.) Iron-carbon alloys with C content between 0.008 and 2.1 wt % are classified as ***steels,*** although in practice the C content is rarely much higher than 1 wt %. Alloys with between 2.1 and 6.7 wt % C (normally less than about 4.5 wt %) are used for ***cast iron*** materials.

With respect to properties, you know from earlier discussion that ferrite and low-C steels in general are softer and more ductile. By contrast, cementite is a hard and brittle material. Since ferrous materials generally consist of a combination of the two phases, their properties will be intermediate. This, of course, points towards a method for tailoring material properties. We will discuss properties and microstructures extensively below.

8.5.2 Equilibrium Microstructures

For the discussion of microstructures, we now focus on the region of the phase diagram around the eutectoid point. As we consider different cooling paths, once again we assume that the cooling takes places so slowly that the sample is essentially in thermal equilibrium at all times.

Path 1 is at the eutectoid composition. At point 1a and down to just above the eutectoid temperature, a single solid phase is present, namely γ or austenite, and the microstructure would show practically featureless grains. Once the temperature is below 727°C, all the γ has to turn into the eutectoid microstructure composed of α and Fe_3C, or ferrite and cementite, according to the reaction specified in Eq. 8.23. The eutectoid microstructure is layered for the same reason the eutectic microstructure of Cu-Ag was layered above. In fact, under the microscope these two structure would look quite alike.

You will notice that the cooling paths 1 in Figs. 8.19 and 8.13 are very similar. The only real difference is that in Fig. 8.13 solidification at the eutectic composition takes place, whereby liquid is transformed at once into two solid phases, as in Eq. 8.14. On the other hand, in Fig. 8.19 the transformation is from a solid phase at the eutectoid composition to two new solid phases, as in Eq. 8.23.

A quick application of the lever rule to the eutectoid microstructure would show that about 7/8 of it is ferrite and 1/8 cementite. Also, given the shape of the phase diagram, this structure would change very little as the temperature is lowered to room temperature. The eutectoid microstructure of the Fe-Fe_3C plays such an important role for ferrous alloys that it has its own name: It is called **pearlite**.

Path 2 is at a **hypoeutectoid** composition. Again, this is similar to path 2 in Fig. 8.13. At point 2a the system is in the single-phase γ region. As the line is crossed at around 820°C, proeutectoid α starts to form. By the time the

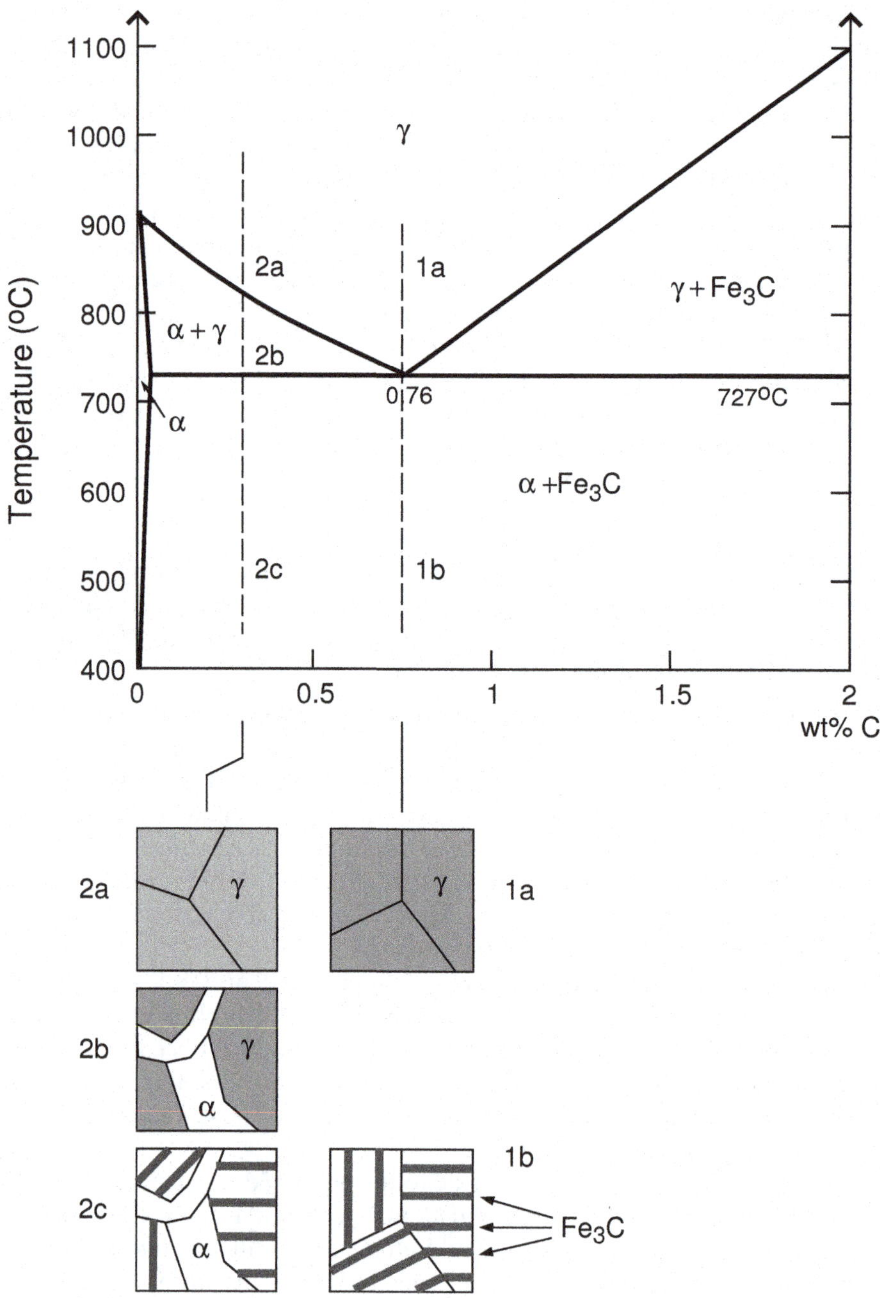

Figure 8.19: Fe-C phase diagram around the eutectoid point with schematic microstructures for two cooling paths.

eutectoid line is reached, about 0.4 of the sample is α already and 0.6 is still γ. As soon as the eutectoid line is crossed, this remaining γ is converted to the eutectoid microstructure. Note that the proeutectoid α tends to nucleate along the grain boundaries of the γ.

It is also quite common that a ferrous material has a composition above eutectoid, e.g. 1 wt % C. If we considered a cooling curve for that case, we would find something similar to path 2 above, except that the proeutectoid phase would be Fe_3C, but once again below the eutectoid line all the remaining γ would be converted to the eutectoid microstructure.

Note that the conversion of austenite to the eutectoid microstructure tends to originate along grain boundaries, and the layers tend to grow in a direction perpendicular to the grain boundaries. The process is illustrated schematically in Fig. 8.20.

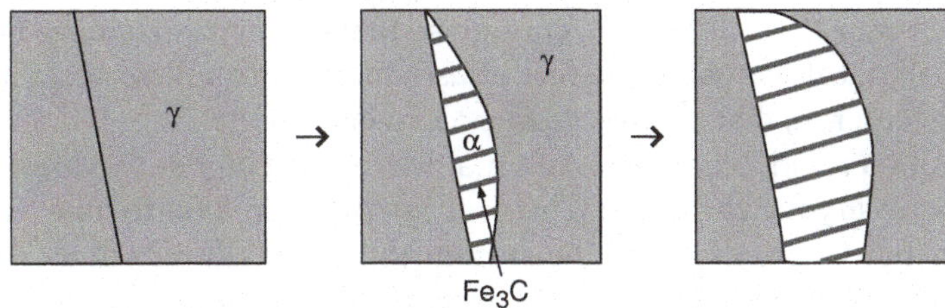

Figure 8.20: Schematic illustration of the growth of the pearlite microstructure.

The first panel in the figure above shows two neighboring grains of γ phase. In the other two panels, you should visualize the C atoms diffusing out of the austenite, away from ferrite, and towards the cementite, as the front of eutectoid material expands.

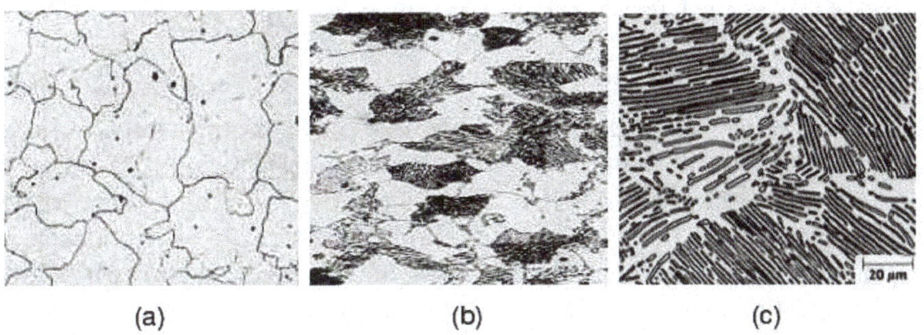

Figure 8.21: Microstructures of ferrite, hypoeutectoid, and eutectoid steel.

Fig. 8.21 displays micrographs of the microstructures of ferrite, a hypoeutectoid steel and a eutectoid steel. The first two panels on Fig. 8.21 are optical micrographs at a lower, unspecified resolution, and the third panel is an electron micrograph with a 20 μm scale bar. In Fig. 8.21a you can see the almost feature-less grain structure of single-phase α ferrite. Fig. 8.21b shows a roughly equal mix of proeutectoid α phase (light) and layered, eutectoid microstructure (dark). The individual pearlite layers are only barely resolved in a few dark areas. Fig. 8.21c shows the fully-developed pearlite microstructure of a eutectoid steel, ferrite being light in color and cementite dark.

8.5.3 Kinetics of Phase Transformation: TTT Diagrams

Previously I emphasized on several occasions that phase diagrams are associated with thermal equilibrium material conditions. That is, a phase diagram describes the state of a material at any point of composition and temperature (X_0, T) when we have waited long enough so that no further changes take place. This implies that if we want to use a phase diagram to follow a change in the state of the material, that change must occur very slowly.

On the other hand, when we cool a material from a higher to a lower temperature, by necessity the change in temperature occurs in a finite time. In practice the process may be too fast for the system to remain equilibrated. Hence we may well end up with the system in a **metastable state**, where it is not truly equilibrated yet not able to change further on a practically relevant time scale.

Moreover, if as a consequence of a change in temperature we observe a phase change, a transformation reaction takes place. This again takes time since the reaction evolves at a finite the rate. All these arguments are meant to show that time must be an important variable in phase transformations.

We encountered some examples of the importance of time in earlier discussions. The issue came up in connection with the solidification of pure substances in Chapter 7 (glass formation; nucleation and growth of crystals (Fig. 7.12)). We also noted briefly an effect of time in the formation of the eutectic microstructure in the Cu-Ag systems (see Fig. 8.12).

Avrami Phase Transformation Kinetics

Now we will have a closer look at the role time plays in phase transformations in general, and we will be particularly interested in transformations in the Fe-C system. First let me point out that the generic process outlined in Chapter 7 for a change of phase, namely the initial formation of independent nuclei of the new phase, followed by the growth of these nuclei, applies to any phase transformation, not just to solidification. For example, when austenite transforms into the pearlite microstructure of ferrite and cementite as the sample is cooled below the eutectoid temperature, nuclei of the new microstructure must form first and then take time to grow and convert the entire sample.

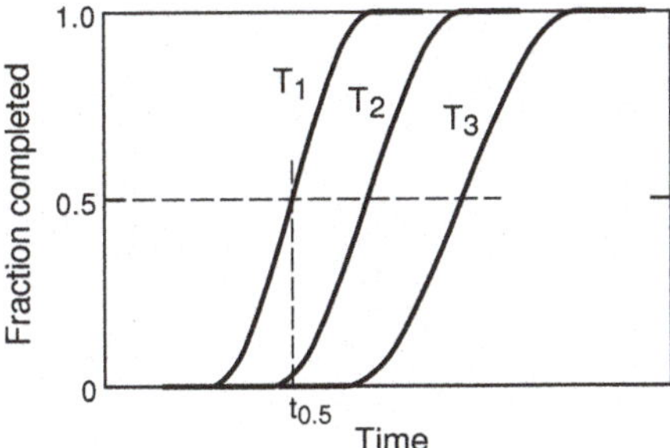

Figure 8.22: Generic time evolution of a phase transformation. Fraction completed vs. time for three different temperatures.

When one follows the progress of this kind of a reaction, a fairly universal behavior is observed (Fig. 8.22). The extent of the reaction typically follows an S-shape as a function of time. The reaction starts almost imperceptibly, with the nucleation phase. This is followed by a time of rapid growth. Finally, the reaction levels off and gradually approaches completion. The shape of this curve and its position on the time axis depend on temperature.

Curves of this type can be expressed conveniently in the form

$$y = 1 - \exp(-kt^n) \tag{8.24}$$

where y indicates the fraction of the transformation completed and k and n are suitable parameters depending on temperature. The parameter n is related to the dimensionality of the problem. In the simplest model of isotropic growth in three dimensions, n is equal to 4. Experimentally, when k and n are treated as empirical fitting parameters, values for n between 1 and 4 have been found. 8.24 is known as the **Avrami equation**.

If you think of the variable y as the amount of reaction product, then the reaction rate would be equal to $\frac{dy}{dt}$. This is displayed qualitatively in the left panel of Fig. 8.23 together with $y(t)$. The right panel of Fig. 8.23 shows the effect of n on the shape of $y(t)$.

You can see that the lower the value of n, the lower is the maximum slope in the middle of the $y(t)$ curve, and thus the lower is the maximum reaction rate. At the same time, for a fixed $t_{0.5}$, the lower the value of n, the longer the reaction takes to go to completion.

From these graphs it is apparent that the most important characteristic of the curve is its midpoint $t_{0.5}$. It is a measure of the time scale over which the transformation occurs. If you set $y = 0.5$ in Eq. 8.24 and take the natural log on both sides, you obtain after some simple algebra:

$$t_{0.5} = (\frac{\ln 2}{k})^{\frac{1}{n}} \tag{8.25}$$

Of course, for a more complete characterization of the process, one would also

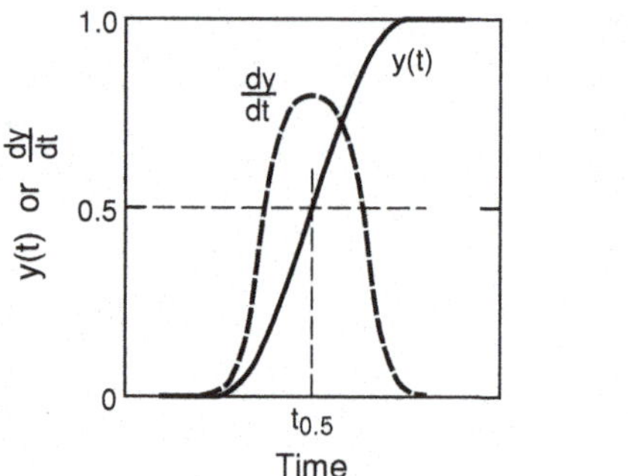

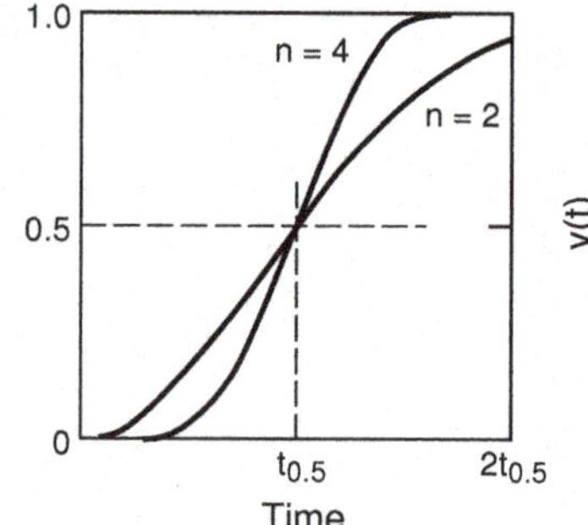

Figure 8.23: Extent of phase transformation, $y(t)$, rate of phase transformation, $\frac{dy}{dt}$, and the effect of the parameter n on the shape of $y(t)$.

need to know, for example, the width of the curve in Fig. 8.22, or the slope at the midpoint.

Generic Time-Transformation Curves

Since the rate of a reaction, in an average sense, is inversely proportional to the time the reaction takes, we will simply identify the inverse of $t_{0.5}$ with *the rate* of the transformation as envisioned in Fig. 7.12:

$$\text{transformation rate} = \frac{1}{t_{0.5}} \tag{8.26}$$

The implication is that if we plotted $t_{0.5}$ vs. T, we would obtain a curve similar to the transformation rate of Fig. 7.12, but flipped upside down (Fig. 8.24). From an experimental point of view, this is more convenient because the extent of a transformation vs. time, i.e. the function $y(t)$, is often easier to observe than the transformation rate.

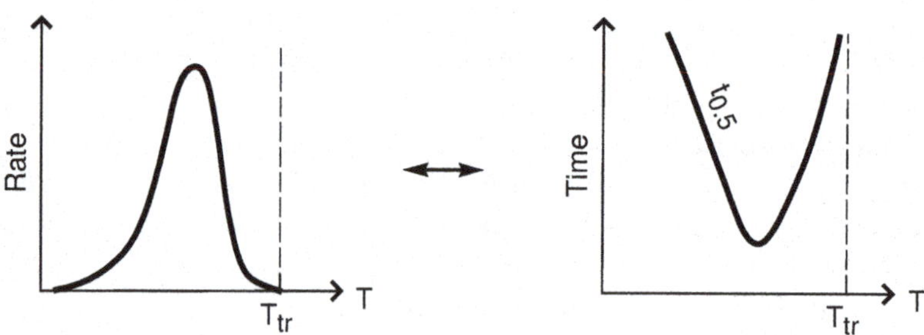

Figure 8.24: Generic phase transformation reaction rate and reaction time curves vs. temperature.

The left panel in Fig. 8.24 reproduces the reaction rate vs. T from Fig. 7.12, and the right panel displays its "inverse", i.e. the time it takes to reach the point of 50% conversion, $y = 0.5$, vs. T. Also note that T_{tr} denotes the temperature below which the transformation can take place.

The form in which such plots are typically used differs slightly from the one displayed in Fig. 8.24. It is common to have the x- and y-axes switched even though the temperature is the real independent variable. In addition, as we are interested in the progress of a transformation with time, we wish to keep track of when the transformation starts and when it ends. These terms are, of course, not very precise. So we will define them with a practical sense: The beginning of the transformation is set to the time when 1 % has been converted, and the end to when 99 % has been converted. Both these conditions can be represented by curves similar to the time curve in Fig. 8.24, as is illustrated in Fig. 8.25:

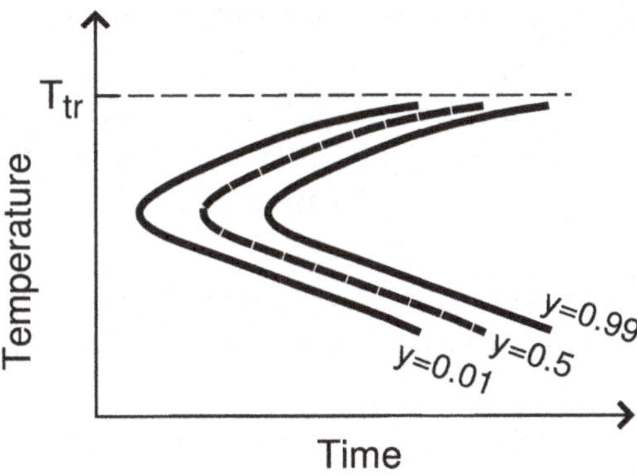

Figure 8.25: Generic time-transformation curves describing the progress of a phase transformation as a function of temperature.

Plots such as the one of Fig. 8.25 are referred to as isothermal transformation diagrams, or **TTT-diagrams** (for time-temperature-transformation). The set of three curves depicts in a somewhat simplified way how a phase transformation proceeds at a given, constant temperature. For example, any point (T,t(T)) on the $y = 0.5$ curve indicates the time t(T) which it takes for the transformation to go to 50% completion at the temperature T.

TTT Diagrams and Microstructures for the Fe-C System

We are now ready to examine the most important TTT-diagram for this book, namely the diagram for the eutectoid reaction of austenite (the γ phase) to the pearlite microstructure made up of ferrite (the α phase) and cementite (the Fe_3C phase). Just as a brief reminder: Austenite is the phase present above 727°C, and we are interested in the alloy with the eutectoid composition of 0.76 wt % C (cf. Fig. 8.19). The TTT-diagram is shown in Fig. 8.26.

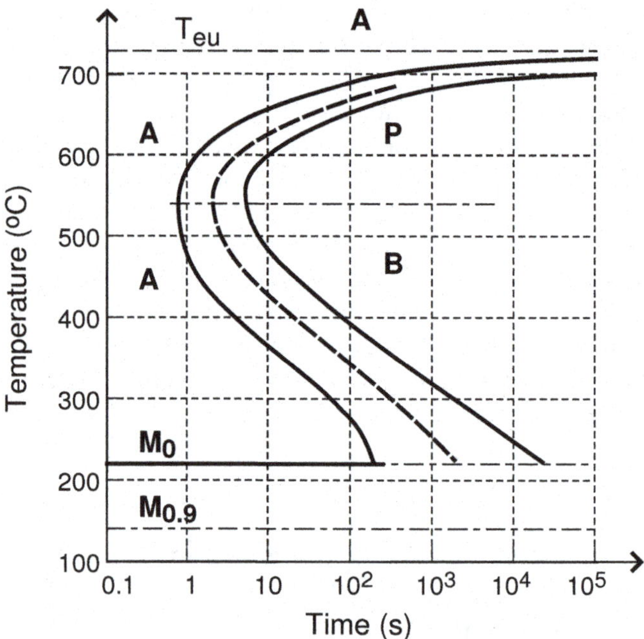

Figure 8.26: TTT diagram for the isothermal transformation of an Fe-C alloy with eutectoid composition from austenite to a low-temperature microstructure.

The main part of the diagram are the three curves indicating the beginning, 50% completion, and the end of the transformation (cf. Fig. 8.25). The curves show a "knee", i.e. a minimum for the time to convert, at around 540°C, where the transformation will be completed in about 7 s. At the eutectoid temperature T_{eu} of 727°C the curves approach infite time, meaning that exactly at T_{eu} no nucleation takes place, and thus no transformation occurs. Nucleation must be initiated by a certain amount of undercooling, just as in the case of solidification in Chapter 7.

Please note the logarithmic scale on the time axis Fig. 8.26. This is chosen so that the figure can display the large range of time necessary to cover this type of transformation.

Fig. 8.26 also incorporates new information regarding this phase transformation. First, the symbol A denotes the austenite phase, and the symbols P, B, and M denote possible microstructures. P stands for ***pearlite***, and the other two will be explained shortly. The dash-dotted horizontal line at the "knee" indicates that only at temperatures of around 540°C or above will the pearlite microstructure P be formed.

The figure is to be read from left to right, at a given temperature. For example, let us assume that a sample has been fully equilibrated at at temperature somewhat above 727°C. The *entire* sample is then **quenched** to T = 600°C and kept at that temperature. (To quench means to lower the temperature very quickly on the time scale relevant for the reaction, say in less than 1 s). This temperature jump starts the clock for the reaction. If you now imagine a horizontal line at 600°C across the graph, you see that it takes about 1.5 s for the transformation to begin, The reaction reaches its midpoint after about 5 s, and it is completed after about 10 s. If you could follow the microstructure of the sample as a function of time, at the beginning it would like panel 1a in Fig. 8.20, and after completion it would like panel 1b.

Fig. 8.26 also implies that the pearlite microstructure can be achieved over a range of temperatures. It is just a matter of time. If the reaction were made to proceed at 700°C, it would take more than 10^2 s to start and more than 10^4 s to go to completion (cooling path 1 in Fig. 8.27). On the other hand, at 560°C the reaction would start in less than 1 s and be completed after about 7 s (cooling path 2 in Fig. 8.27).

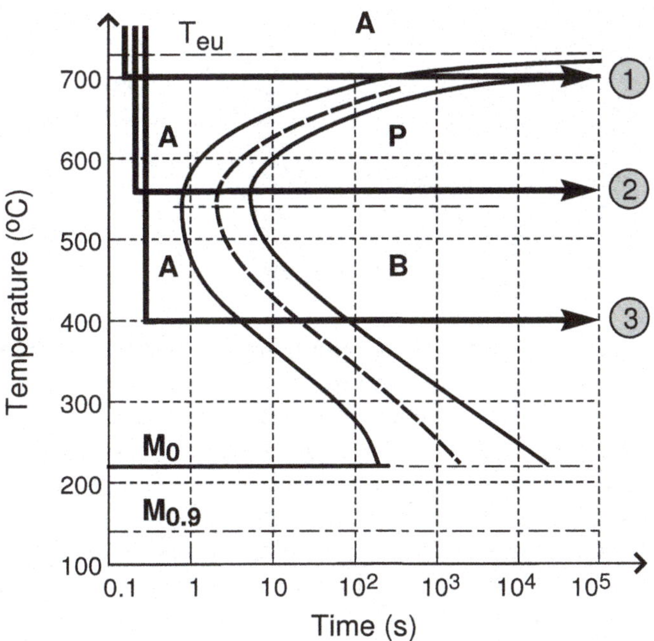

Figure 8.27: Fe-C TTT diagram with three cooling paths.

The difference between these two situations can be inferred from our previous discussion on nucleation and growth in solidification. We know that, simply put, just slightly below the transformation temperature nucleation is slow but growth is fast. Thus, at 700°C we would expect a microstructure with fewer, but larger features, which in the present case translates to a eutectic layered structured with fewer but wider layers. By contrast, at a much lower temperature, say at 560°C, nucleation will be faster but growth slower. This would give rise to a eutectic layered structured with more but narrower layers. For this reason, microstructures obtained at temperatures in the upper part of the 720–540°C range are referred to as **coarse pearlite**, and those obtained in the lower part of this range as **fine pearlite**. (An example of this same effect was displayed earlier, in Fig. 8.12, for the CuAg eutectic system).

If we quench the sample rapidly enough to a temperature lower than 540°C while avoiding the "knee"entirely, no pearlite is formed. For example, quenching to 400°C as in path 3 in Fig. 8.27, and then keeping the sample at 400°C for at least 100 s, will produce an entirely new microstructure known as **bainite** (symbol B in Fig. 8.27). The key feature of bainite is that it is not layered, but has fine Fe_3C particles embedded in an α-Fe matrix. Overall the bainite microstructure is finer than pearlite. An example is displayed in Fig. 8.28.

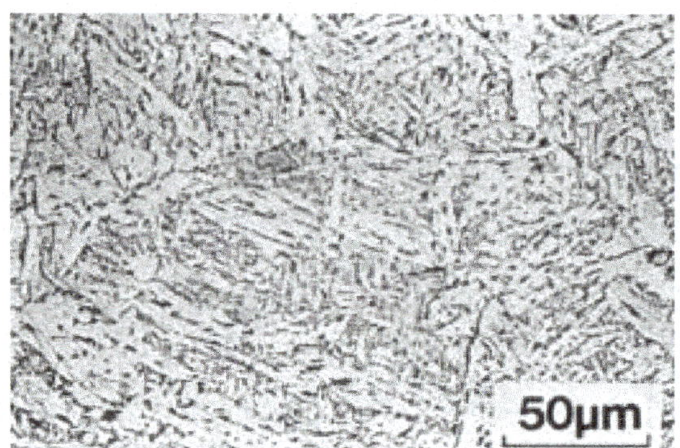

Figure 8.28: Bainite microstructure

With respect to the transformation temperature and the difference between pearlite and bainite, it should be clear that there cannot be a sharp separation or boundary between the two regimes. In fact, different authors quote somewhat different temperatures in this regard. However, above about 560°C a eutectoid layered structure appears, and below about 500°C the new structure without layers appears. Still, both types of microstructures have some important things in common: 1) They both result from diffusional processes: C atoms move in and out of austenite regions, and 2) in both cases the phases giving rise to the microstructures are the same, namely ferrite and cementite. The two phases just aggregate in different fashion, depending on the transformation temperature.

For bainite, too, there is a distinction made sometimes depending on the temperature at which it has been formed. In the upper temperature range, say around 500°C, one may speak of upper bainite, and in a lower range of around 350°C, one may speak of lower bainite. The most important thing to remember, though, is not so much the names but that all else being the same, the lower the temperature of the transformation, the finer the microstructure.

It is also possible to obtain a mixed microstructure, for example containing both pearlite and bainite as in path 4 of Fig. 8.29. After the initial quenching step, the sample is held at 580°C for 3 s, after which time half of the sample has been converted to pearlite and the rest is still austenite. The sample is then quenched again, to 360°C. This resets the clock, and after another 300 s or so the entire remaining amount of austenite has been converted to bainite. Note that during this second isothermal period nothing happens to the previously created pearlite, since pearlite is a stable equilibrium structure. Thus the final microstructure would be made up of 50% pearlite and 50% bainite.

With all the cooling paths discussed so far, paths 1 - 4 in Figs. 8.27 and 8.29, it is important for you to realize that if at the end of a path the temperature were lowered to 100°C or to room temperature, nothing further would happen. Once a path has crossed the 99 % completion line, the sample has reached a stable microstructure that does not change in going to a lower temperature. The only way a stable microstructure can change is if its temperature is raised and kept at a higher temperature for a sufficient period of time.

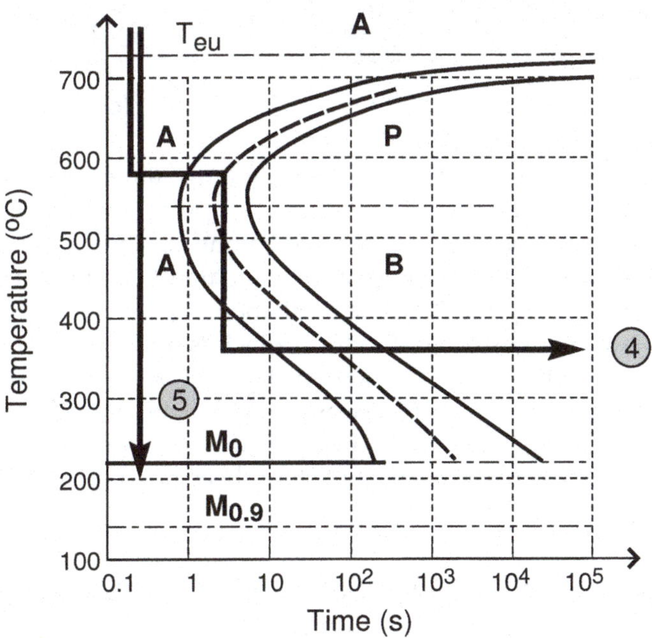

Figure 8.29: Fe-C TTT diagram with two additional cooling paths.

If we quench an austenite sample so fast that we avoid the "knee" and cool down to a temperature of about 200°C or lower (path 5 in Fig. 8.29), yet another microstructure called **martensite** arises (symbol M in Figs. 8.27 and 8.29). Martensite is different in several ways from both pearlite and bainite. First, it is a complex structure with its own body-centered tetragonal unit cell (a unit cell with 90° angles, one unit vector different than the other two). This structure comes about by a distortion of the FCC austenite lattice. Hence, in contrast to the microstructures we have encountered before, the martensite transformation is diffusion-less: It does not require any diffusion of C atoms and does not involve ferrite and cementite at all. Yet martensite is still a supersaturated solution of C in Fe, and the C atoms reside in interstitial sites of the tetragonal lattice. Furthermore, the extent of the completion does not depend on time, but only on the temperature: The transformation begins slightly above 200°C and goes to 90% completion at about 130°C (Fig. 8.29)).

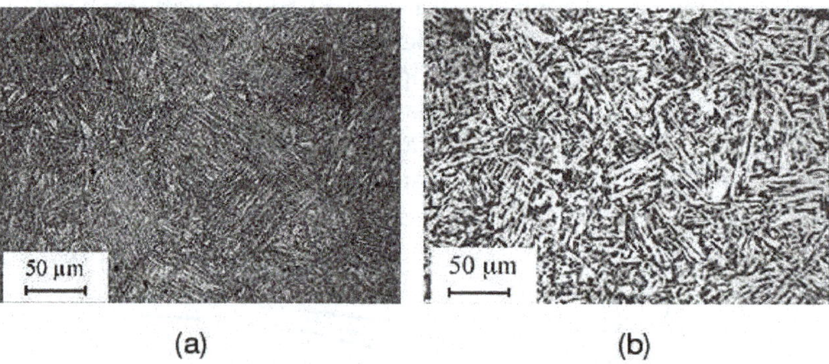

Figure 8.30: Comparison of a) martensite, and b) bainite microstructures.

Electron micrographs of martensite and bainite are shown in Fig. 8.30. You will note the finer structure in martensite compared to bainite. The grains in martensite are very fine and often needle-shaped.

Because of its unique crystal structure, which is different from all the other phases in the Fe-C phase diagram, martensite is not only a microstructure, but is also considered its own phase. However, it is a metastable phase, and therefore is not in the equilibrium phase diagram. If its temperature is raised, after the fact, to a point where diffusion of C atoms becomes appreciable, it will decompose into ferrite and cementite.

8.6 Comments on TTT Diagrams and Microstructures

So far, we have concentrated on eutectoid Fe-C alloys and on isothermal transformations. Now we will add some comments regarding alloys with other compositions, more realistic cooling curves, as well as the effect of all these processing details on the properties of the materials.

Non-eutectoid Fe-C alloys

We have mentioned before on several occasions that carbon steels are Fe-C alloys which cover a range of composition, from a small fraction of 1 % C to more than 1 % C. If there are no other impurities present, such materials are called

plain carbon steels. On the other hand, a ferrous material may incorporate a number of other elements at various concentrations, especially other metallic elements, in which case it is called an **alloy steel**.

Such alloys can undergo transformations similar to the one for eutectoid Fe-C, and can form similar microstructures, but in principle each alloy has its own TTT diagram. For example, the completion curves may differ somewhat in the exact location of the "knee". We will not go into these details, except to make a comment about **hypo-** and **hypereutectoid** Fe-C alloys, i.e. materials containing either less than or more than 0.76 wt % C.

In those cases the phase diagram indicates that the transformation will start at a temperature higher than 727°C, and thus will produce some pro-eutectoid phase: either pro-eutectoid ferrite if the composition is < 0.76 wt % C, or pro-eutectoid cementite if the composition is > 0.76 wt % C. The pro-eutectoid phase would be embedded in the microstructure formed later at the eutectoid temperature from the remaining austenite. (See Fig. 8.21b for a microstructure example of a hypoeutectoid composition).

Continuous cooling curves

In the cooling curves examined so far, we have always assumed that the conversion took place at a constant temperature. This is rather hypothetical. In practice cooling will be continuous.

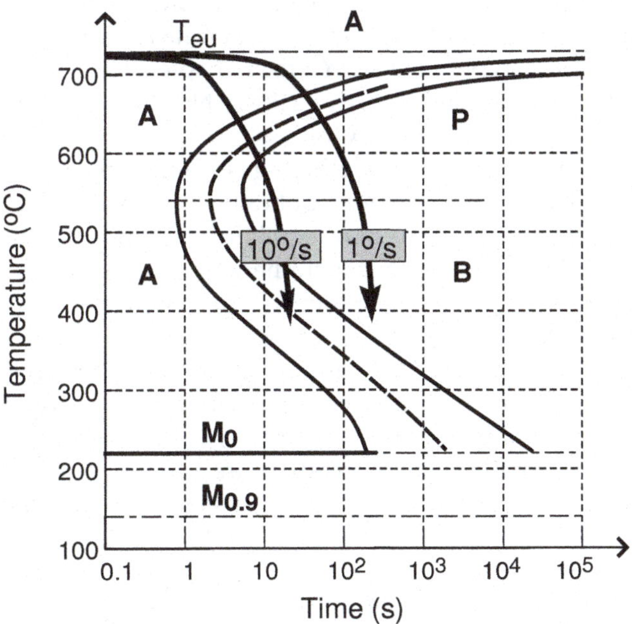

Figure 8.31: TTT diagram with continuous cooling curves.

This means that the sample temperature will decrease continuously during the entire cooling process. Even if the sample temperature on the outside could be kept constant during cooling, the temperature on the inside would be different and changing with time.

We can visualize the effect of continuous cooling by taking a normal TTT diagram and drawing in cooling curves that correspond to specific cooling rates, as illustrated in Fig. 8.31. The graph shows two cooling curves with constant cooling rates of 1°/s and 10°/s. The complication, compared to the isothermal case, is that here the beginning of the reaction and the end occur at different temperatures. The difference between these two temperatures is the larger, the greater the cooling rate.

It should also be pointed out that Fig. 8.31 includes a minor simplification: The conversion curves as drawn are for isothermal reactions. In practice, it is found that the 1 % and 99 % conversion curves themselves depend somewhat on the cooling rate. They turn out to be shifted slightly to larger times, i.e. the transformation is slightly delayed in continuous cooling.

Microstructure of alloy steels

When dealing with alloy steels, i.e. Fe-C alloys that contain additional metal impurities, again similar phenomena are observed but the details may vary. For example, both the phase diagram and the TTT diagram (the conversion curves) may depend on the alloy content.

A dramatic example of the dependence of the microstructure on composition beyond plain carbon steels is offered by **stainless steels**. We will discuss the properties of these types of steel in a later chapter. Here we will introduce them by a simplified definition, namely as Fe-Cr-Ni alloys containing much less than 0.76 wt % C, usually 0.15 wt % C or less.

You note, of course, that once two alloying elements are involved, we are not dealing with a binary system anymore, but rather with a ternary system. Thus, the phase diagram would have to show equilibrium phases as a function of two alloying components. We will be content here by putting things as simply as possible: The Cr makes these steels stainless, i.e. non-corroding, whereas the Ni controls the microstructure.

More specifically, Cr is typically around 16 wt % in these materials. If Ni is then close to 0 wt % (i.e. with very little Ni and very little C), the stable microstructure at room temperature is ferritic, i.e. the normal BCC. However, if Ni is > 8 wt % or so, the stable microstructure at room temperature remains austenitic, i.e. FCC! In other words, *enough Ni prevents the conversion of austenite to ferrite.*

Microstructure and mechanical properties

The main reason why microstructure is so important for these materials is because of its profound effect on mechanical properties. It is by the control of microstructure that the materials engineer can tailor the material properties to meet specific requirements. A very simple overview of this effect is shown in the table below:

Table 8.2: Fe-C Microstructure and Mechanical Properties

Microstructure	T.S. (MPa)	Ductility	HV
Ferrite	300	0.5	130
Pearlite	950	0.15	300
Bainite	1200	0.15	400
Martensite	1800	0.06	500

Please note in the table that ferrite is a phase and could only be identified with a microstructure if it were pure (i.e. if it had a very low C concentration) and did not contain any cementite phase. Also, HV refers to Vickers hardness, and as always you should consider the numbers to be average and representative. The mechanical properties of stainless steels also vary greatly, depending on their microstructure.

In interpreting the numbers in this table, keep in mind that ferrite is a relatively soft and ductile phase, whereas cementite is hard and brittle. Thus the eutectoid microstructure of pearlite will have in-between properties. Furthermore, the correlation between feature size and properties is manifest again here: Fine pearlite turns out to be stronger and less ductile than coarse pearlite, whereas bainite with its even finer microstructure is still stronger and less ductile. Martensite is in a class of its own, not being a mixture of ferrite and cementite. Pure martensite which has not been treated any further tends to be very hard and brittle, in fact so much so as to be not practically very useful.

Heat treatments

You learned earlier in the section on recrystallization in Chapter 5 that heat treatment can change the microstructure, in that case by reforming and growing new grains in a work-hardened material, and thus can have a significant effect on the mechanical properties. The same procedure of controlled reheating can be applied to certain materials that were produced originally with a metastable microstructure. We will look at two cases in some detail.

a) Tempered Martensite

Martensite can be made more ductile, without losing much of its strength, by a process called tempering. This involves heating the martensite gently, to a temperature where some diffusion of C atoms takes place, typically 300-

500°C. This causes very fine, uniformly dispersed cementite particles to form in a ferrite matrix. As a consequence, the material's strength is reduced but its ductility and toughness increased. The extent of this effect, of course, depends on the temperature and the time of the treatment.

b) Precipitation Hardening

Another heat treatment that allows for adjustment of the strength/hardness of certain materials is known as precipitation hardening. It involves a clever application of the phase diagram to an alloy which had been solidified under non-equilibrium conditions. This is also an example that a metastable microstructure can be obtained in systems other than Fe-Fe_3C.

Our example will be the precipitation hardening of Al, actually an Al-Cu alloy, a process used widely in engineering practice.

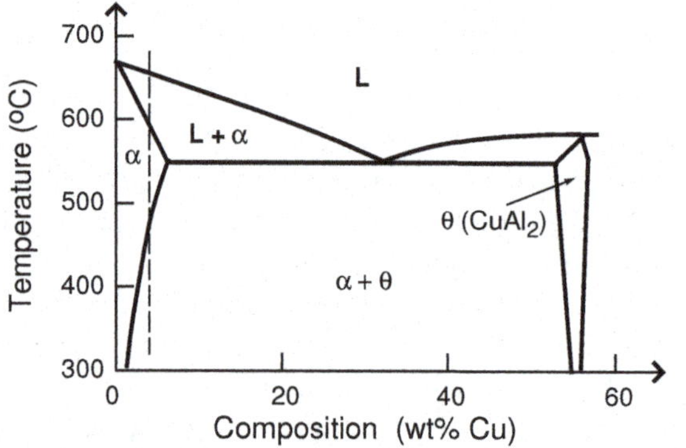

Figure 8.32: Partial phase diagram of the Al-Cu system. Note the intermetallic compound $CuAl_2$.

Cu has limited solubility in Al, thus forming a eutectic system with phases α (solid solution of Cu in Al) and θ (intermetallic compound $CuAl_2$). We prepare the starting material by equilibrating an alloy with 4 wt % Cu at 500°C, i.e. in the single-phase α region and then quenching it to room temperature. If we do this quickly enough, the single phase material will be preserved, as there is not enough time for it to form the equilibrium eutectic microstructure. At that point we have a metastable phase consisting of supersaturated solution of Cu in Al. This is very similar to quenching austenite prior to the beginning of the eutectoid reaction in the Fe-C system.

In a next step, the temperature of the material is raised again, to a point near, but below, the solvus line. This enables some diffusion of Cu atoms to occur in the Al matrix such that very small θ phase ($CuAl_2$) particles can nucleate and begin to grow. If this is done in a controlled manner, i.e. for the right period of time at the right temperature, the yield strength and tensile strength of the Al-alloy will increase due to the presence of the small precipitate particles. However, if the particles grow to be too large, then the strength of the material decreases again.

8.7 Summary: Phase Diagrams and Phase Transformations

We have discussed extensively in this chapter some of the central issues of materials science related to phase diagrams and transformations. In particular we have examined binary, mostly metallic, alloys in which the two component exhibit either complete or limited miscibility.

A two-component system with complete miscibility such as Cu-Ni is quite similar to a pure metal. The main difference is that in the two-component system melting occurs "gradually", in the sense that the liquid and the solid solution can coexist over a range of temperature rather than just exactly at the melting point. The microstructure of the solid is usually fairly simple, consisting of grains of the solid solution whose average size depends on how rapidly the liquid was solidified.

When the miscibility of the two components is limited, the behavior of the material is more complicated. In the simplest case of a eutectic system, as long as only liquid or a liquid and a solid phase are present, the system resembles the case above with complete miscibility. However, in a eutectic system such as Cu-Ag there exists a range of compositions such that below a certain temperature known as the eutectic temperature, in thermal equilibrium a single solid phase cannot exist, but liquid is transformed by the eutectic reaction into two coexisting solid phases. This is only the simplest of many possible cases, and important materials such as brass exhibit even more complex behavior.

Perhaps the most important two-component system is the Fe-C system. It is more complex than a simple eutectic system but can be understood along similar lines. The characteristic feature of its phase diagram most relevant for practical ferrous materials is the eutectoid point and the phase transformation that takes place there. This so-called eutectoid reaction is analogous to a eutectic reaction except that it involves the transformation of one solid phase into two new solid phases which are arranged in a characteristic, layered configuration called pearlite.

Moreover, the configuration, or microstructure, which the two new phases assume depends not only on the composition of the material but also on the kinetics, i.e. the time history, of the transformation. If it takes place much below the eutectoid temperature, a different microstructure results called bainite. The point to note is that pearlite and bainite look totally different under the microscope, yet both are made up of the same basic building blocks, namely small regions of the ferrite and cementite phases.

The most difficult concepts to grasp for students, in my experience, are the concept of a solid solution, the concept of a solid phase, particularly *different solid phases*, and the concept of a microstructure. I have tried to illuminate these concepts from various angles and with many examples, but perhaps it is worth making an attempt at providing a shortcut to contrast these differences one more time. This seems worthwhile especially in light of all the names ending in "-ite" we have come across in connection with the Fe-C system.

So here are the bare essentials of phase diagrams, TTT diagrams, and microstructures:

Phases, phase diagrams, TTT diagrams, and microstructures

1. If it is in the phase diagram but not in the TTT diagram, it is a phase.
2. If it is not in the phase diagram but in the TTT diagram, it is a microstructure.

But as with every simple rule, it is not only simple but also slightly simplified, and of course you know better by now. As regards Point 1, the phase γ-Fe, also known as austenite, does show up as the starting point in the TTT diagram. Of course, it has a microstructure of its own, but that is of no concern. What matters is the transformation of one solid phase into two other solid phases, and the possible microstructures that may be the outcome.

Image sources

Title picture: reproduced with permission by ASM International
`http://www1.asminternational.org/asmenterprise/apd/help/intro.aspx`

Fig. 8.12: reproduced with permission by Elsevier
M. Erol, U. Böyük, T. Volkmann, D.M. Herlach, J. Alloys Compounds 575 (2013) 96-103
`http://www.sciencedirect.com/science/article/pii/S0925838813009092`

Fig. 8.21: reproduced with permission by ASM International
`http://www1.asminternational.org/asmenterprise/apd/help/intro.aspx`
`http://practicalmaintenance.net/?p=1315`

Fig. 8.28: reproduced with permission by TWI
`http://www.twi.co.uk/technical-knowledge/faqs/material-faqs/faq-what-are-the-microstructural-constituents-austenite-martensite-bainite-pearlite-and-ferrite/`

Fig. 8.30: reproduced with permission by C. Rey, LMSSMAT, ECP, France
`http://archive.mssmat.ecp.fr/Etudes-experimentale-et-numerique,3377`

Fig. 8.33:
http://metallurgie.iehk.rwth-aachen.de/moodle/mod/resource/view.php?id=77&lang=de_utf8

Fig. 8.37
http://www.onera.fr/en/dendrites-gallery

References

Both Shackelford, and Callister and Rethwisch discuss similar topics at the same level, but a bit more expansively. Guy goes into more detail and deeper, but the relevant topics are not all in one place. Smith has more information on practical engineering aspects.

For a more extensive introduction to phase diagrams, visit:
http://www1.asminternational.org/asmenterprise/apd/help/intro.aspx

Software has been developed to calculate alloy phase diagrams from thermodynamic data. For some examples see:
http://www.calphad.com

For stunning pictures of metal alloy microstructures visit:
http://www.onera.fr/en/dendrites-gallery
Some of these microstuctures show eutectic parts, mixed with pro-eutectic dendrites.

A website with an atlas of microstructure micrographs is at:
http://pwatlas.mt.umist.ac.uk/internetmicroscope/micrographs/microstructures.html

For a quick overview of stainless steels, go to:
http://www.aksteel.com/markets_products/stainless.aspx

We have concentrated our discussion of microstructures on the Fe-C system. For similar information on the metallurgy of Cu-based alloy, go to:
http://www.copper.org/resources/properties/703_5/703_5.html

For a beautiful example of directional eutectic solidification, and the effect of cooling rate on microstructure in a system other than Fe-C, visit:
In-situ observation of eutectic growth in Al-based alloys by light microscopy, V.T. Witusiewicz, U. Hecht, S. Rex, J. Cryst. Growth 372 (2013) 57-64
http://www.sciencedirect.com/science/article/pii/S0022024813001656

Exercises

1. Take a NiO-MgO sample at 50 wt% MgO and 2500°C.
   What are the phases present, their compositions, and their amounts ?

2. What phases are present, and what are their fractional amounts, for a Cu-Ag system with 80 wt % Ag at room temperature ? What can you say about the nature of these phases, for practical purposes ?

3. Verify that the molecular composition of mullite, 3 Al_2O_3 - 2 SiO_2, is consistent with a composition by weight of about 28 wt % SiO_2.

4. Take a sample of Al_2O_3 - SiO_2 at 60 wt % SiO_2.
   a) What are the phases present and their fractional amounts just below the horizontal line at 1587°C ?
   b) What are the phases present and their fractional amounts just above the horizontal line at 1587°C ?
   c) As you go from a) to b), you cross from the m+c to the m+L phase region. So what did actually melt in the process ? All of the c phase, some of the c phase, or something else ?

5. In the Fe-C phase diagram, find a system point (X_0, T) where only the δ phase exists.

6. What phases are present, and what are their fractional amounts, for equilibrated Fe-C alloys with
   a) 0.2 wt % C at 800°C;
   b) 0.8 wt % C at 800°C;
   c) 1.4 wt % C at 800°C ?

7. What is the composition of eutectoid Fe-C in atom percent?

8. Check the statement in the text about the amounts of ferrite and cementite in the eutectoid microstructure with the lever rule.

9. The color of the γ phase in panel 1a in Fig. 8.19 is a little darker than the color of γ in panel 2a, whereas the colors of γ in panels 1a and 2b are the same. What is this meant to indicate ?

10. Critique the following statement from
    `http://wiki.answers.com/Q/What_is_the_difference_between_pearlite_and_cementite`

    Question: What is the difference between pearlite and cementite?
    Answer: "Therefore, the difference between pearlite and cementite is that pearlite is a composition of steel, and cementite is a composition of pearlite."

11. What is at least unorthodox or, some might say, even wrong in the Fe-C phase diagram of Fig. 8.33 ?

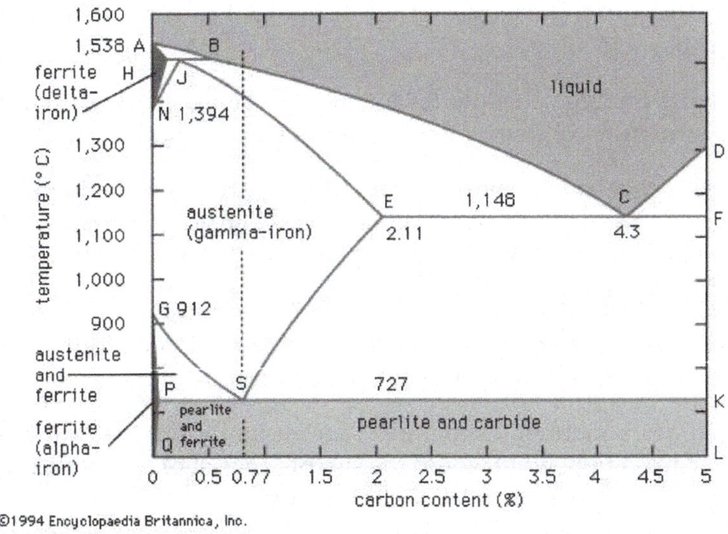

Figure 8.33: Fe-C phase diagram from the Encyclopedia Britannica.

12. What are the microstructures from the two cooling paths in Fig. 8.34 ?

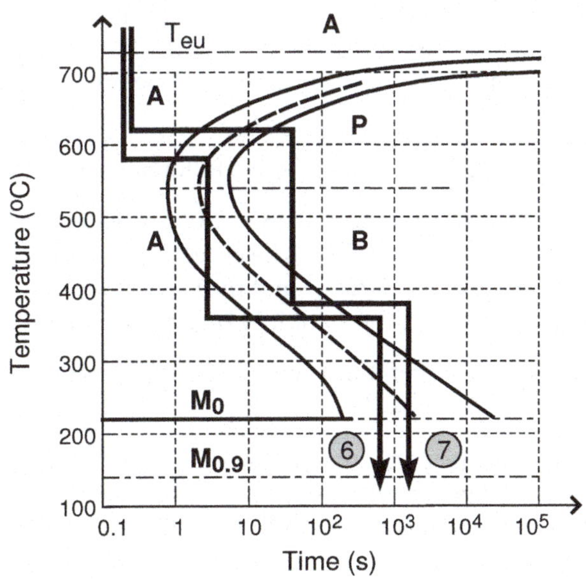

Figure 8.34: Two more cooling paths for the eutectoid Fe-C TTT diagram.

13. If an Fe-C alloy with 0.2 wt % C were fully equilibrated at 730°C, then quenched to 400°C and held there for 150 s, what would be the resulting

microstructure ? Assume that the pertinent TTT diagram is the same as for a eutectoid alloy.

14. Design a cooling schedule for a eutectoid Fe-C sample so that the final microstructure consists of $\frac{2}{3}$ fine pearlite and $\frac{1}{3}$ bainite.

15. It is said that bainite consists of fine Fe_3C particles embedded in an α-Fe matrix. Why is it that way, and not the other way around (i.e. why not α-Fe particles in an Fe_3C matrix) ?

16. If you wanted to "avoid the knee" in the eutectoid Fe-C TTT diagram, i.e. avoid the formation of any pearlite, what should your cooling rate be, at a minimum, in a continuous cooling process ?

17. If you did continuous cooling on an austenite sample with the eutectoid composition, what would the sample consist of at room temperature, qualitatively speaking, for a cooling rate of
a) 10°C/s , or   b) 50°C/s ?

Problems

1. You heat a sample of mullite with 27 wt % silica from 1200 to 1830°C, at which point you are just above the horizontal line in the phase diagram. Describe, and comment on, what happens in the process in as much detail as you can.

2. In a study of the CeO_2-CoO system, the following results were obtained:

   A eutectic reaction was found at 1645 ± 5°C. The eutectic point was calculated to be at 82 ± 1.5 mol % CoO. The eutectic phases were the CeO_2-rich phase (containing <5 mol % CoO) and the CoO-rich phase (containing 0.5 mol % CeO_2). At 1580°C, the solubility of CoO in CeO_2 was 3 mol %.

   Based upon this information, draw a phase diagram for the CeO_2-CoO system.

3. Consider the ceramic compounds NaCl, KCl, and RbCl. You can make several binary systems from them. Enumerate these systems. Comment on what types of systems they would be with respect to their phase diagrams, and why.

4. The figure below shows a salt-water phase diagram. A solution of salt in water is called brine.

   a) What type of a phase diagram is this ? Label the empty regions with the appropriate phases.

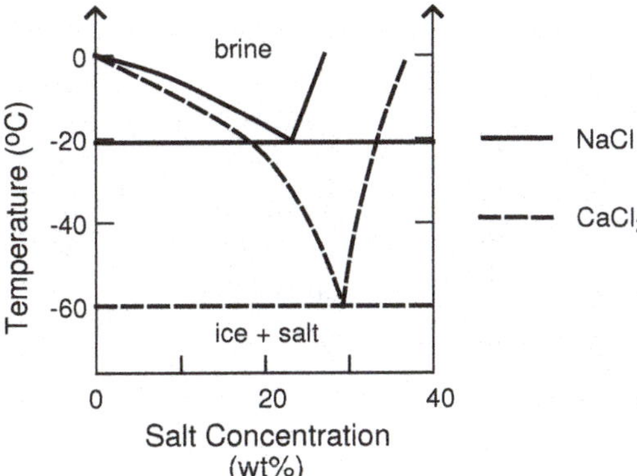

Figure 8.35: Salt-water phase diagram.

b) What would be the lowest temperature at which the NaCl-water system is suitable for de-icing ?
c) What are the advantages of using $CaCl_2$ over NaCl for de-icing ?
d) Are there any disadvantages to using $CaCl_2$?

5. Why is the solubility for C in ferrite about 100× lower than in austenite ?

6. Does the Avrami equation apply to the austenite-to-pearlite transformation and, if so, what are the values of the parameters k and n, at the following temperatures:
a) 560°C ? b) 640°C ?

7. The eutectoid microstructure of the Fe-C system is sometimes said to have layers with relative widths in the ratio of about 8:1. Is this correct, and if so, why ?

8. Devise a semi-quantitative argument as to why the austenite-to-martensite transformation at a temperature of around 200°C cannot be diffusion-driven. State your assumptions clearly, and give good reasons for them. (Tip: Review what we discussed earlier about the diffusion of C in Fe.)

9. Come up with a plausible argument as to why Ni at >about 8 wt % in stainless steel inhibits the austenite-to-ferrite conversion.

10. Can an Sn-rich Pb-Sn alloy be strengthened by precipitation hardening? If not, why not? If so, under what conditions, and with what kind of a heat treatment ?

11. Is the diagram in Fig. 8.36 below a possible phase diagram? If so, label all regions of the diagram, and interpret them. If not, explain why not.

12. Research lead-free solders. List some examples, with properties, phase diagrams, and applications.

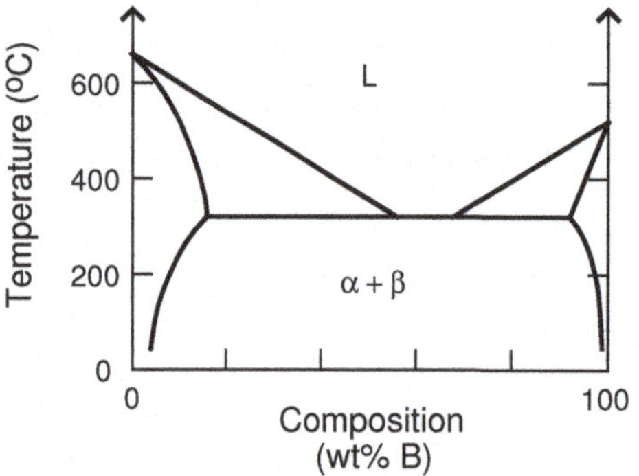

Figure 8.36: A potential phase diagram with components A and B.

13. The picture below shows the microstructure of a Nb-Si alloy. It is Nb-rich and said to consist of Nb (light) and Nb_3Si (dark) phases. Note the dendrites embedded in eutectic microstructure. The scale bar is 50 μm.

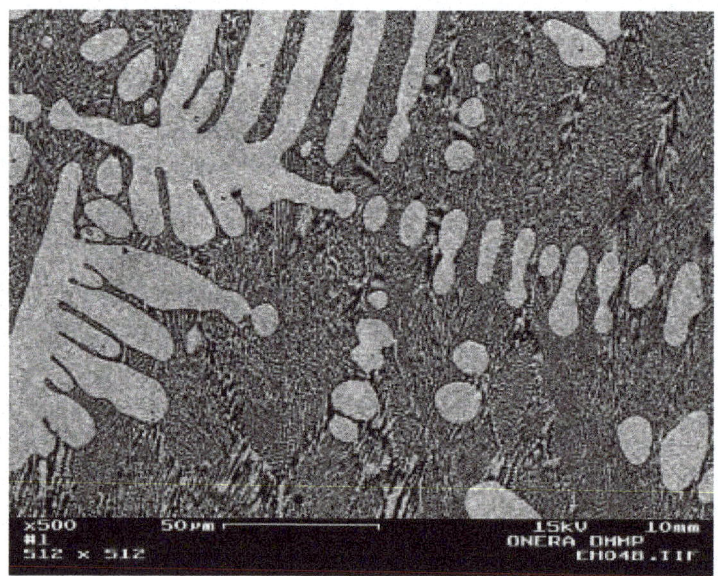

Figure 8.37: Nb-Si microstructure

a) Look up the Nb-Si phase diagram and comment on the statement above about the phases involved.
b) Figure out what phase the dendrites are.
b) Make an educated guess what the overall composition of this alloy was.

Chapter 9
Metal Alloys

Overview

Metals are generally used in the form of alloys in structural applications. This is due to the fact that alloying, in connection with suitable heat treatments, allows for a great deal of control over the alloy microstructure and mechanical properties, in particular tensile and yield strength vs. ductility and toughness. In this chapter a brief overview of these issues will be given for Fe-, Cu- and Al-based commercially important alloys. The focus will be on various alloying elements and their effects on phase diagrams and microstructures, and ultimately on mechanical properties.

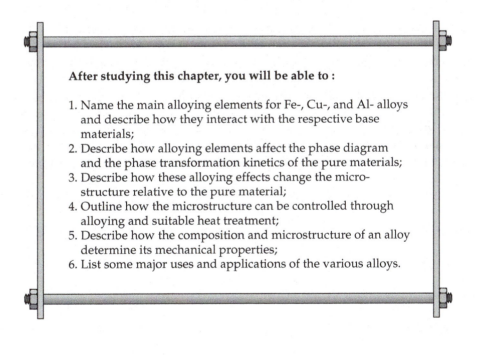

After studying this chapter, you will be able to :

1. Name the main alloying elements for Fe-, Cu-, and Al- alloys and describe how they interact with the respective base materials;
2. Describe how alloying elements affect the phase diagram and the phase transformation kinetics of the pure materials;
3. Describe how these alloying effects change the microstructure relative to the pure material;
4. Outline how the microstructure can be controlled through alloying and suitable heat treatment;
5. Describe how the composition and microstructure of an alloy determine its mechanical properties;
6. List some major uses and applications of the various alloys.

The Stonecutters Bridge in Hong Kong

The pictures show the Stonecutters Bridge in Hong Kong. It is a so-called cable stayed structure with a total length of 1596 m and a main span of 1018 m. Two 298 m high tapered pillars at each end support a 50 m wide deck. The lower sections of the pillars are a tubular reinforced concrete construction and carry three sets of cables. The upper sections of 115 m carry 25 sets of cables and are a composite of an outer skin, made of duplex stainless steel, a tubular reinforced concrete core, and inner stainless steel cable attachment boxes. The design life of the bridge is 120 years.

The materials science issue of interest is the special grade of stainless steel used. It was chosen for its combination of high strength and excellent corrosion resistance.

9 Metal Alloys

In the next three chapters we will re-examine the major classes of materials - metals, ceramics and glasses, and polymers - and add further details regarding their microstructure and properties. This expanded view will focus especially on engineering materials and will be informed by our previous discussion of microstructures. The present chapter will deal with structural metal alloys, primarily those based on iron (ferrous alloys), and on copper and aluminum (non-ferrous alloys).

9.1 Ferrous Alloys

9.1.1 Overview

Ferrous alloys have iron as their base metal and include steels and cast irons (Fig. 9.1). As I mentioned in Chapter 8, the distinction between these two broad groups is made on the basis of the Fe-C phase diagram (Fig. 8.18): Steels have less than 2.1 wt % C and cast irons have more than 2.1 wt % C. This amounts to saying that steels are materials that, because of their composition, do not undergo the eutectic transformation at 1127 °C.

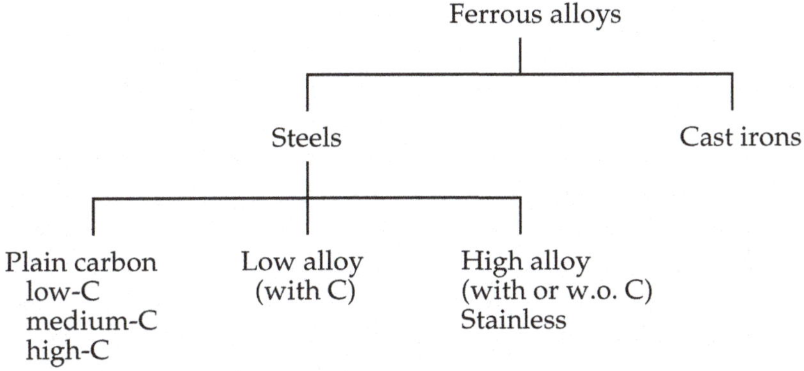

Figure 9.1: Classification of ferrous alloys

There are different ways of categorizing steels. We will distinguish three broad categories: **plain carbon**, **low alloy** and **high alloy steels**. Plain carbon refers to alloys without major alloying elements other than carbon. Low alloy refers to a total content of alloying elements of up to about 5 wt % including some carbon (this number is not exactly fixed), and high alloy to more than about 5 wt %. The most important alloying elements besides C are: Ni, Cr, Mo,

V, W, Mn, and Si, but others may also be used. The high alloy steels of interest to us are primarily the stainless steels, which contain only an insignificant amount of carbon.

The plain carbon steels fall into three general categories: **low-carbon, medium-carbon**, and **high-carbon steels**. Low-carbon designates a C-content of up to about 0.25 wt %, medium-carbon 0.25-0.60 wt %, and high-carbon 0.60-1.4 wt %, the latter being an approximate upper limit of C-content used in practice for steels.

As mentioned, plain carbon steels are material that contain no significant alloying elements other than C. Tool steels are mostly alloyed high-carbon steels. The element responsible for imparting corrosion resistance to stainless steels is Cr, typically at a concentration of at least 11 wt %. Corrosion resistance may be enhanced by the addition of Ni and Mo.

Table 9.1: AISI/SAE Designations of Plain Carbon and Low Alloy Steels

AISI/SAE	Mn	Ni	Cr	Mo	Other
10xx	0.75				
13xx	1.75				
2xxx		3.5-5			
3xxx		1.25-3.5	0.6-1.6		
4xxx		1-3.5	0.3-1.5	0.1-0.3	
5xxx			0.3-1.5		
6xxx			0.6-1		V 0.1
7xxx			0.75		W 1.75
8xxx		0.5	0.5	0.2	
92xx	0.75		0-0.65		Si 1.4-2

The numbering system in Table 9.1 is named after the AISI (American Iron and Steel Institute) and its successor SAE International, two organizations concerned with standards. According to the AISI/SAE system, a plain carbon or low alloy is designated by a four-digit number abxx. The number a identifies the alloy class according to the major alloying element, b gives additional information about other alloying elements, and xx indicate the wt % C multiplied by 100. Tool and stainless steels carry different designations. Tool steels are identified with a letter and a single number, stainless steel with three numbers and possibly an additional letter.

A unified numbering system (UNS) is used for labeling all alloys, not just the ferrous ones. The UNS number has 6 characters and consists of a letter for the alloy family, the AISI numbers and one or two additional numbers. There is also an equivalent European system using the EN steel number.

Table 9.2: Examples of Steels, with Composition, Properties, Common Uses

AISI Nr.	Comp.	T.S. (MPa)	Y.S. (MPa)	Ductility	Typical Uses
1010	0.1 C 0.75Mn	320	210	0.5	car body, nails, screws, rebars
1040	0.4 C 0.75Mn	600	400	0.25	crank shafts, gears
1080	0.8 C 0.75Mn	950	620	0.18	hammers, chisels, piano wire
1095	0.95 C 0.75Mn	950	610	0.15	springs, saws, knives
4140	0.9Mn 1Cr 0.2Mo	620	410	0.26	axles, shafts, valves, sprockets
4340	0.7Mn 1.8Ni 0.8Cr 0.25Mo	710	480	0.23	transmission gears, aircraft landing gear
304	19Cr 10Ni	590	240	0.55	chemical plants, food processing
316	17Cr 12Ni 2.5Mo	590	240	0.60	construction, welding
430	18Cr 1Mn 1Si	520	300	0.30	automotive, appliances

Table 9.2 gives typical data and properties for some examples of the different classes of steels: plain carbon, low alloy, and stainless. Both data and properties are somewhat simplified. In the plain carbon steels, Mn is also present at a level of typically about 0.75 wt %. In the low alloy steels, C is present at a level of typically 0.4 wt % in addition to the other alloying elements. The mechanical

property parameters are representative for hot rolled materials. These parameters can easily vary by ± 20% or so depending on the state of annealing or hardening of the material. Note that the 304 and 316 stainless are austenitic, i.e. have the FCC austenite microstructure, due to the high Ni content, whereas the 430 stainless steel is ferritic, i.e. has the BCC ferrite microstructure. Stainless steels generally have very low carbon content. Note also the large ductility of the austenitic stainless steels compared to other steels with similar strength.

9.1.2 Plain Carbon Steels

Plain carbon steels are the simplest and cheapest types of steel. They contain C as the most relevant impurity. Commercial materials range in C content from 0.1 wt % to over 1 wt %. They exemplify most clearly the trade-off between low strength/high ductility and high strength/low ductility. However, their strength depends non only on the C content, but just as much on the microstructure produced by the processing. For example, many of these steels are available hot rolled (implying, with relieved stress and slightly annealed) or cold drawn. In the cold drawn state they are about 10-15% stronger than hot rolled but correspondingly less ductile. On the other hand, if they were fully annealed, they would be about 10% less strong than hot rolled.

Medium-carbon and high-carbon steels are often heat-treated. This means that their desired strength cannot be achieved by simply letting them cool at the appropriate rate. Rather, they are produced originally as martensite (i.e. at a very high cooling rate), then reheated for tempering. This provides more control over the microstructure and properties.

9.1.3 Low Alloy Steels

Plain carbon steels have some limitations with regard to their properties: Some of these limitations are:

1. Plain medium carbon steels are difficult to quench rapidly enough so as to achieve a martensite microstructure. In addition, rapid quenching may lead to distortion and cracking. (Recall that the transformation to martensite is a lattice distortion, and thus accompanied by shear stresses).

2. Plain carbon steels have low resistance to corrosion.

3. Plain carbon steels loose their toughness (impact resistance) at low temperature.

These problems may be alleviated by alloying plain carbon steels with other elements, primarily Mn, Cr, Ni, or Mo. Additional elements may be used as well, such as V, W, Si and B. The alloying causes the materials to be heat-treatable more easily and results in stronger materials at a given ductility. These types of steel span the range of 13xx to 92xx (see Table 9.1, and Table 9.2 for two examples) and are sometimes referred to as high-strength low-alloy (HSLA) steels. High-carbon alloyed steels represent the special category of so-called tool steels and carry their own designation.

9.1.4 Stainless Steels

Stainless steels are materials formulated specifically for being resistant to oxidation in air and corrosion in an aqueous medium. In the simplest instance this means that they are rust-free at ambient conditions. However, corrosion resistance may also be of interest under more severe conditions, such as high temperature, high moisture, or a chlorine-containing environment.

Corrosion resistance is achieved by alloying the steel with at least about 12 wt % of Cr. The mechanism by which Cr imparts corrosion resistance to the steel is that in an oxidizing ambient the Cr forms a very thin, self-limiting Cr-oxide film at the surface, which protects the rest of the material from further attack.

There are three ***main types of stainless steel***, corresponding to their microstructures:

1. ferritic: containing no Ni; 0.01-0.05 wt % C.
2. austenitic: containing at least 6.5 wt % Ni, usually 8-12 wt % Ni; 0.05-0.06 wt % C (unless formulated as low-C, with 0.02 wt % C, e.g. in 304L) .
3. martensitic: containing no Ni; 0.14-0.4 wt % C.

Ferritic stainless steel is named in reference to the normal microstructure of a steel, consisting mostly of ferrite (i.e. the BCC α phase). Martensitic refers to a very rapidly quenched, then tempered, steel which has largely maintained its martensite microstructure. Note that neither of these two types contains any Ni. Austenitic stainless steel typically contains upward of 8 wt % Ni. This is because Ni stabilizes the austenite (i.e. the FCC γ) phase by extending its phase region all the way down to room temperature.

Ferritic and martensitic stainless steels are numbered 4xx, possibly with an additional L or H, for low or high C content. Austenitic steels are numbered 3xx, also with a possible L or H. Three very common materials are listed above in Table 9.2.

If you hear about a stainless steel in colloquial terms as 18/8, or you see this as a mark on your cutlery, it will refer to a grade with 18 wt % Cr and 8 wt % Ni, one of the most often used grades and similar to 304 (Table 9.2). The Cr-Ni steels are the ones in widest general use. The Arch in St. Louis is clad in plates of 304 stainless steel.

Protection against corrosion is not perfect in stainless steels. The material can be attacked if the protective Cr-oxide is damaged. Moreover, depending on the processing, the stainless steel may become sensitized, meaning that Cr-carbide has formed in grain boundaries. This can deplete regions near the grain boundaries of Cr, thus making the material more susceptible to attack.

A chlorine-containing environment is an especially severe test for the corrosion resistance of a stainless steel, as chlorine ions tend to attack and penetrate the protective surface layer. Such attack may lead to pitting of the material. The extent to which this happens depends on the chlorine ion concentration. 316 type steels are more resistant to this type of corrosion than 304 types, and Mo as an alloying element tends to improve this corrosion resistance.

9.1.5 Cast Irons

Although I have chosen not to focus very much on cast irons, here are a few simple facts. First, as you recall from the Fe-C phase diagram, Fig. 8.18, cast irons are materials with 2.1 wt % or more of carbon, usually up to 4 wt %. In the phase diagram, this means materials for which the single γ-phase region is avoided entirely and the C content is less than at the eutectic point. Upon cooling, such a material will evolve from L + γ to γ + Fe_3C and then to α + Fe_3C.

Under some conditions, graphite may also be formed during cooling, due to the decomposition of the cementite, Fe_3C, into Fe and C. The formation of graphite is favored for high C content and slow cooling, and it can be enhanced by the presence of Si in the material.

Types of cast irons:

1. **White Cast Iron**
   Contains C mostly in the form of cementite. Composition: typically 2.5-3.0 wt % C, 0.5-1.5 wt % Si. High solidification rate.

2. **Grey Cast Iron**
   Carbon present largely as graphite flakes. Composition: typically 2.5-4.0 wt % C, 1-3 wt % Si. Slow solidification rate.

3. **Malleable Cast Iron**
   Produced from white cast iron by heat treatment. This causes cementite to decompose, forming graphite.

4. **Ductile Cast Iron**
   Similar to gray cast iron, but with extensive heat treatment. This transforms the graphite flakes into spherical nodules in a ferritic matrix.

Both white and gray cast iron are rather weak and brittle, whereas ductile and malleable cast iron are much stronger and show some ductility. Cast irons are useful because in general they are cheap to produce and they can be formed into shapes that might be difficult to achieve any other way.

We will wrap up this section with two microstructures that should help you reconnect with things seen before. The figure below shows on the left a 4140 low alloy steel (cf. Table 9.2) and on the right a white cast iron.

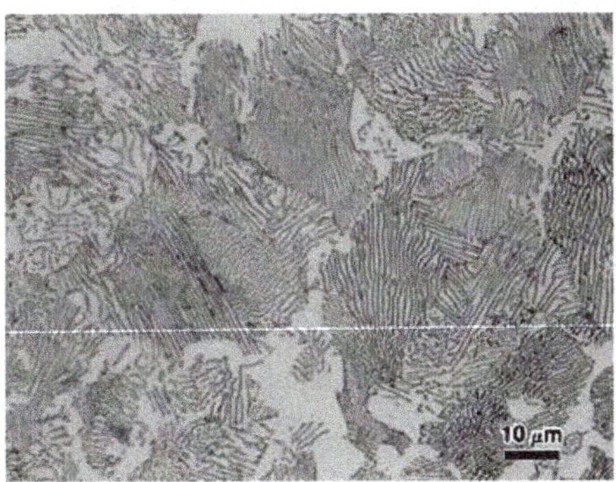

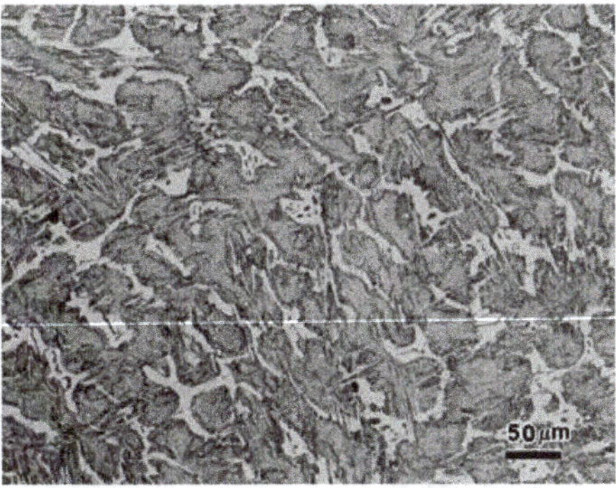

Figure 9.2: Electron micrographs of 4140 low alloy steel and white cast iron.

In the panel on the left, what you see is proeutectoid ferrite surrounded by pearlite, and in the panel on the right cementite (light) and pearlite (dark).

9.1.6 Alloying Elements and Microstructure

In Chapter 8 we examined the Fe-C phase diagram and its phase transformations. Strictly speaking, these arguments apply without modification only to plain carbon steels. Once we add another alloying element, the system becomes ternary. Moreover, many steels of practical interest contain more than one such element. Hence the question is what we can say about the microstructure in these more complex systems. We will not engage in a full discussion of this issue here, but we will examine some hints as to how what you know already can be adapted to more complex alloyed materials.

First, the main effects of some of the most common alloying elements can be summarized very briefly as follows:

Chromium
Chromium is the main element used for improving corrosion resistance.

Nickel
Nickel enlarges the temperature range in which austenite (γ-Fe) is stable.

Molybdenum
Molybdenum improves corrosion resistance; increases strength at high T.

Manganese
Manganese promotes the stability of austenite.

Silicon
Silicon improves strength and acts as deoxidizing agent binding oxygen.

In the simplest case, we need to consider two alloying elements at the same time. For example, if these were C and Cr, what we would have to do is fill in the three-dimensional phase diagram in Fig. 9.3 as follows:

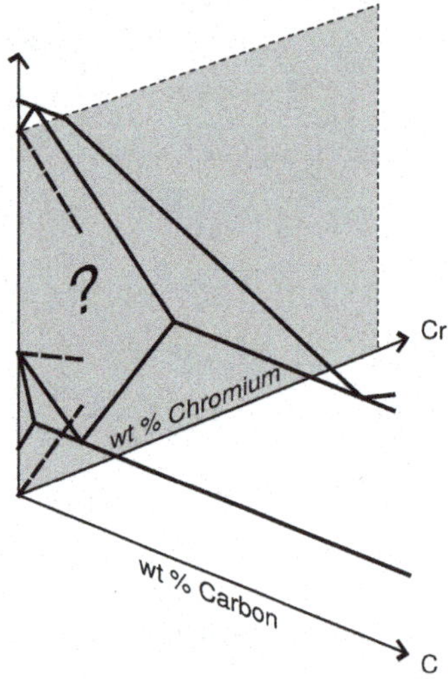

Figure 9.3:
Combined phase diagrams of Fe-C (in front) and Fe-Cr (in gray plane in back). The ternary system would include the entire region between the two two-component phase diagrams.

Think of Fig. 9.3 as combining the Fe-C and the yet unknown Fe-Cr phase diagrams with a common temperature axis. The ternary diagram would then include points in-between the two 2D coordinate systems T(C) and T(Cr). We will outline below how to fill in a few small parts of this puzzle.

The most critical feature of the Fe-C phase diagram is the eutectoid line. Thus, we will focus first on the effect of alloying on the eutectoid temperature. Fig. 9.4 shows that this effect depends on the alloying element. Cr and Si increase the eutectoid temperature, and Ni and Mn decrease it.

We can also ask what the effect of alloying will be on the carbon composition at the eutectoid point. You recall that in the Fe-C system the eutectoid point occurs at 0.76 wt % C. We do not need a quantitative answer here. Simply put: All alloying elements tend to reduce the carbon composition at the eutectoid point.

For our further discussion we will keep in mind primarily stainless steels, i.e. the alloying elements Cr and Ni, but we will also mention others. From the verbal descriptions above and the curves in Fig. 9.4, we may conclude that the alloying elements with a tendency to stabilize austenite are also the ones which lower the eutectoid temperature.

Partial binary phase diagrams for Fe-Cr and Fe-Ni are shown in Fig. 9.5. In comparison to the Fe-C diagram (Fig. 8.18), you realize that the T-axis (i.e. pure Fe) is identical in all three cases. In particular, on the T-axis the γ-phase exists between 912 and 1394 °C. But how the diagram extends to small amounts of the alloying element differs greatly between C, Cr, and Ni.

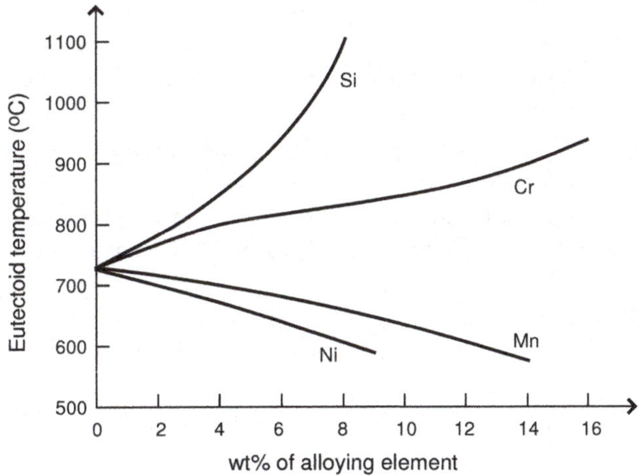

Figure 9.4: Effect of alloying on the eutectoid temperature in the Fe-C phase diagram

For small amounts of Ni, the Fe-Ni diagram looks similar to Fe-C: You can see a γ- and an α-region, depending on the temperature. At larger Ni content and low temperatures there is a region where α- and γ-phases can coexist, adjoining a γ-region at even larger Ni content. Of course, α- and γ-phases mean the same as for Fe-C, namely BCC and FCC structure. Most noteworthy is the fact that ***for about 18 wt % of Ni or more, the γ-phase (austenite) is stable at room temperature***! So Ni greatly extends the phase region for austenite.

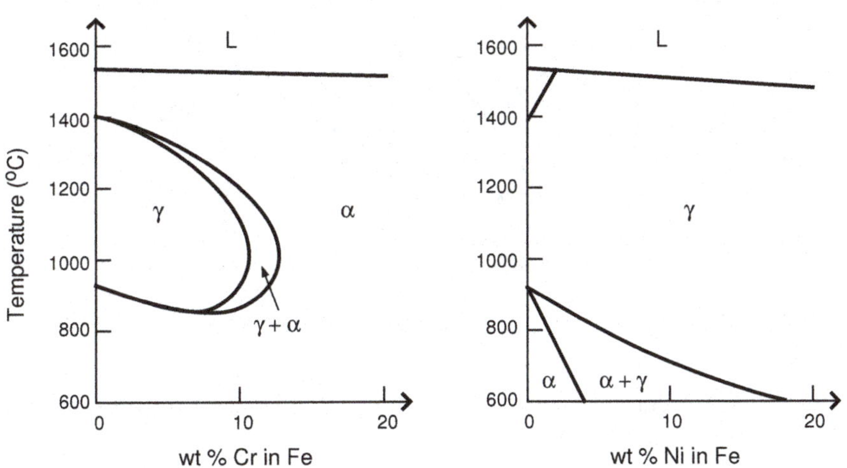

Figure 9.5: Binary Fe-Cr and Fe-Ni phase diagrams.

The Fe-Cr diagram also looks somewhat similar to Fe-C for small amounts of Cr. But the γ-region is now much more limited, and ***for about 12 wt % of Cr or more, no γ-phase exists at any temperature***! In this case one say that Cr promotes the occurrence of ferrite.

But what about Cr and Ni together, and possibly combined with other alloying elements? It has been found that the effect of all these elements can be taken into account at least qualitatively based upon whether they tend to promote ferrite as does Cr, or austenite as does Ni. Two parameters named Cr_{eq} and Ni_{eq} have been defined as follows:

$$Cr_{eq} = Cr + Mo + 1.5\,Si + 0.5\,Nb \tag{9.1}$$

$$Ni_{eq} = Ni + 30\,C + 0.5\,Mn \tag{9.2}$$

In equations 9.1 and 9.2 the elements are understood to represent alloy content in wt %. With these two parameters in mind, it is possible to use a qualitative plot called Schaeffler diagram (Fig. 9.6) to determine what phases are present at room temperature depending on the composition of the steel.

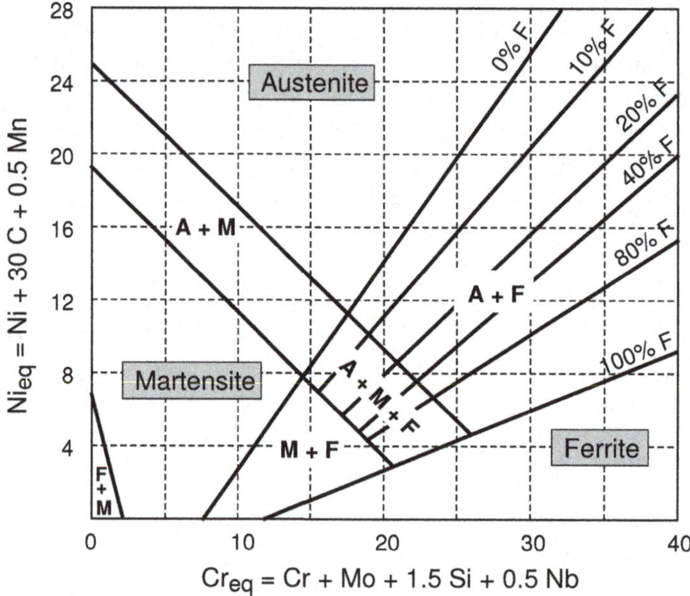

Figure 9.6: Schaeffler diagram indicating what phases are present in a multi-component Fe alloy.

9.1.7 Steels at Low Temperature: The Ductile-to-Brittle Transition

Our previous working assumption with respect to the mechanical properties of metals has been that they do not depend strongly on temperature, certainly not when compared to inorganic glasses and polymers. We will now discuss an effect that represents a major exception to this rule, namely the so-called ductile-to-brittle transition.

You will recall from Chapter 5 that, loosely speaking, strong metals typically have low ductility, and weak metals have high ductility. Most desirable are materials with both good strength and high ductility. The pertinent figure of merit is the toughness, i.e. the area under the $\sigma(\epsilon)$ curve (Fig. 5.6).

For steels in particular, one finds that many of them undergo a marked transition from high ductility/toughness to much lower ductility/toughness as the temperature is decreased to a low enough value. The common method to evaluate this transition is the so-called Charpy test. The test is performed on a material sample of standardized geometry and measures how much energy is dissipated in the sample as it fractures upon impact. This type of impact stress, or equivalently very high strain rate, is thought to present a worst case condition with respect to resistance to fracture. The dissipated energy, called impact energy or impact strength, as a function of temperature is used as an indicator for the ductility or brittleness of a material.

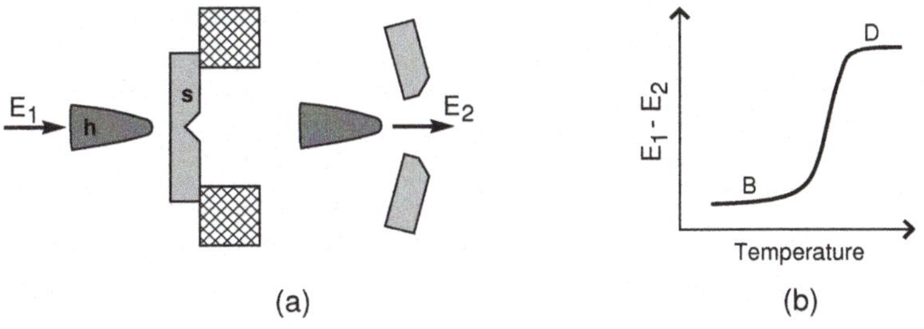

Figure 9.7: Schematic of Charpy test, and impact energy E_1-E_2 vs. T.

Fig. 9.7a shows a schematic of a Charpy test. The hammer h hits the sample s with kinetic energy E_1, causing it to fracture. The energy of the hammer after impact is E_2. The impact energy of the material is defined as the difference E_1 - E_2, and is assumed to have been dissipated in the sample during fracturing. A generic example of the impact energy of a material with a ductile-to-brittle transition is shown in Fig. 9.7b. At high temperature, the impact energy is high due to plastic deformation of the sample, indicating a ductile material. At low temperature the impact energy is low, indicating a brittle material.

Low- and medium-C plain carbon steels display a ductile-to-brittle transition in dramatic fashion (Fig. 9.8). The transition temperature, taken as the

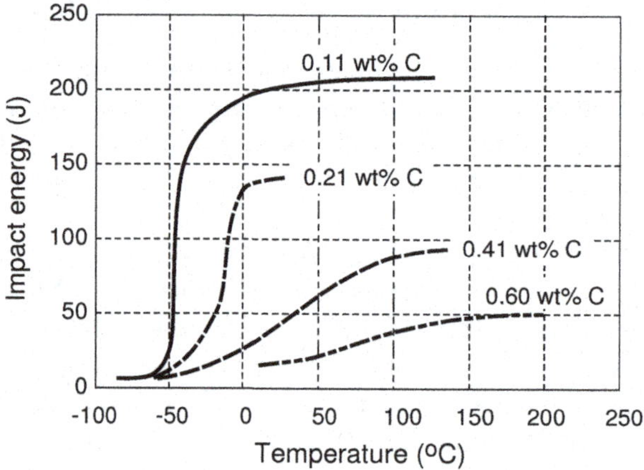

Figure 9.8: Fracture toughness of plain low- and medium carbon steels.

midpoint of the drop-off of the curve, is around -50 °C for a 0.11 wt% carbon steel and increases to around 80 °C for a 0.6 wt% carbon steel. Note also that the difference between the ductile and brittle side is much less pronounced for the medium-C steels, which is not surprising given that they have a lower maximum ductility to begin with.

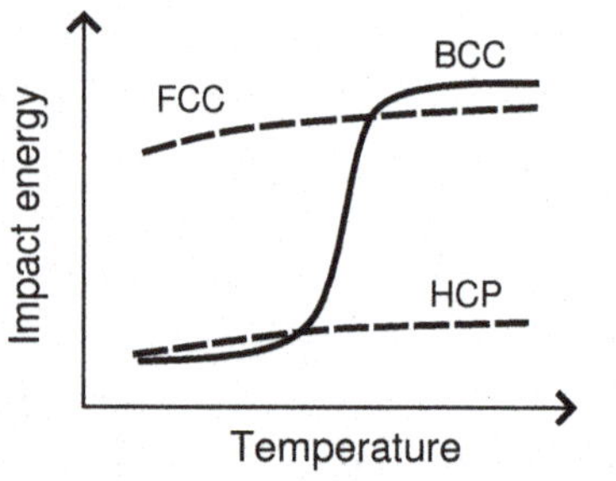

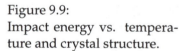

Figure 9.9:
Impact energy vs. temperature and crystal structure.

The impact energy depends on many factors other than temperature. The most significant one is crystal structure (Fig. 9.9). BCC metals do exhibit a ductile-to-brittle transition, whereas FCC and HCP metals do not. (Of course, HCP metals are brittle around room temperature to begin with). This is of great practical importance for steels. It indicates that ferritic and martensitic steels generally will be subject to such a transition, whereas the austenitic stainless steels, being FCC, will not. In fact, austenitic stainless steels will maintain their ductility down to -200 °C. Hence, such steels are excellent materials for service in low- or even cryogenic temperature conditions.

In addition, the impact energy of steels depends on material parameters other than crystal structure. Some of these parameters are listed below.

Parameters affecting impact energy:

1. Alloying elements,
2. Impurities,
3. Grain size,
4. Microstructure,
5. Strain rate.

This suggests that it may often be difficult to compare data quantitatively for nominally equal materials. Note also that impact energy correlates with, but is not the same as, fracture toughness. Impact energy as tested in a Charpy configuration involves stress applied as a brief pulse, i.e. a very high strain rate, whereas fracture toughness refers to the condition for crack propagation under a static load.

A famous instance with catastrophic consequences, where impact energy played a pivotal role, is the ***sinking of the Titanic*** on her maiden voyage on April 12, 1912, because of a collision with a huge iceberg off of Greenland. In 1996 parts of steel from the hull of the ship were recovered and subject to metallurgical analysis. This analysis showed that the steel used for the Titanic was similar to a hot-rolled ASTM A36 steel. A36 is a low carbon steel similar to AISI 1020, but with higher manganese content. On the other hand, the Titanic steel had higher S content and lower Mn content than A36. This is significant because S in Fe tends to form sulfides at grain boundaries which make the material more brittle. The role of the Mn is to bind with S, which turns out to counteract the formation of iron sulfides.

The general conclusion drawn was that the steel used in the Titanic was not suited for service at low temperature. You can see from Fig. 9.8 that this type of steel should be expected to loose ductility below 0 °C. In fact, Charpy tests on steel from the Titanic showed a ductile-to-brittle transition temperature of around 50 °C. The sea water temperature at the time of the collision of the Titanic was reported to have been -2 °C.

9.2 Non-Ferrous Alloys

Non-ferrous alloys comprise all metal alloys not based on Fe. Here we will concentrate on copper and aluminum alloys, even though many others are of interest, e.g. magnesium, titanium, or nickel alloys.

So why would one want to use a non-ferrous alloy rather than a type of steel? A quick answer would be: All else being the same, because Al and Cu are inherently less prone to corrosion, and Al, Mg, and Ti are lighter (actually, less dense) than Fe. The important qualification, of course is: All else being the same. For one, when it comes to combining strength and ductility, it is difficult for other types of materials to compete with steels. Also, steels are cheap overall, compared to the other types of materials.

A brief view of the price history of the main metals is given in Table 9.3 below.

Table 9.3: World Market Cost of Metals

Metal	Density	Cost (U.S.$/lb)		
	(g/cm^3)	1989	2010-07-31	2013-07-31
Mg	1.74	1.63	1.35	1.25
Al	2.70	0.96	0.95	0.80
Ti	4.54	5.50	11.00	10.00
Zn	7.14	-	0.95	0.82
Cr	7.19	-	-	4.00
Fe	7.87	0.20	0.30	0.30
Ni	8.90	7.00	10.00	6.00
Cu	8.96	1.45	3.50	3.00

The price figures for 1989 in Table 9.3 are from the book by W.F. Smith. The table clearly shows that Fe is the cheapest base metal by far for alloys. Also, it is evident that Cr and Ni are the main contributors to the price of the metals in stainless steel. Moreover, all these prices have changed greatly over time, and not always in one direction. For example, between July 2010 and July 2013, the price for Ti went as high as about $ 12.70 in the latter part of 2011 and Ni was almost twice in early 2011 what it is now.

9.2.1 Copper Alloys

Copper is another important base metal for alloys. In its pure form, it is especially suited for electrical applications due its very high electrical conductivity. It also has a very high thermal conductivity. When alloyed, it has useful mechanical properties, combined with excellent behavior in corrosive environments.

The table below lists a few commercial copper alloys, one from each major group, with their most important mechanical property parameters. The figures displayed are representative for annealed alloys. The yield and tensile strengths would be much higher, and the ductilities much lower, for hardened materials.

Table 9.4: Examples of Copper Alloys, with Composition, Properties, Common Uses

UNS Nr.	Comp.	T.S. (MPa)	Y.S. (MPa)	Ductility	Typical Uses
C10200	99.95Cu	220	70	0.5	electrical wires, gaskets, tubing
C23000	15Zn	280	90	0.47	plumbing, fittings, nipples
C26000	30Zn	350	130	0.55	ammunition, home fixtures
C17200	1.9Be 0.2Ni	490	165	0.35	bellows, springs
C51000	5Sn 0.25P	400	175	0.5	bellows, springs
C71500	30Ni	385	125	0.36	piping in corrosive environment
C75700	23Zn 12Ni	430	200	0.35	jewelery, musical instruments

The compositions given in Table 9.4 list the most import alloying elements, with the rest being copper. C10200 is pure copper for practical purposes. The brasses (items 2 and 3 in the table) can be work-hardened, whereas the bronzes (items 4 and 5 in the table) can be precipitation-hardened. This is due to the fact that Cu-Zn is a single-phase FCC material up to about 35 wt% Zn, whereas the alloying elements in the Cu bronzes have very limited solid solubility. The last item in Table 9.4 is commonly known as nickel silver, German silver, or nickel brass, due to its silvery appearance.

9.2.2 Aluminum Alloys

Aluminum is the third important base metal for alloys. In some ways it it similar to copper, in that it also has high electrical conductivity (about 30 % lower than Cu) and high thermal conductivity. When alloyed, it can be made strong enough to be useful for structural uses, in combination with good resistance to corrosion in a number of environments.

Similarly to steels, aluminum alloys are also classified by four digits, in a system devised by the Aluminum Association (AA). The AA number may be combined with a letter and additional digits to indicate the state of hardening of the alloy. The corresponding UNS number consists of A9 plus the AA number for a wrought alloy, and A0 plus the AA number for a cast alloy.

The table below lists several commercial aluminum alloys, one from each major group, with their most important mechanical property parameters and some typical applications.

Table 9.5: Examples of Aluminum Alloys, with Common Uses

AA Nr.	Comp.	T.S. (MPa)	Y.S. (MPa)	Ductility	Typical Uses
1100	0.12Cu	90	35	0.4	food and chemical equipment
2024	4.4Cu 1.5Mg 0.6Mn	490	350	0.18	plumbing, fittings, truck wheels
3003	1.3Mn	155	140	0.12	kitchen utensils, storage tanks
4043	5Si	200	-	-	welding filler
5052	2.5Mg	265	220	0.12	pressure vessels, tubes, marine applications
6061	1Mg 0.5Si	310	275	0.13	couplings, fittings, hydraulic pistons
7075	5.5Zn 2.5Mg 1.5Cu	580	510	0.11	aircraft structural parts

In Table 9.5, compositions are indicated by listing the major alloying elements. The values for mechanical properties are for hardened alloys, except for 1100 which is essentially pure Al and is annealed. The 1xxx, 3xxx, and 5xxx alloys are hardened by cold work, whereas the 2xxx, 6xxx, and 7xxx alloys are hardened by precipitation hardening heat treatment.

A new generation of Al alloys containing Li is being developed for the aerospace industry. These alloys are given an AA Nr. 8xxx. They are especially light-weight (i.e. low density) and have an excellent strength-to-weight ratio.

Image sources

Title pictures: Stonecutters Bridge in Hong Kong
http://www.arup.com/Projects/Stonecutters_Bridge/Stonecutters_Bridge_1.aspx
http://www.worldstainless.org/Files/issf/non-image-files/PDF/Stonecutters_Bridge_Case_Study-2.pdf

Fig. 9.1: adapted from Callister and Rethwish (Fig. 13.1).

Fig. 9.2: reproduced with permission by G. Vander Voort
http://www.georgevandervoort.com/metallography-articles/894-microstructure-of-ferrous-alloys.html

Fig. 9.4: reproduced with permission by ASM International
http://www.metalpass.com/metaldoc/paper.aspx?docID=272

Fig. 9.5: reproduced with permission by C.W. Bale (CRCT)
http://www.crct.polymtl.ca/FACT/documentation/SGTE/SGTE_Figs.htm

Fig. 9.6: adapted from ASM International, with permission
http://www.asminternational.org/content/ASM/StoreFiles/05231G_Sample.pdf

Fig. 9.8: reproduced with permission by ASM International
http://news.alibaba.com/article/detail/metalworking/100191045-1-metals-knowledge%253A-contribution-fracture-mechanics.html

References

Both Shackelford, and Callister and Rethwisch go into more details about classes of alloys and individual materials. Smith also has additional information on practical aspects such as metal and alloy fabrication.

SGTE Phase diagrams
http://www.crct.polymtl.ca/FACT/documentation/SGTE/SGTE_Figs.htm An amazing repository of about 300 binary metal phase diagrams.

Steel Manufacturing

How A Blast Furnace Works
http://www.steel.org/Making%20Steel/How%20Its%20Made/Processes/How%20A%20Blast%20Furnace%20Works%20larry%20says%20to%20delete.aspx

Steel-making Processes
http://www.keytometals.com/Articles/Art2.htm

General Steel References

Introduction to the SAE-AISI Designation System
http://www.keytometals.com/page.aspx?ID=CheckArticle&site=kts&NM=333

Plain Carbon Steels

http://www.engineersedge.com/materials/carbon-steel-properties.htm
http://www.tf.uni-kiel.de/matwis/amat/mw1_ge/kap_8/backbone/r8_4_1.html
They have an Fe-C phase diagram with a temperature-color scale.

Stainless Steels

A quick overview of commercial stainless steels is at:
http://www.aksteel.com/markets_products/stainless.aspx
The file Stainless_Steel_Comparator.pdf has a good summary of the effects of individual alloying elements.

Stainless Steels - Introduction To Grades, Properties and Applications
http://www.azom.com/article.aspx?ArticleID=2873
There are many links here to detailed properties.

Stainless steel in architecture
http://www.nickelinstitute.org/~/Media/Files/TechnicalLiterature/TimelessStainlessArchitecture_11023_.pdf
Discusses surface finish for stainless steel, many examples of buildings.

A History of Stainless Steel
http://www.stainlesssteelcentenary.info/StainlessHistory

Copper Alloys

Effects of Composition, Processing, and Structure on Properties of Copper and Copper Alloys
http://www.keytometals.com/Article98.htm

Microstructures of Copper and Copper Alloys
http://www.copper.org/resources/properties/microstructure/homepage.html

Aluminum Alloys

Aluminum Alloys Effects of Alloying Elements
`http://www.keytometals.com/Article55.htm`

A comprehensive account of the mechanical properties of Al alloys:
`http://www.alueurope.eu/wp-content/uploads/2012/01/AAM-Materials-4-Microstructure-and-properties.pdf`

Exercises

1. What are the phases present and their amounts in 1020 and 1080 plain carbon steel?
2. What are the important differences between a low carbon steel such as 1020 and a high carbon steel such as 1095 with respect to their equilibrium microstructure?
3. Why are stainless steels usually low-carbon steels? On the other hand, why might you want to increase the C content in stainless steel?
4. Given that the lowest carbon content for cast iron is 2.1 wt % C, what is the lowest amount of Fe_3C in equilibrated cast iron? How much of that Fe_3C is proeutectoid?
5. Why can there be pearlite in cast iron?
6. Is the statement in the text true that the alloying elements are mainly responsible for the price of stainless steel?
7. Look up which stainless steels are magnetic (i.e. ferromagnetic) and which ones are not. Check this out on some stainless steel implements in your kitchen.
8. If you were to advise a manufacturer of tableware with regard to what grade of stainless steel to use, what would you recommend for forks, spoons, and knives?
9. Why is the Cu-Ni alloy in Table 9.4 harder than pure copper?
10. What kind of a microstructure would you expect for the last alloy in Table 9.4, so-called nickel silver?
11. Formulate a general criterion describing when a binary metal alloy can or cannot be strengthened by precipitation hardening (as opposed to e.g. work hardening).

Problems

1. Why is Ni effective in stabilizing the austenite microstructure in stainless steel, but not Cr?

2. Are there any castable stainless steels? If so, what are their properties and applications?

3. If Cr is the alloying element responsible for the stainless property, why is the Cr content in stainless steel generally close to 18 wt % and not, say, 25 wt % or 30 wt % ? In other words, why wouldn't more Cr be even better for corrosion resistance?

4. In our discussion of the microstructure of stainless steels, we were silent about a factor we had found important in other, related circumstances. What is this factor, and what difference does it make for stainless steels?

5. Research the topic of duplex stainless steels. What are they? What is special about their microstructure? What are they used for?

6. In Chapter 8 we saw that Al containing a small amount of Cu can be precipitation-hardened. Does this also work for a small amount of Al in Cu? If so, describe a suitable process. If not, explain why not.

7. Consider a Cu-Cr alloy.
   a) Test your understanding of binary alloys by quickly guessing what the effects would be of adding a small amount of Cr to Cu.
   b) Look up the Cu-Cr phase diagram and discuss how good your quick guess for a) was.

Chapter 10
Ceramics and Glasses

Overview

Ceramics exhibit phenomena analogous to metals when it comes to alloys, phase diagrams, and phase transformations. Binary mixtures of alkali halide ceramics can be anything between completely miscible and not miscible at all, depending on the similarity of the components. Systems with limited miscibility are eutectic, and intermediate compounds are often observed. This type of behavior gives rise to corresponding microstructures, including layered eutectics. Silicates can solidify to form a crystalline or a glassy material, depending on the cooling rate. Engineering ceramics are used primarily where high temperature stability is required. Control of their microstructure facilitates control of material properties, especially mechanical properties, in similar ways as for metals.

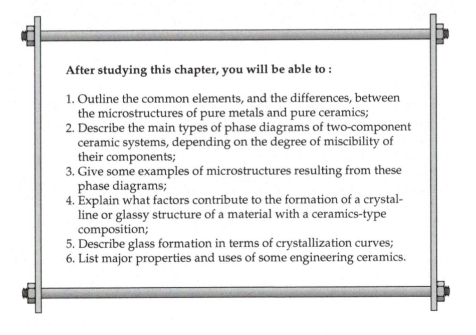

After studying this chapter, you will be able to :

1. Outline the common elements, and the differences, between the microstructures of pure metals and pure ceramics;
2. Describe the main types of phase diagrams of two-component ceramic systems, depending on the degree of miscibility of their components;
3. Give some examples of microstructures resulting from these phase diagrams;
4. Explain what factors contribute to the formation of a crystalline or glassy structure of a material with a ceramics-type composition;
5. Describe glass formation in terms of crystallization curves;
6. List major properties and uses of some engineering ceramics.

Microstructures in a Eutectic CeO_2-CoO Ceramic Alloy

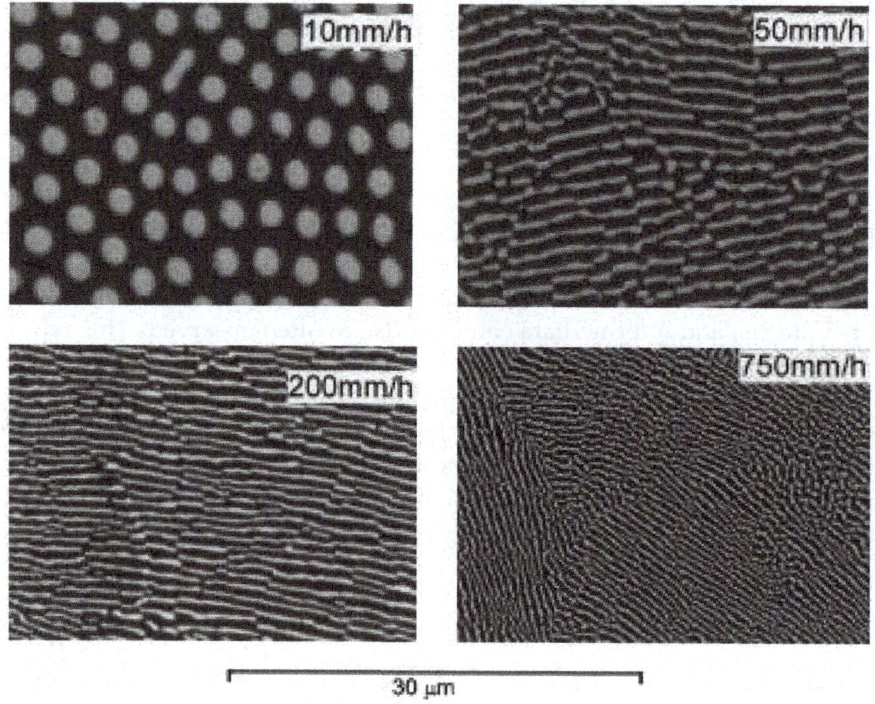

The micrographs in the picture above display eutectic microstructures of a CeO_2-CoO alloy. This material is of interest for solid state fuel cells. The panels show structures obtained in small ceramics rods which were heated locally in a narrow zone by a laser. The heated zone was moved along the rod at the speeds indicated so as to effect directed crystallization of the material.
Note the fibrous microstructure in the first panel, at the lowest speed of crystal growth, and the usual layered eutectic structures at the higher speeds. Also note how the width of the layers decreases as the speed of growth increases.

10 Ceramics and Glasses

In this chapters we will re-examine ceramics and inorganic glasses with a perspective informed by phase diagrams and phase transformations. We will also shed further light on the two classes of materials from the point of view of crystallization. We will conclude this chapter with a discussion of various engineering ceramics and their properties.

10.1 Phase Diagrams and Microstructure of Ceramics

In Chapters 2 and 3 we introduced ceramics as metal-nonmetal compounds, i.e. roughly speaking, combinations of elements from the left and right sides of the periodic table. Typical examples are metal halides such as LiF and CaF_2, metal oxides such as NiO, MgO, and Al_2O_3, and we also included SiO_2 in this category. The main characteristic of these types of materials is the strong ionic character of the atom-atom bonds. This in turn is the major determinant of their crystal structure. The structure of many of these compounds can be understood as the close packing of positive and negative ions.

For the microstructure of ceramics, basically the same principles apply as with metals and metal alloys. The equilibrium microstructure depends first of all on the number of components, and can be deduced form the phase diagram. We gave two examples of such phase diagrams before, namely in Fig. 8.16 for NiO-MgO as a binary system with complete **miscibility**, and in Fig. 8.17 for SiO_2-Al_2O_3 as a much more complex binary system.

Just as with metals, the simplest microstructure is observed for a pure, one-component ceramic (Fig. 10.1). However, even this relatively simple microstructure is somewhat more complicated than its metallic counterpart.

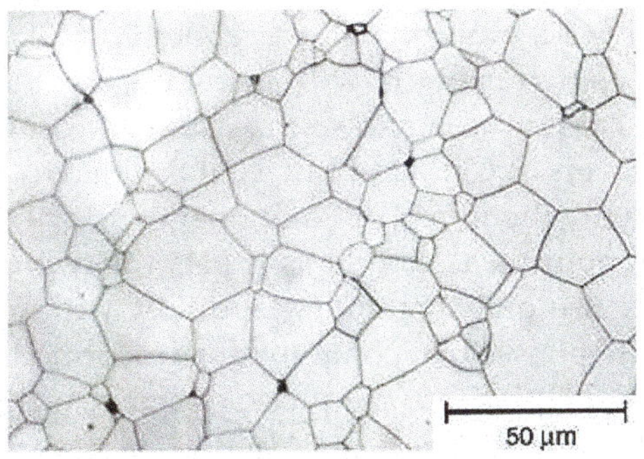

Figure 10.1: Microstructure of polycrystalline Al_2O_3.

In Fig. 10.1 you can see individual grains, as in metals, but there are also a few dark spots visible. These are small voids which are related to the way ceramic materials are usually prepared.

Since ceramics are not ductile and thus difficult to machine, they must generally be prepared close to the shape in which they are needed. The raw material for a ceramic is often a powder, which is then pressed into a form and sintered. As noted in Chapter 5, sintering refers to a high-temperature treatment similar to annealing, but at a high enough temperature so that ions near the surfaces of the powder particle can diffuse around, equilibrate the particle shapes, bind particles together, and fill empty spaces between particles. During this process of sintering, the material becomes more dense, but it may not reach its full, theoretical density. That means that occasional small voids (the ellipsoidal dark areas mentioned above) may remain in the material. In addition, the boundary areas between the grains may not be fully crystalline but may consist of less dense, amorphous material.

In two-component ceramics, the microstructural phenomena you are familiar with can be seen again. The simplest cases are two-component metal-halide systems. They range from completely **miscible** (as for NiO-MgO in Fig. 8.16) to eutectic and completely **immiscible**, as e.g. in the CaF_2-LiF system:

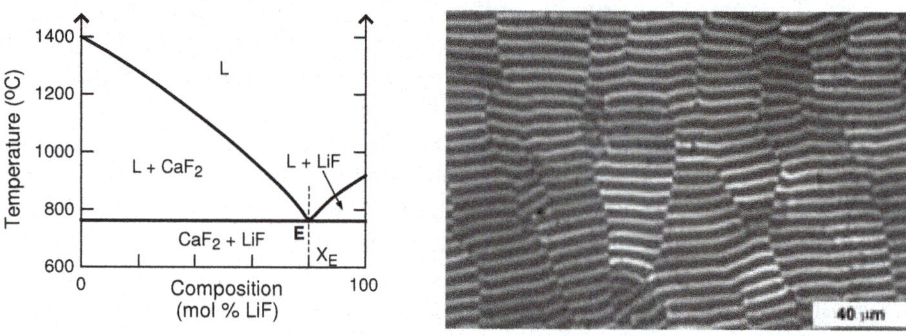

Figure 10.2: CaF_2-LiF phase diagram (left panel) and eutectic microstructure (right panel).

The CaF_2-LiF phase diagram shows a eutectic point at about 20 mol % of LiF and 767 °C (note that the composition is not in wt % here, but in mol %). The microstructure for the eutectic composition shows the usual lamellar appearance, consisting of alternating layers of CaF_2 and LiF. That is, the two low-temperature solid phases are essentially the two pure components.

A system with just one eutectic point is as simple as it gets for two components with limited solubility. It is also quite common, especially with metal oxides, that one or more intermediate phases (i.e. compounds) are formed. An example is SiO_2-Al_2O_3 shown previously, where the phase diagram includes the intermediate compound mullite, with composition 3 Al_2O_3 - 2 SiO_2, at about 28 wt % SiO_2 (cf. Fig. 8.17).

An even simpler example is displayed in Fig. 10.3 below with the phase diagram for the $CaCl_2$-KCl system:

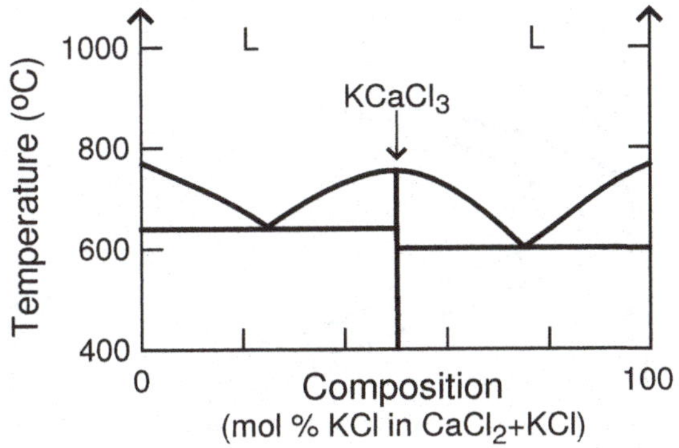

Figure 10.3: Phase diagram for the $CaCl_2$-KCl system.

This phase diagram has an intermediate compound at 50 mol % KCl with composition $KCaCl_3$ and two eutectic points. Again, note that the composition is not given in wt % here, but in mol %.

Another interesting example is the CeO_2CoO system, a material under investigation for potential application in solid fuel cells (see the title picture at the beginning of the chapter). This system shows limited solubility in the solid state and has a eutectic point at around 1645 °C and 82 mol % CoO. In a directional crystallization experiment, the material showed various eutectic microstructures depending on the growth rate (i.e., basically, the time allowed for growth to occur). At very low growth rate the structure was fibruous: One solid phase had the form of fibers embedded in the matrix of the other solid phase. At higher growth rates the microstructure took on the layered shape typical for eutectics, which you have seen before with metals.

10.2 Ceramics, Glasses, and Crystallization

Given that both ceramics and inorganic glasses are in essence compounds of metals and non-metals, a question of great fundamental as well as practical interest is why some of these compounds tend to crystallize whereas others do not. Based upon what you learned in Chapter 8 on solidification, you realize of course that this is not so much a question of "either or", but rather "under what circumstances".

In Chapter 8 our tacit assumption was that the liquid would crystallize with a sufficient amount of undercooling, once nucleation got initiated. We can refine this view now with help of what you learned in Chapter 9 on the kinetics of phase transformations. Consider Fig. 10.4 below, which displays two generic time-transformation curves for crystallization vs. glass formation (cf. Fig. 8.25).

The parameter y once again denotes the extent of the transformation. The first of the time-transformation curves represents the beginning of crystallization and the other curve the completion. (When discussing crystallization, it is

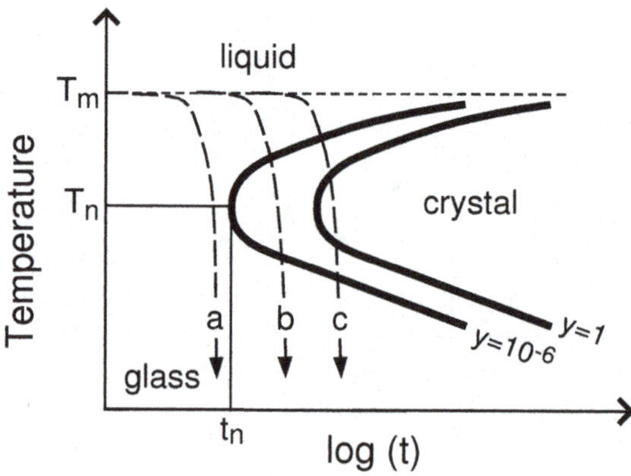

Figure 10.4: Generic time-transformation curves for crystallization and glass formation (solid lines), and three continuous cooling curves a, b, and c, corresponding to different cooling rates (dashed lines).

common to identify the beginning of the transformation with the point at which 10^{-6} of the total volume has crystallized). Thus, curve (a) among the three cooling curves indicates a cooling schedule which avoids any crystallization and thus leaves the sample as a glass. For curve (b) about half of the sample would end up crystalline, and for curve (c) the entire sample would have crystallized.

You can also see from Fig. 10.4 that if the objective is to avoid crystallization, then a decent estimate for the critical cooling rate can be obtained by taking the ratio T_n/t_n where T_n is the temperature of the "nose", and t_n is the time at the tip of the "nose".

The key to applying these arguments to an real system is, of course, to know the actual values of T_n and t_n, and these are often not known. What one can say is that among the ceramic compounds we have discussed, the alkali halides and their alloys are just as hard not to crystallize as are metals. On the other hand, silicates especially (i.e. SiO_2-based compounds), and also some other oxides, are much easier to synthesize in glassy, amorphous form, even though with enough time crystals can certainly be prepared.

One clue as to why this is so can be gleaned from the viscosities of these types of materials in the liquid state, as listed on Table 10.1. The point of these data is that molten salts such as KCl and molten metals have similar, and very low, viscosities just above their melting points. A glass does not have a melting point but becomes practically molten over a temperature range. Still, at a viscosity where the glass flows much like a liquid, say at an η of the order of 1, or equivalently at a temperature much above the working point (Fig. 7.8), the typical viscosity for the glass is about a factor 10^3 higher than the viscosity of a molten metal or molten salt.

Table 10.1: Viscosities of liquid metals, alkali halides, and silicates

Material	Viscosity (Pa s), T	Melting Point
water	0.001 at 20 °C	100 °C
Al	0.003 at 700 °C	660 °C
Sn	0.0013 at 400 °C	232 °C
K	0.0005 at 100 °C	63 °C
KCl	0.001 at 800 °C	772 °C
soda-lime glass	10 at 1470 °C	1470 °C
boro-silicate glass	10 at 1560 °C	1560 °C

If you go back to Chapter 7 and look over our discussion of the viscosity there, you will realize that viscosity has to do with the strain rate a material can sustain. Thus it is a measure of how quickly the material can rearrange its internal structure. The viscosity data tell us that this is easy for a molten salt, but much more difficult for a silicate glass. This difference is fairly straightforward to understand from an atomic standpoint.

A molten alkali halide salt is similar to a molten metal in that both are essentially made up of hard spheres in a slightly disordered arrangement, although in the case of the salt the spheres are also charged. In solidification, these hard spheres try to become as close-packed as possible, which is fairly easy, as it does not require a very substantial atom rearrangement compared to the liquid state. On the other hand, a silicate glass consists of SiO_4 tetrahedra joined together at their corners into an irregular 3D network. With an increase in temperature, the glass network just becomes gradually looser, as some of the bonds between SiO_4 tetrahedra are broken. On decreasing the temperature, the disordered network simply reforms, unless the material is given the very long time it takes for the large scale rearrangements to occur that will align the tetrahedra into a crystalline array.

In our discussion of ceramics and glasses, so far we have focused on alkali halides and silicates. However, there are many more materials in the general class of metal-nonmetal compounds, e.g. other oxides, nitrides, carbides, and beyond those a vast number of three- and multi-component materials. Some of these may form glasses, but others will not. So, the final question we want to address in this section is whether there are general rules with regard to material composition which govern glass formation, i.e. rules describing which kinds of materials it will be difficult to produce in crystalline form.

In the 1930's a man by the name of W.H. Zachariasen did extensive studies on oxide materials and formulated some rules as to when an oxide tends to form a glass and when not. His conclusions were that compounds with the generic formulae A_2O and AO will not form a glass, A_2O_3 will if triangles are formed, and AO_2 and A_2O_5 will if tetrahedra are formed. Eventually it was found that these rules do not cover all materials but have exceptions. However, a number of important compounds do fall into these categories, as you can appreciate from the representative list below:

- Important glass formers
  B_2O_3, SiO_2, GeO_2, P_2O_5

- Non-glass formers
  Li_2O, Na_2O, K_2O, CaO, MgO, PbO

The key to a compound being a glass former is that it is able to build a 3D network. You will also recognize the non-glass formers as what we termed, in Chapter 3, network modifiers. These do not enter the network themselves but rather serve to balance charges arising from bonding imperfections in the network itself. In doing so, they have a marked effect on the flow properties of the glass (i.e. on the viscosity vs. temperature), and on the mechanical properties overall.

In addition, when one examines compounds other than oxides with regard to the nature of the interatomic bonding and its connection with the tendency to form a glass, it appears that an intermediate type of bonding is important for glass formers. If the bonding is strongly covalent or strongly ionic, a compound tends not to form a glass.

10.3 Glass-Ceramics

An essential feature of a glass is that it does not contain any crystalline grains. In fact, crystals in glass are highly undesirable from the point of view optical transparency, since crystalline grains will have a tendency to scatter light and thus to reduce the transparency of the material.

On the other hand, it is possible to produce a glass with a controlled amount of crystalline grains in it. In this case the material is called a glass-ceramic. The advantage of this type of material is typically that it has a very low coefficient of thermal expansion. This gives it great mechanical stability under variations of temperature and greatly reduces its susceptibility to thermal shock.

A glass-ceramic can be realized from a glass with the proper materials composition and the proper heat treatment, as you might expect from the related processes we discussed with metal alloys (e.g. precipitation hardening). The principle is similar: The material is prepared first in a low-temperature, metastable form, i.e. as a glass, and then a certain amount of crystallization is allowed to take place by reheating the material for the appropriate time.

The actual process is somewhat more complicated. The starting material usually contains a nucleating agent such as TiO_2 in an alumino-silicate glass (Li_2O-Al_2O_3-SiO_2). The heat treatment occurs in two steps. In a first step, at a lower temperature, crystalline nuclei are formed at the desired concentration. In a second step at higher temperature, the growth of the initial nuclei to the desired size is performed.

Commercial products in which this type of process has been applied are known as Corning Ware® (used for kitchen dishes), ZERODUR® by Schott, and CLEARCERAM® by Ohara. The latter two are materials with zero thermal expansion coefficient and are used for precision optics and as mirror blanks for telescopes. The coefficient of thermal expansion is specified as $0 \pm 0.10 \times 10^{-6}/K$ for ZERODUR® and regular CLEARCERAM® and $0 \pm 0.10 \times 10^{-7}/K$ for CLEARCERAM®-Z EX.

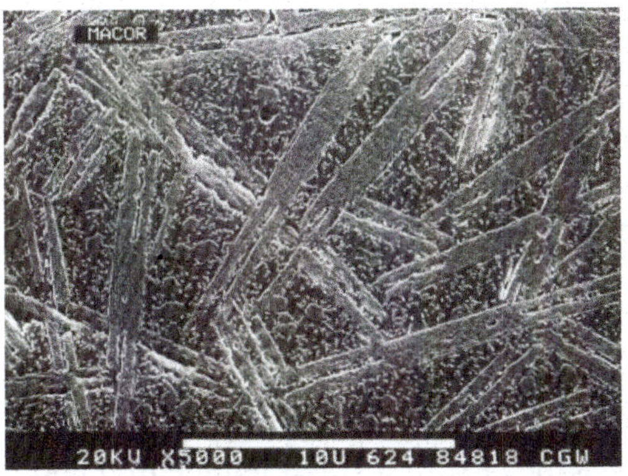

Figure 10.5: Microstructure of MACOR® . Note the sheet-like mica crystallites embedded in a glassy matrix.

A related product of interest is MACOR® (Corning) because it is a glass-ceramic that is machinable with ordinary machine tools and can be polished to an excellent finish. MACOR® starts out as a fluorine rich glass with a composition similar to fluorphlogopite mica ($KMg_3AlSi_3O_{10}F_2$). Upon cooling from the melt, the glass undergoes a phase separation into two glassy phases, with one consisting of fluorine-rich droplets. The subsequent heat treatment crystallizes the fluorine rich droplets. The result is a microstructure with randomly oriented small sheets of mica crystals within an alumino-borosilicate glass matrix. The volume fraction of the crystalline phase is about 0.55, and the average length of the crystallites is about 20 μm (Fig. 10.5).

The uniform distribution of randomly oriented mica sheets within the glass matrix gives rise to the unique material characteristics. Mica itself is a layered crystalline material with the property of cleaving easily, i.e. fracturing easily, along a plane in a layer. This causes the composite glass-ceramic to fracture locally, at the microscopic scale, in a controllable fashion when being drilled or milled.

10.4 Engineering Ceramics

10.4.1 General Background

Engineering ceramics are mixtures of two or more basic ceramic compounds such as SiO_2, Al_2O_3, MgO, and others. They have rather complex microstructures, not only because they contain several components, but also because with respect to microstructure they may not be fully equilibrated. A large and important group is based on oxides, but there are other important materials as well, e.g. SiC, Si_3N_4, BN, etc.

The raw materials for producing engineering ceramics are minerals. One such class are **clays**, which are layered layered aluminosilicate materials. One of the most common types is Kaolinite, with chemical formula $Al_2(Si_2O_5)(OH)_4$. It is made up of planar silicate layers where SiO_4 tetrahedra are joined together in O-corners, so that the repeating unit is $(Si_2O_5)^{2-}$:

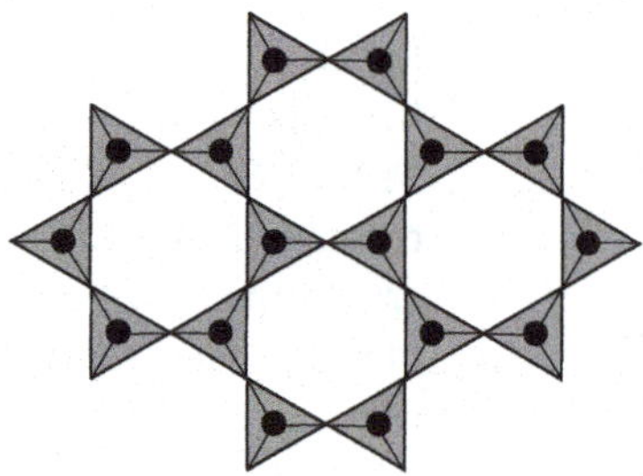

Figure 10.6: Planar layer of SiO_4 tetrahedra. The repeating unit is $(Si_2O_5)^{2-}$.

In Fig. 10.6, each triangle denotes the base of an SiO_4 tetrahedron and the black dot represents a Si atom. The top O atom of the tetrahedron is not drawn. You have encountered this structure before, in the left panel of Fig. 3.28:

In the Kaolinite clay, this silicate layer is joined to another planar layer of $Al_2(OH)_4^{2-}$. The structure can take up water between these two layers to become moldable with the typical consistency of clay. Hence, clays can be shaped easily and are often used as mined.

Other raw materials come from rock-type minerals. For example, a major source of SiO_2 is quartz from sand, the most abundant mineral in the Earth's crust. Quartz is also one of the three main constituents of granite, the other two being **feldspar** and **mica** as important mineral sources of their own. The mineral feldspar occurs in different varieties, with formulae $KAlSi_3O_8$, $NaAlSi_3O_8$, or $CaAl_2Si_2O_8$. Mica refers to a large group of layered silicates, also termed phyllosilicates, containing the same kind of parallel sheets of silicate tetrahedra noted above, in combination with $(OH)_2$ or $(OH)_4$ groups and various metals.

The desired ceramic part is first formed into what is sometimes referred to as the "green body" either from clay, for a traditional ceramic, or from ceramic powder or particles with an added binder, for **refractory** ceramics. The green part is then dried if necessary, and finally fired at a temperature in the range of 900-1400 °C in a kiln or furnace. During firing a fusion or sintering process takes place, possibly involving the formation of a small amount of liquid phase, which binds the ceramic particles together. The final microstructure can be very complex, combining several phases (Fig. 10.7).

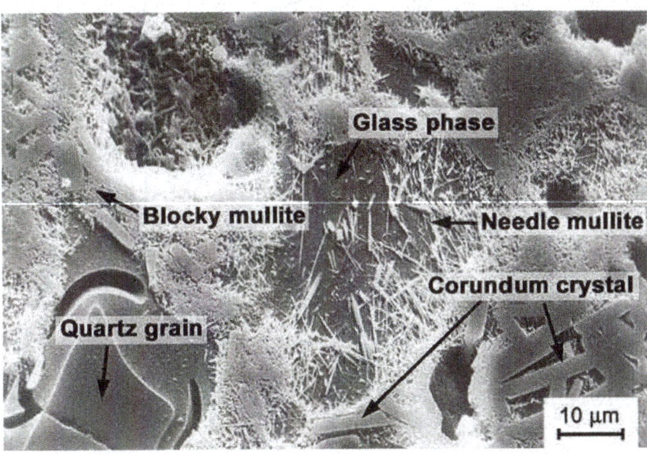

Figure 10.7: Electron micrograph of the microstructure of porcelain. The sample was etched slightly to highlight the various phases. Corundum means crystalline Al_2O_3. See the SiO_2-Al_2O_3 phase diagram in Fig. 8.17.

The end product depends on the composition of the starting material and on the firing temperature. For certain applications, it may also be provided with a glassy glaze. The shape of the final product may be something as simple as a brick or tile, or it may be fairly complex. One manufacturer goes as far as calling their products kiln furniture. Products you may be familiar with include whiteware (porcelain, pottery, tableware), kitchen appliances, bathroom fixtures, structural products (bricks, tiles, sewage pipes) and refractories for lining metal-processing furnaces.

10.4.2 Oxide Ceramics

In this section, for the purpose of illustrating principles, we will concentrate mainly on aluminosilicate materials. This is not to say that there are no other important oxide ceramics. For example, MgO- and ZrO-based materials are also used for structural applications, and in the area of electronic ceramics a broad range of other materials are of interest for their electrical properties, e.g. so-called piezoelectrics such as $BaTiO_3$ or $SrTiO_3$. You may also remember that the first high-temperature superconducting materials were in fact ceramics.

The range of aluminosilicate ceramics can best be appreciated by looking again at the phase diagram of SiO_2-Al_2O_3 (see also Fig. 8.17). There are four types of materials, depending on their composition. They are labeled as bricks

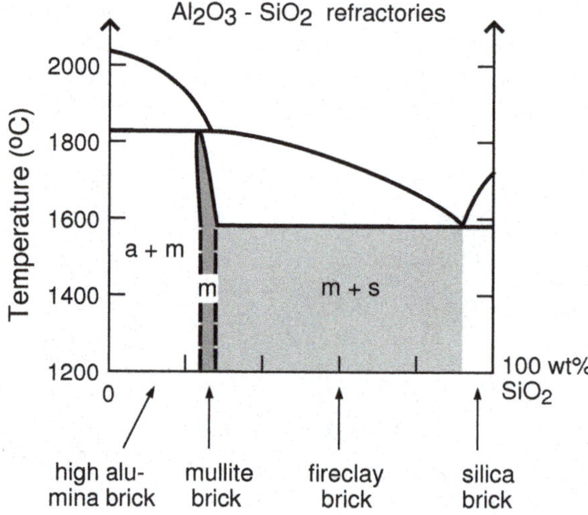

Figure 10.8: Main types of aluminosilicate refractory materials.

in Fig. 10.8 because bricks are an important application of them. High-alumina bricks have a higher Al_2O_3 content than mullite. Firebricks range in composition from the mullite boundary to the eutectic composition at 92 wt % SiO_2.

Some common refractory engineering ceramics are listed in Table 10.2 below, together with the traditional ceramic of porcelain for comparison.

Table 10.2: Compositions of common engineering ceramics

	Content in wt %				
Material	SiO_2	Al_2O_3	MgO	Other	Typical uses
fireclay brick	70-50	25-45		5	furnace and kiln linings, hot-metal ladles
high-alumina brick	5-45	90-50		5	furnace linings
silica brick	96			2.2 CaO, 1.8 other	chemical reactor linings
magnesia brick		95-78	3-20	2	furnace and kiln linings
porcelain	60	32		6 K_2O, 2 other	whiteware, electrical insulators

An example of the use of a refractory lining is displayed in Fig. 10.9. The figure shows a design drawing of a ladle for transporting molten iron from the furnace to the steel plant.

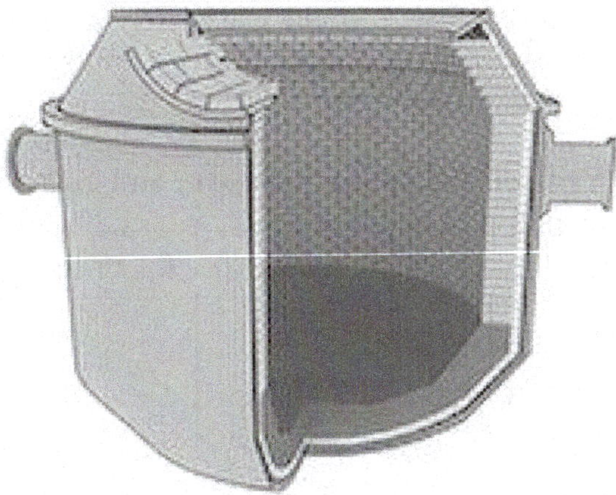

Figure 10.9: Cross section of a ladle for molten iron. Note the lining made with refractory ceramic bricks.

The choice of ceramic is dictated by the conditions for its application, primarily the use temperature and the chemical environment. Hence refractories can also be classified according to their chemical reactivity, namely as acidic, basic, or neutral, especially relative to the so-called **slag** produced in iron and steel furnaces.

Slag is the nonmetallic byproduct of steel production, floating on top of the molten iron. It is siphoned off before the molten iron is drained from the furnace. In the production of iron, the blast furnace is charged with iron ore (a mixture of iron oxides, silica, and alumina), fluxing agents such as limestone ($Ca(CO)_3$) and dolomite ($CaMg((CO)_3)_2$), and coke as fuel and reducing agent. From this mixture molten iron and molten slag are formed. Slag consists primarily of calcium, magnesium, and aluminum silicate. Hence it is not only a by-product but also a resource for the minerals left in it.

Aluminosilicates are examples of acidic refractories. They are chemically resistant to attack by acidic materials, but are attacked by basic materials. Examples of basic refractories are CaO, MgO, and dolomite. These are resistant to attack by basic materials, but are attacked by acidic materials. Graphite, silicon carbide (SiC), chromite ($FeCr_2O_4$), and zirconia (ZrO_2) are neutral refractories and resist attack by both acidic and basic materials. Thus a further important factor, other than the use temperature, in the choice of material for the lining of a furnace is the nature of the slag produced, since erosion of the lining by the slag has to be minimized.

10.4.3 Non-Oxide Ceramics

Properties of a few non-oxide ceramics, together with ZrO_2 and the most common oxide ceramics for comparison, are shown in the table below. T_m in this table is the melting point. The parameter HK refers to the Knoop hardness test with a load of 1 kg (see Fig. 5.4), and the parameter values are listed in GPa. The parameter K_{Ic} is the fracture toughness (see Chapter 5) and is obtained with a notched sample in a flexural test.

Table 10.3: Properties of some non-oxide and oxide ceramics

Material	T_m (°C)	HK (GPa)	K_{Ic} (MPa $m^{1/2}$).
SiC	2760	28	4
WC	2870	13-21	6
AlN	2250	12	3.5
Si_3N_4	1900	22	5-8
BN	2730	0.25-2	2.6
ZrO_2	2700	16	13
Al_2O_3	2050	20	4.5
mullite	1820	14.5	2.0
quartz	1720	8.2	-
fused silica	1720	5.2	-

A quick glance at the table suggests that the carbides and nitrides have very desirable properties compared to the common oxides: They tend to be more refractory (i.e. have a higher melting point), have comparable or higher hardness and generally better fracture toughness. Other properties of interest are resistance to mechanical wear and to thermal shock. The drawback of these materials is that they are more difficult to process and therefore more expensive.

Nevertheless, carbides and nitrides are under development for mechanical types of applications at high temperature, especially when large amounts of ceramic material are not required. For example, SiC can be used in furnace heating elements due to an interesting aspect of its high temperature behavior: The Si at the surface becomes oxidized to SiO_2, which has a protective function for the material underneath. SiC is also used as an abrasive, or in low-friction, low-wear mechanical seals.

In principle, pure carbon, i.e. graphite, would also be a very refractory material, since at atmospheric pressure it sublimes around 3600 °C (its triple point is said to be at around 11 MPa and 4300 °C). However, in an oxygen-containing atmosphere carbon will quickly be oxidized at high temperature, and thus it is only stable in a chemically inert environment or in vacuum. Under those conditions, it has been used as a thermal insulation material.

AlN has proved to be of interest as an electronic substrate in the semiconductor industry primarily because its thermal expansion coefficient matches closely that of Si.

WC is used most often as a ceramic phase in a metal-ceramic composite material known as a cermet. It has found application in metal cutting and forming tools. Si_3N_4 is a material with strong covalent bonding, similar to SiC, and has found use as a high-temperature structural ceramic due to its superior heat resistance, strength and hardness.

A vivid illustration of the difference an increase in fracture toughness makes for a ceramic is provided in the following figure:

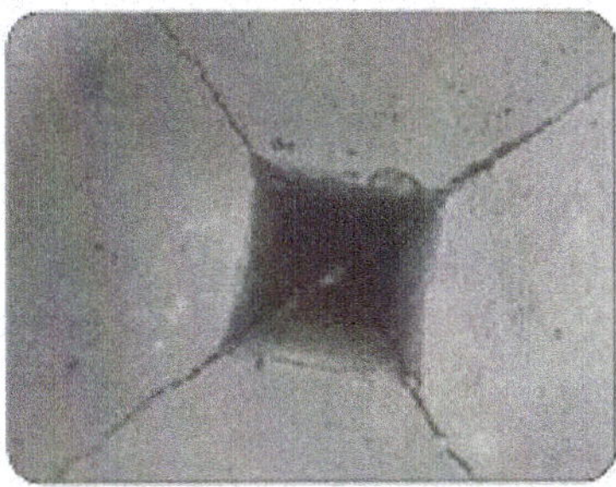

Figure 10.10: Vickers hardness tests on BN with K_{Ic} of about 3 MPa $m^{1/2}$ and toughened ZrO_2 with K_{Ic} of about 12 MPa $m^{1/2}$.

You note that BN on the left, with the lower fracture toughness, shows cracks in the corners of the indentation, as a sign of the onset of brittle fracture, whereas ZrO_2 on the right, with a four times higher fracture toughness, shows no cracks and looks as if it had been deformed plastically.

As a final reminder, it should be pointed out that especially the mechanical properties of ceramics are quite variable, and probably more so than for metals. This has to do on the one hand with the many components, thus the slightly different compositions, which may make up a ceramic material. On the other hand, the processing of ceramics, which often starts out with a powder, may give rise to variable porosity in the end product. Many commercial ceramic materials are available with varying purities, or with a range of porosities but nominally the same composition.

Image sources

Title pictures: reproduced with permission by Elsevier
Microstructures in a Eutectic CeO_2-CoO Ceramic Alloy
L. Ortega-San-Martin et al., J. Eur. Ceram. Soc. 31 (2011) 1269-1276
`http://www.sciencedirect.com/science/article/pii/S0955221910003043`

Fig. 10.1: reproduced with permission by ASM International
Preparation and Microstructural Analysis of High-Performance Ceramics
U. Täffner, V. Carle, and U. Schäfer
`www.asminternational.org/content/ASM/StoreFiles/06044G_Chapter_Sample.pdf`

Fig. 10.2: reproduced with permission by R.I. Merino and S.G. Pandalai
Melt grown composite ceramics obtained by directional solidification: structural and functional applications
R.I. Merino et al., Recent Res. Devel. Mat. Sci. 4 (2003) 1-24
`http://fmc.unizar.es/people/rmerino/papers/merino_review03.pdf`

Fig. 10.5: courtesy of L. Rubin (Corning)
`http://accuratus.com/macorfab.html`

Fig. 10.7:
`http://sembach.com/porcelain.html`

Fig. 10.9: reproduced with permission by Rath
`http://www.rath-group.com/en/industries/liquid-steel/transport-ladle/`

Fig. 10.10: reproduced with permission by C.Marsden (CoorsTek)
`http://www.azom.com/article.aspx?ArticleID=3314`

Figure for homework Problem 5:
A Simplified Model for Glass Formation
D.R. Uhlmann, P.I.K. Onorato, and G.W. Scherer
`http://adsabs.harvard.edu/full/1979LPSC...10..375U`

References

Shackelford, Smith, and Callister and Rethwisch cover similar topics but in more detail. Smith is particularly strong on engineering materials processing.

For excellent tutorials on various aspects of ceramics, you may consult

A Brevier of Technical Ceramics
`http://www.keramverband.de/brevier_engl/brevier.htm`

Overview of Refractory Materials
`http://www.pdhonline.org/courses/m158/m158content.pdf`

A vast repository of ceramics phase diagrams is at:
FACT Oxide Phase Diagrams (about 120)
`http://www.crct.polymtl.ca/FACT/documentation/FToxid/FToxid_Figs.htm`
FACT Salt Phase Diagrams (about 280)
`http://www.crct.polymtl.ca/FACT/documentation/FTsalt/FTsalt_Figs.htm`

For data sheets on commercial ceramics, go to:
`www.ceramicindustry.com/ext/resources/files/CCD-MP-Charts-4.pdf`
`http://www.coorstek.com/resource-library/library/cerprop2.pdf`
`http://global.kyocera.com/prdct/fc/index.html`

Exercises

1. With reference to Fig. 10.3, which phases are present, and in what amounts,
   a) at 700 °C and 40 mol % KCl ;
   b) at 500 °C and 40 mol % KCl?

2. The title picture with microstructures of eutectic CeO_2-CoO indicates that feature sizes are smaller at more rapid growth rates.
   a) What is the analogous phenomenon in the Fe-C system?
   b) The experimental trend can be summarized by saying that the system apparently prefers to make larger features if possible, i.e. if given enough time. How do you explain this trend?

3. Are the following glass formers or non-glass formers? Explain.
   a) SrO, b) As_2O_3, c) TiO_2, d) Li_2O.

4. Are SiO_2 and Al_2O_3 soluble in each other. If so, under what conditions?

5. The graph below shows the curve indicating the start of crystallization vs. temperature for GeO_2.

   a) What is the melting point of GeO_2 ?
   b) What is your quick evaluation for the critical cooling rate to prevent crystallization?

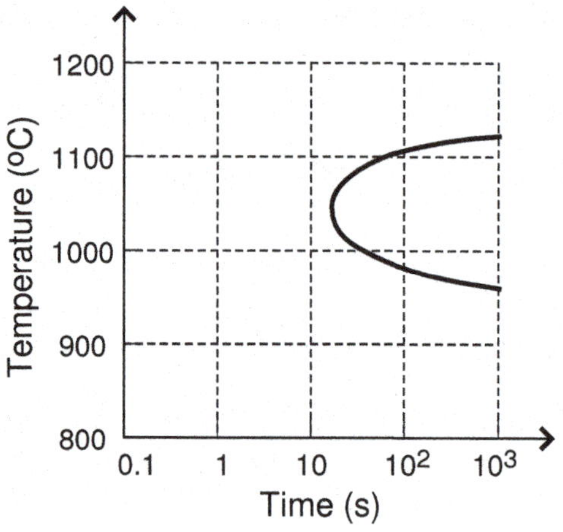

Figure 10.11: Crystallization curve for GeO_2 .

c) Determine a better estimate for the critical cooling rate by trial and error, using graphical curves for various cooling rates.

6. What is the highest service temperature of a typical fireclay brick containing 40 wt % Al_2O_3 ?

7. Illustrate the fact that clays are aluminosilicates containing water by writing the composition of kaolinite, $Al_2(Si_2O_5)(OH)_4$, as a combination of SiO_2, Al_2O_3, and H_2O.

8. Given what you know about the effects of various types of ingredients in forming ceramics or glasses, what role do you think feldspar plays in the formulation of engineering ceramics?

9. How can an aluminosilicate ceramic contain all three solid phases of the SiO_2 - Al_2O_3 phase diagram plus a glassy phase, as in Fig. 10.7 ? Doesn't something violate the Gibbs Phase rule here?

10. Why does slag float on top of molten iron?

11. The data listed in Table 10.3 for ZrO_2 are for a material containing a small amount of yttria (Y_2O_3). This addition tends to stabilize the cubic solid phase of ZrO_2.
    Briefly discuss the mechanical properties of ZrO_2 in comparison to the other oxide ceramics. What characteristics make ZrO_2 particularly attractive as a structural ceramic?

Problems

1. Compare the calculated phase diagram for CaF_2-LiF with experimental data. Comment on the accuracy of the calculation.

2. Look up the LiBr-LiI phase diagram. Describe its important features and how they are similar to, or differ from, other ceramic two-component phase diagrams you are familiar with.

3. Devise a set of rules for the miscibility of binary ceramic alloys which is similar to the Hume-Rothery rules for metal alloys. (Hint: Alternatively, think about the conditions under which two component cannot mix.)

4. You have decided that for reasons of chemical stability, in a certain process you need to use a silica brick refractory. The service temperature may reach as high as 1600 °C. At that temperature the material will contain a small amount of liquid, which is tolerable.
   What would you have to consider with respect to the composition of your material to assure its ability to withstand that temperature?

5. All other things being equal, so to speak, what would be the point of using MgO as a ceramic raw material as opposed to alumina or an aluminosilicate?

6. The graph below shows TTT-curves for lunar breccia 70019, a glass identified in lunar rock.

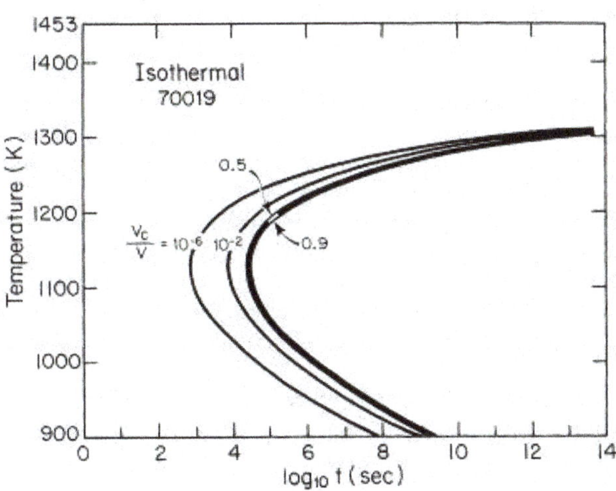

   a) What is the critical cooling rate for this material ?
   b) What kind of a glass is this ? What is its composition ?
   Ref: `http://adsabs.harvard.edu/full/1979LPSC...10..375U`

Chapter 11
Polymers

Overview

Polymers can be classified in various ways, for example according to their macroscopic structure or their mechanical and thermal behavior. At the molecular level, polymer properties depend on chemical structure, molecular geometry, and the length of the polymer molecules. The main mechanisms for polymerization are addition and condensation. These typically give rise to a material with a fairly wide distribution in molecular size, or length. Many material properties show a systematic dependence on the average molecular weight. The properties of a number of important thermoplastic polymers will be presented, and their main characteristics and common applications will be discussed.

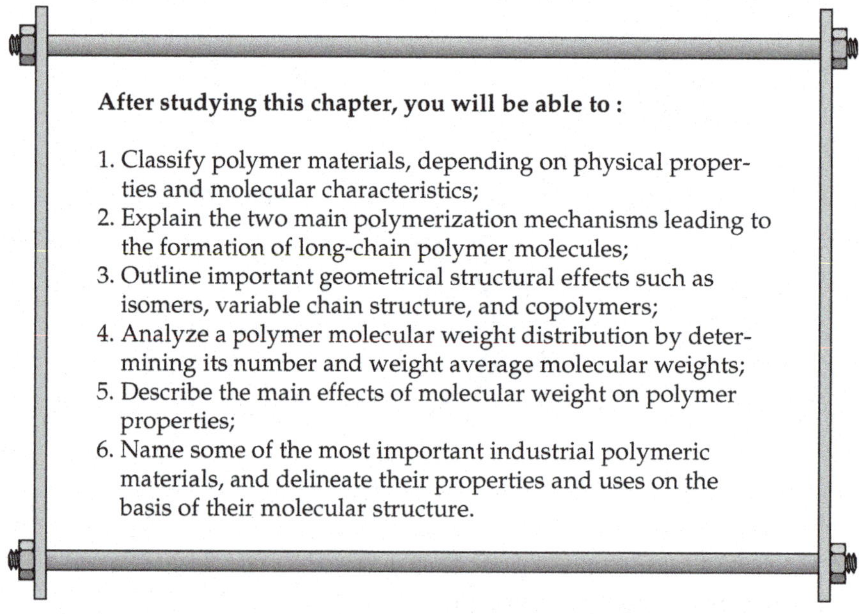

After studying this chapter, you will be able to :

1. Classify polymer materials, depending on physical properties and molecular characteristics;
2. Explain the two main polymerization mechanisms leading to the formation of long-chain polymer molecules;
3. Outline important geometrical structural effects such as isomers, variable chain structure, and copolymers;
4. Analyze a polymer molecular weight distribution by determining its number and weight average molecular weights;
5. Describe the main effects of molecular weight on polymer properties;
6. Name some of the most important industrial polymeric materials, and delineate their properties and uses on the basis of their molecular structure.

Microstructure of Poly(methyl methacrylate) – Polyethylene-Polypropylene Copolymer Blended Polymer

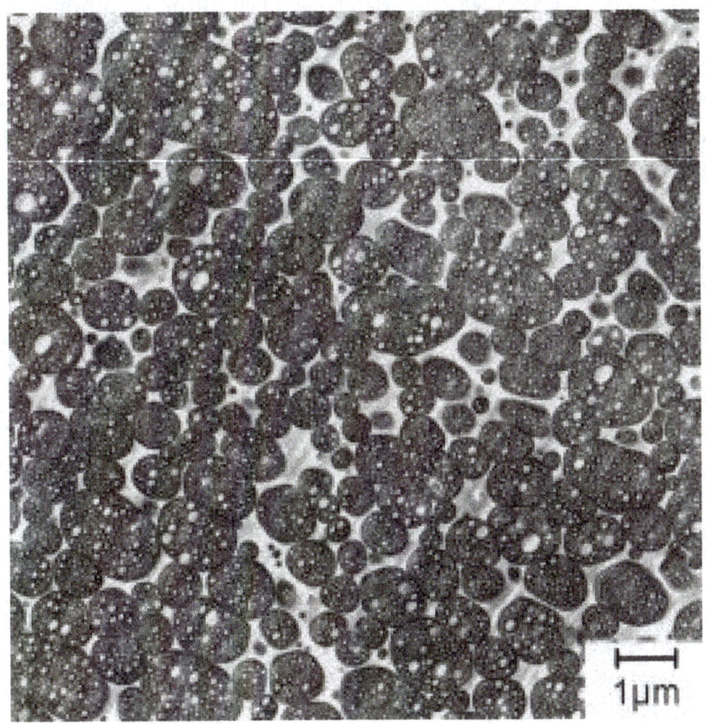

The picture displays a transmission electron micrograph of a blend of poly(ethylene-*co*-propylene) copolymer (EP) and poly(methyl methacrylate) (PMMA), obtained by polymerization of methyl methacrylate (MMA) in an EP/MMA solution. The microstructure shows dark EP domains in a white PMMA matrix. The dark domains have white inclusions themselves. The point of this is that PMMA is a strong but rather brittle material. The EP/PMMA blend has much improved toughness.

11 Polymers

In our prior discussion we have touched on the subject of polymers in two places. In Chapter 3 we focused on structure, crystalline or otherwise, in relationship to the basic building blocks of polymers, which are long chain molecules. In Chapter 5 we discussed mechanical properties of polymers as they are determined by the structure of the material. The purpose of this chapters is two-fold. First, we wish to fill in a few details and add a few new points with respect to various aspects of the structure of polymer molecules and materials. And second, we will point out some engineering applications of polymeric materials.

11.1 Classification of Polymers

Since polymers represent a class of materials with widely varying properties, they can be classified in many ways. A traditional classification is according to their macroscopic behavior and appearance. Another way would be according to the chemical type of monomer which provides the basic building block of the polymer chain molecules. These two points of view have been combined in the following graph:

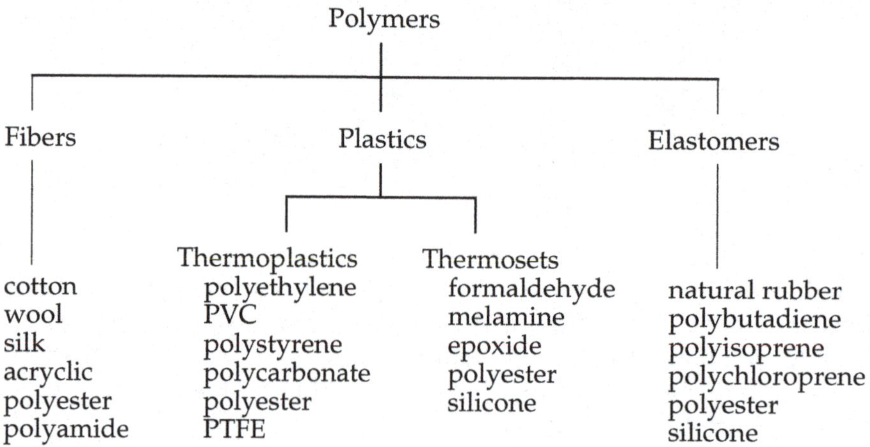

Figure 11.1: Traditional classification of polymers into three groups, with examples differing by chemical structure.

The three main categories are fairly self-evident. Fibers are polymers which have been processed such that they their length dimension is greater by several orders of magnitude than their dimension across. This gives rise to very anisotropic properties along their length and in cross-section.

Plastics can be processed easily into a suitable shape and are able to maintain that shape. For **thermoplastics**, the processing usually requires heating to soften the material and pressure to shape it. Moreover, this sequence of steps can be repeated many times without any deterioration of the material properties. By contrast, **thermosets**, after an initial forming process, undergo chemical reactions such that they cannot be softened and reformed later on.

Elastomers have the unique property that they their range of elastic deformation is very large at around room temperature. They can be stretched to many times their original length, yet they are able to recover completely their original dimensions.

You will note in Fig. **??**, in the category of fibers, that both natural polymers (the first three) and man-made ones are listed. Also, a given chemical type can show up in more than one category, for example polyesters or silicones. The reason is mainly structural, as you will be able to appreciate having studied Chapters 3 and 5: Thermoplastics are linear and lightly branched polymers, thermosets are heavily cross-linked to the point of being amorphous 3D networks, and elastomers are lightly crosslinked. (We will describe the chemical details of some of the listed polymers later in this chapter).

Our discussion here will follow an alternative classification scheme which is based on molecular characteristics of the polymer chain molecules. The scheme is outlined in Fig. **??**:

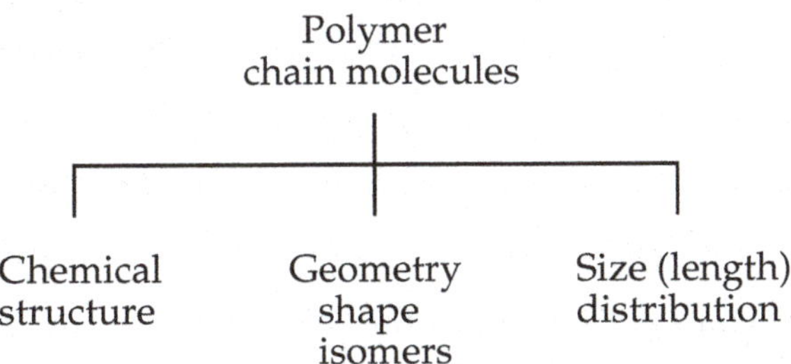

Figure 11.2: Classifying polymers according to characteristics of the chain molecules.

We will examine the various molecular features listed in Fig **??** and how they affect polymer properties, focusing on a few illustrative examples, but without being exhaustive. It should also be clear that the features listed in Fig **??** are not entirely independent. For example, the chemical structure of a polymer molecule often has an important effect on its geometrical properties.

11.2 Chemistry of Polymer Molecules

We introduced a few polymer molecules in Chapters 3 and 5 on structure and mechanical properties. Now we will add a few details about how such molecules are formed and how the polymerization process may inform the structure of the resulting molecules.

11.2.1 Polymerization by Addition: Polyethylene

The simplest such case is the formation of polyethylene from ethylene by polymerization by addition. Even this case is fairly complicated: It is illustrated schematically in Fig. ??

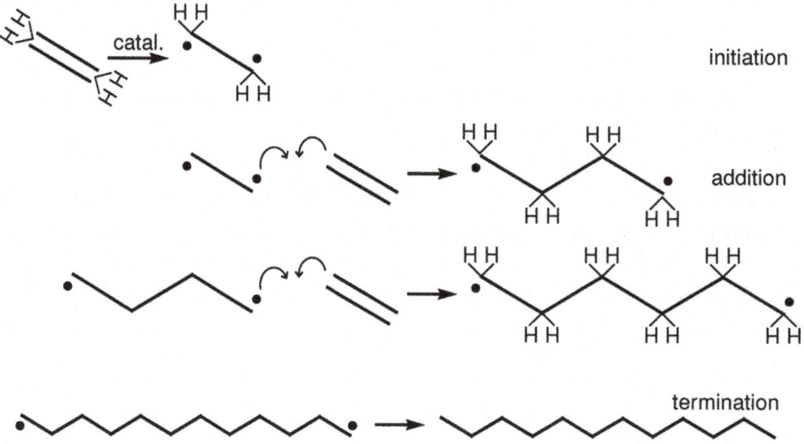

Figure 11.3: Schematic of the three-step reaction mechanism for forming polyethylene from ethylene. Note that all H atoms are not shown, especially in the molecules for the last step.

The reaction begins with the ***initiation step***, in which a catalyst causes one of two bonds in the double bond of ethylene to break open, thus producing a reactive $\cdot\, C_2H_4 \cdot$ intermediate. In the next ***step of addition***, this intermediate pulls an electron from the double bond of another ethylene molecule to form a single bond between the two C_2H_4 species, thus creating a new reactive intermediate, $\cdot\, (C_2H_4)_2 \cdot$, consisting essentially of two ethylene molecules. This addition process can go on for many steps, until eventually it undergoes a ***termination step*** by addition of H atoms to the ends of the (very long) reactive intermediate. The end result is a polyethylene molecule with the structure $CH_3(C_2H_4)_{n-1}CH_3$ and the formula $C_{2n}H_{2n+2}$.

The reaction chemistry of this sequence of events is quite complicated. What is notable for us is that there are about ten different catalysts used for the initiation step, and there are several ways for termination to occur. The catalyst used has important consequences for the reaction process and the final material, as is indicated briefly in Table 11.1:

Table 11.1: Main industrial processes for making polyethylene

	LDPE	**HDPE**
Temperature (°C)	200	60
Pressure (atm)	2000	a few
Catalyst	peroxide	Ziegler-Natta, metal
Molecular structure	some branching	~ no branching

The original process involved a peroxide catalyst and high pressure and temperature. It also allowed for long enough intermediates to fold back onto themselves so that a reactive chain end could abstract an H atom from somewhere along the chain, thus creating a reactive locus in the middle. This enabled a certain amount of branching to occur.

In the Ziegler-Natta process, invented by K. Ziegler and G. Natta in the 1950s and honored with the Nobel prize in Chemistry in 1963, the reaction occurs at much milder conditions and, most importantly, produces very long molecules with almost no branching. This facilitates crystallization and thus higher density which, as you know, has benefits for material properties.

Other linear polymers, such as polypropylene, poly(vinyl chloride), and polystyrene, are also made by addition polymerization.

11.2.2 Polymerization by Condensation: Polyester

Another important method of polymerization is polymerization by condensation. This is used, for example, in making polyester. In the simplest case, the method consists of joining two molecules by abstraction of an H_2O molecule from them. It begins with the step illustrated schematically in Fig. **??**:

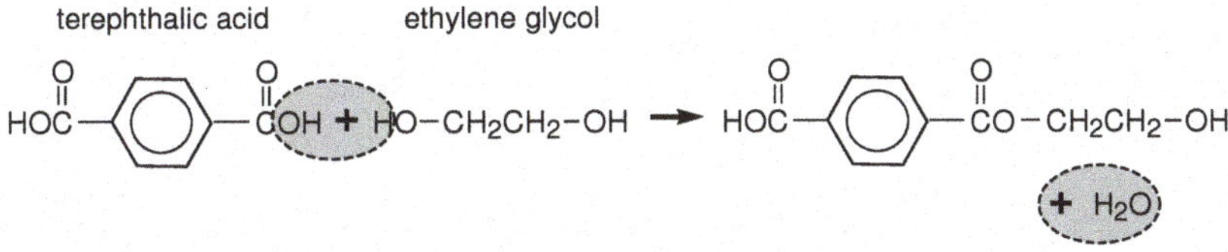

Figure 11.4: Simplified mechanism for forming an ester from ethylene glycol and terephthalic acid.

A water molecule is removed as indicated on joining a terephthalic acid molecule A and ethylene glycol B. The result is an ester molecule A-B, ethylene terephthalate, which has a COOH group at one end and an ethylene with an OH group at the other end, thereby combining functionalities of ethylene glycol and terephthalic acid. As a next step, two ester molecules combine, again by removal of an H_2O, to form a longer molecule A-B-A-B, etc. The resulting polymer molecule is called poly(ethylene terephthalate), or PET for short.

The process above yields linear molecules. These are the basis of polyester fibers. On the other hand, if one starts the polymerization with an alcohol with three rather than just two OH groups as in ethylene glycol, it is possible to make cross-linked PET. This is the raw material for soft drink bottles.

A second type of polymer produced with a similar condensation mechanism are the polyamides, commonly known as nylon. For this type of molecule the reaction begins with an acid A a COOH group at each end and a diamine B with an NH_2 group at each end. Condensation connects the acid and the diamine to form A-B via removal of H_2O (H from the NH_2 + OH from the COOH). Units A-B can then add in the same fashion as in the case of the polyester to form long chains -A-B-A-B-A-B- .

11.3 Geometrical Aspects of Polymer Molecules

An important geometrical effect comes arises with molecules such as polypropylene, where the monomer has a unit other than H atoms attached to the C atoms. In analogy to polyethylene, you can visualize the polymerization of propylene as in Fig. **??**.

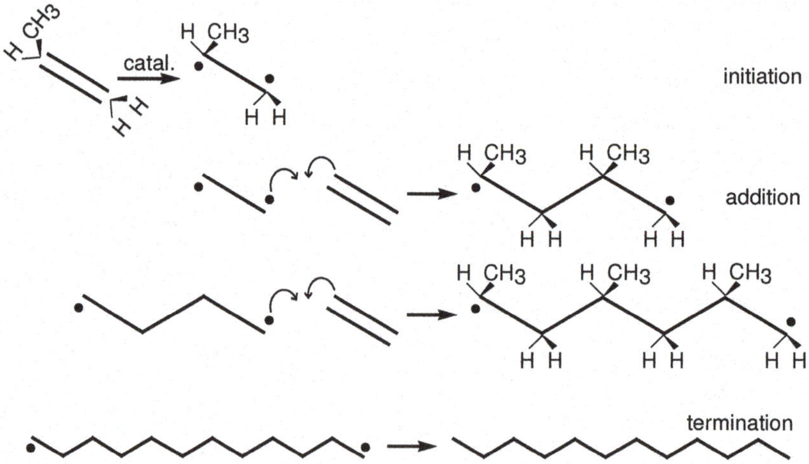

Figure 11.5: Schematic of three-step reaction mechanism for forming polypropylene. All H atoms and CH_3 groups are not shown.

Think of the figure as displaying side views of the polymer chains. Now it is important to realize that the CH_3 groups of polypropylene are shown in a particular configuration, namely all pointing towards the viewer. In other words, the CH_3 groups are ***in front*** of the zigzag plane of the polymer backbone, whereas the associated H atoms are ***behind*** the zigzag plane.

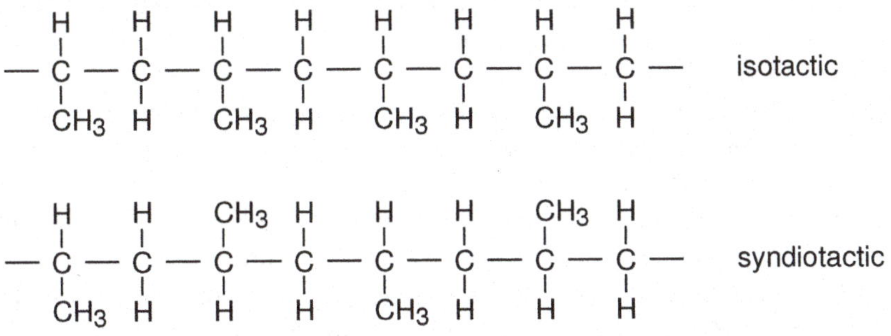

```
   H     H     H     H     H     H     H     H
   |     |     |     |     |     |     |     |
 — C  —  C  —  C  —  C  —  C  —  C  —  C  —  C —      isotactic
   |     |     |     |     |     |     |     |
  CH3    H    CH3    H    CH3    H    CH3    H

   H     H    CH3    H     H     H    CH3    H
   |     |     |     |     |     |     |     |
 — C  —  C  —  C  —  C  —  C  —  C  —  C  —  C —      syndiotactic
   |     |     |     |     |     |     |     |
  CH3    H     H     H    CH3    H     H     H
```

Figure 11.6: Top view of two possible configurations of polypropylene.

Another view of polypropylene molecules, from above, as if they were flattened onto a plane, is shown in Fig. **??**. The two configurations are named **isotactic** and **syndiotactic**. Isotactic is the most regular arrangement, in that all CH_3 groups are on the same side of the backbone. (The prefix "iso" means "the same", as it does in other expressions such as isotropic or isotherm. The word "tactic" refers to "arrangement"). Syndiotactic means that the CH_3 groups alternate with respect to which side of the backbone they are on. There is a third possibility named **atactic**, not shown, where the order of the CH_3 groups is random along the backbone. (Here the prefix "a-" means "without", as in amorphous or asymmetric).

The two configurations in Fig. **??** clearly look quite different. However, there is only a subtle difference in the way they are put together (Fig. **??**):

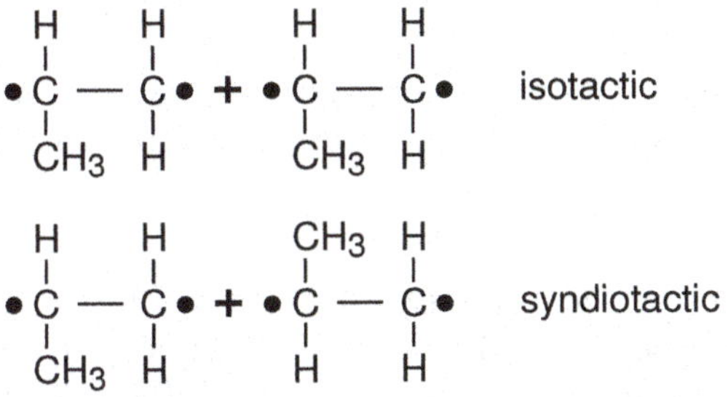

```
   H     H        H     H
   |     |        |     |
 • C  —  C • + •  C  —  C •     isotactic
   |     |        |     |
  CH3    H       CH3    H

   H     H       CH3    H
   |     |        |     |
 • C  —  C • + •  C  —  C •     syndiotactic
   |     |        |     |
  CH3    H        H     H
```

Figure 11.7: Configuring isotactic and syndiotactic polypropylene.

The only difference between the two reactions is that in the syndiotactic case the second mer is rotated 180° around the molecular axis, but it is really the same radical CH_3CHCH_2 as the other three. The three forms of polypropylene – isotactic, syndiotactic, and atactic – are called **isomers** of polypropylene: They consist of identical mers, but the mers are put together differently.

The most interesting thing about these isomers is that they exist at all! You might think that in the addition step of the polymerization the orientation of the mers would be random, and thus the prevailing geometry of the polymer would be atactic. The magic of achieving something regular lies with the catalyst. In fact, the right catalyst allows to form mostly either the isotactic or the syndiotactic form.

The distinction between the three isomers is not just academic: They have rather different properties. First, isotactic polypropylene crystallizes quite easily, but atactic polypropylene does not crystallize at all, which of course gives these materials rather different mechanical properties. In addition, the melting points of the three isomers are about 165 °C, 130 °C, and <0 °C.

Another example of various degrees of geometrical order is observed with so-called **copolymers**, i.e. polymers consisting two or more chemically different mers (Fig. **??**):

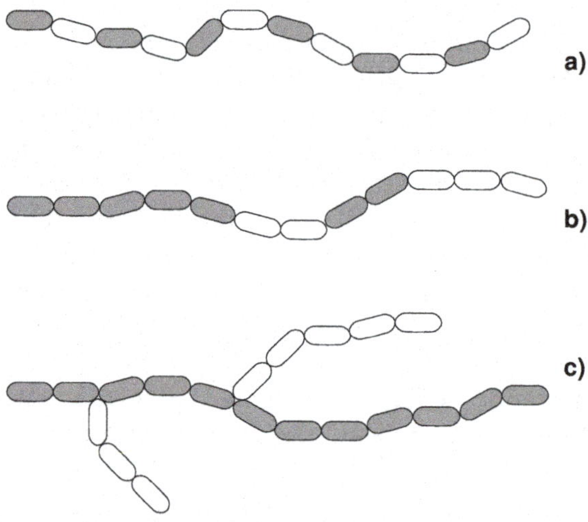

Figure 11.8: Various geometrical configurations of copolymers. Grey and white ovals represent two different mers. The configurations are: a) alternating, b) block, and c) graft copolymers.

Again the question is in what order the mers are put together. The configurations displayed in Fig. **??** are: a) alternating, b) block, and c) graft copolymers. And again, there is a fourth configuration, not shown, in which the two mers are arranged randomly. These copolymers may have different microstructures, depending on how the different parts of the molecules interact with each other. Many synthetic rubbers are examples of copolymers.

11.4 Size Distribution of Polymer Molecules

When you consider the mechanism of polymerization by addition, as with polyethylene, it should be evident that shortly after the initiation of the reaction, reactive intermediates of various sizes, i.e. various lengths, are present. These can combine with unreacted monomers or with each other randomly, independent of size. As a result, after termination the material will contain molecules with a wide variety of sizes (i.e. lengths). This implies that a full characterization of the final material requires knowing the size distribution of the polymer molecules. In fact, this is how commercial materials are typically sold: with a specified size distribution.

The size distribution of a typical polyethylene material is shown in Fig. ??:

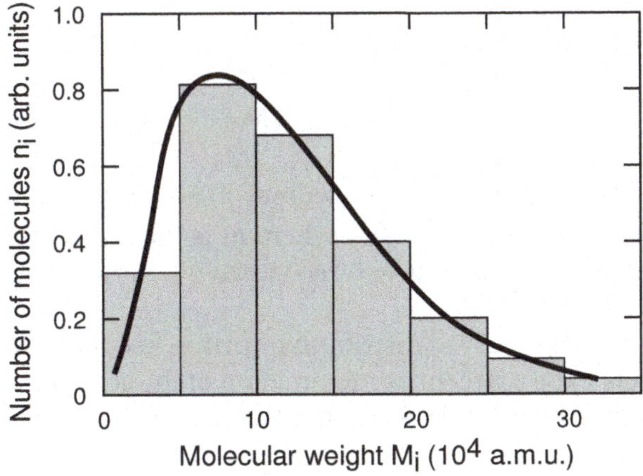

Figure 11.9: Typical size distribution of a polyethylene material. The bar graph is an approximation of the continuous curve.

Please be aware that the distribution is quite broad, extending over about an order of magnitude for the molecular sizes. (In general, it would be difficult and expensive to produce material with a much narrower size distribution, either during the reaction itself, or after the fact by a size selection method.)

The size distribution is defined as the number of molecules, n_i, as a function of their molecular weight, M_i (really, molecular mass). The index i designates a specific size, and you can think of it as the number of mers in the molecule. You will note that, strictly speaking, n_i is a discrete variable, although the function $n_i(M_i)$ in Fig. ?? has been drawn as a continuous curve.

If we knew $n_i(M_i)$ explicitly, that would be a sufficient description of our sample, but often we may not know $n_i(M_i)$. In that case, and also for the purpose of a quick comparison, a first useful parameter to know is the average molecular weight $\overline{M_n}$, defined as follows:

$$\overline{M_n} = \frac{\sum_i^\infty n_i M_i}{\sum_i^\infty n_i} \tag{11.1}$$

The index i runs over all molecule sizes. (The limits of the summation will always be the same in equations similar to Eq. **??**. So in the future we will not specify them.) Eq. **??** represents the usual way of defining an average over a distribution. In the present context we will call $\overline{M_n}$ the **number average** molecular weight.

It is also convenient to define the number fraction, x_i, of molecules i among the total number of molecules as

$$x_i = \frac{n_i}{\sum n_i} \tag{11.2}$$

which allows us to rewrite Eq. **??** as

$$\overline{M_n} = \sum x_i M_i \tag{11.3}$$

because by definition, $\sum x_i = 1$. Sometimes is is also useful to speak of the **degree of polymerization,** $n_n = \frac{\overline{M_n}}{m}$, where m is the molecular weight of the mer of the chain molecules. You will recognize n_n as the average number of mers in a polymer chain.

In order to evaluate the size distribution further, we will use the bar graph in Fig. **??**. It seems to be a decent approximation of the continuous curve. From the bars we can read off the following numbers:

Table 11.2: Data for bar graph form of the size distribution

n_i	x_i	$10^{-4}M_i$	$10^{-4}x_iM_i$	w_i	$10^{-4}w_iM_i$
0.32	0.125	2.5	0.31	0.026	0.06
0.82	0.320	7.5	2.40	0.198	1.49
0.68	0.266	12.5	3.32	0.274	3.43
0.40	0.156	17.5	2.73	0.226	3.95
0.20	0.078	22.5	1.76	0.145	3.27
0.10	0.039	27.5	1.07	0.089	2.44
0.04	0.016	32.5	0.51	0.042	1.36
			$\overline{M_n} = 12.1 \times 10^4$		$\overline{M_w} = 16.0 \times 10^4$

From the numbers in the first column, $\sum n_i = 2.56$. This in turn determines the numbers in column 2, as per Eq. ??. $\overline{M_n}$ is the sum of the numbers in column 4, and is equal to 12.1. You can convince yourself that the numbers in column 2 are proper fractions, in that their sum is equal to 1.

You will also have noticed two additional columns in Table 11.2. It turns out that Eq. ?? is not the only way to take an average. As its name says, $\overline{M_n}$ is the number average of the molecular weight. This is because in the calculation of $\overline{M_n}$, the average is taken according to the number of molecules in each bar of the distribution. It is also possible to calculate another average, $\overline{M_w}$, called weight average molecular weight, by using the weights of the molecules in each bar instead of their number.

To this end, one can define weight fractions w_i as

$$w_i = \frac{x_i M_i}{\sum x_i M_i} = \frac{x_i M_i}{\overline{M_n}} \tag{11.4}$$

and then the **weight average** molecular weight $\overline{M_w}$ as

$$\overline{M_w} = \sum w_i M_i \tag{11.5}$$

If you compare Eqs. ?? and ??, you see that their structure is exactly the same, namely that of taking an average. What is different is the "weighing factors": They are either number fractions x_i or weight fractions w_i. Incidentally, you can check that the w_i are a legitimate set of fractions by verifying that their sum is also equal to 1, just as for the x_i.

From Table 11.2 it is apparent that $\overline{M_w} > \overline{M_n}$. This turns out to be true no matter what exactly the shape of the distribution is. The ratio of $\overline{M_w}$ / $\overline{M_n}$ is known as the **polydispersity index** PDI. It provides a rough indication of the width of the distribution. In our case, we find that $\overline{M_w}$ / $\overline{M_n}$ = 1.32.

The PDI approaches the lower limit of 1.0 for special polymers with very narrow molecular weight distributions. The PDI is equal to 1.0 if the distribution consists of a single bar (see homework). For typical commercial polymers obtained by a step reaction (condensation) mechanism, the PDI is said to be ~2. Chain reactions (addition) are said to yield PDIs from 1.5 - 20.

As a final note, you can easily check that $\overline{M_w}$ can be written in a form equivalent to Eq. ??, namely as

$$\overline{M_w} = \frac{\sum x_i M_i^2}{\sum x_i M_i} \tag{11.6}$$

The number average molecular weight $\overline{M_n}$ is sometimes referred to as the first moment of the distribution, and the weight average molecular weight $\overline{M_w}$ as the second moment. This is related to the power with which M_i appears in the corresponding formulae, Eqs. ?? and ??.

11.5 Material Properties and Applications

At this point you may well ask what this business of polymer geometry, size distributions and average molecular weights has to do with real materials. We did touch briefly upon some of these connections earlier, when we noted that mechanical properties of polymers do depend on the size of the molecules and on the material's crystallinity, which itself depends on the geometry of the molecules. Here we will fill a few more details of interest.

11.5.1 Molecular Weight Effects

The first point to note is that a wide variety of polymers show a universal qualitative behavior of certain mechanical properties with respect to the size and size distribution of the molecules in the material.

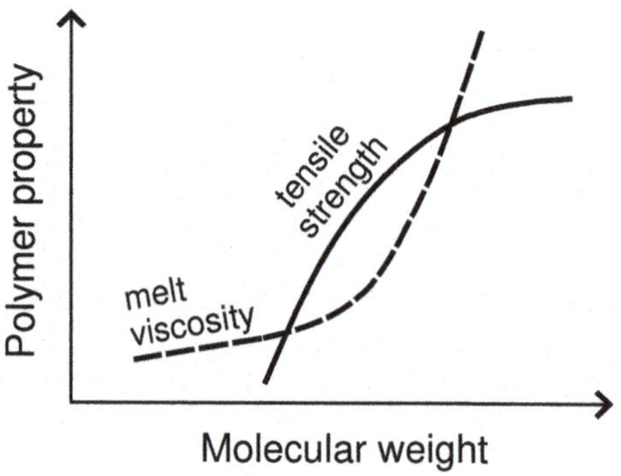

Figure 11.10: Qualitative dependence of mechanical property parameters on average molecular weight.

Properties such as the elastic modulus and the tensile strength tend to show a marked increase in a certain range of intermediate molecular weights and level off at very large $\overline{M_n}$. On the other hand, the melt viscosity measured at low strain rate increases only modestly at low molecular weights, but more rapidly above a certain threshold value of $\overline{M_n}$.

This behavior has immediate practical consequences for polymer processing. A lot of this requires flow and forming operations of the material at elevated temperature, where the viscosity is an important parameter. The figure suggests that the larger $\overline{M_n}$, the more difficult a material will be to process. In addition, when it comes to parameters related to material strength, it appears that there is a point beyond which increasing $\overline{M_n}$ brings little benefit.

The curve of viscosity η vs M_n has an interesting shape in itself. For many different polymers it has been found that there is a critical value M_n such that

$$\eta = K M_n \quad \text{for} \quad M_n < M_c \tag{11.7}$$

$$\eta = K' {M_n}^{3.4} \quad \text{for} \quad M_n > M_c \tag{11.8}$$

In other words, below M_c the viscosity η varies linearly with M_n, and above M_c as a power law with an exponent of 3.4. (This work usually involved material with a narrow size distribution, so that $M_n = \overline{M_n}$).

The physical reason for this behavior is that the rapid rise of tensile strength and η are due to strong entanglement of the polymer chains. Entanglement limits the motion of polymer chains relative to each other, and in order for entanglement to take place to an appreciable extent, the chains have to have a certain minimum length.

Another parameter which shows interesting behavior as a function of molecular weight is the glass transition temperature T_g. This should not come as a surprise, since T_g is also related to the ease of chain motion, and below T_g such motion is not possible. Data of two investigative groups for polystyrene are displayed in the figure below:

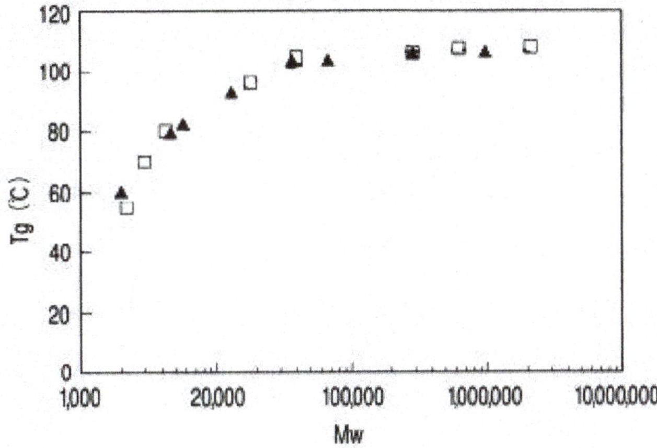

Figure 11.11: T_g of polystyrene as a function of molecular weight.

The data are for practically monodisperse polymer samples, with a PDI of around 1.1. Note the logarithmic scale for M_w, which under these conditions is essentially the same as $\overline{M_w}$ and $\overline{M_n}$. The function $T_g(M_w)$ has the same shape as the mechanical strength properties in Fig. ?? above: T_g rises in the low-M_w range and levels off at high M_w.

For materials with a narrow size distribution, Fox and Flory showed that the behavior of $T_g(M_n)$ data was described quite well by a simple expression, which is now known as the Fox-Flory law:

$$T_g = T_{g,\infty} - k/M_n \tag{11.9}$$

In Eq. ?? the parameter k is an empirical constant characteristic of a particular material, and $T_{g,\infty}$ is the limiting value of the glass transition temperature for very large M_n. (You will figure out in a homework how good this expression is.)

11.5.2 Material Properties and Uses

Polyethylene

We begin by examining some properties of polyethylene, the simplest polymer. This material is produced in different grades. The two main ones are: **LDPE** for low-density polyethylene, and **HDPE** for high-density polyethylene. LDPE consists of shorter polymer chains with a considerable amount of branches which themselves may be branched; HDPE consists of longer chains with almost no branching. There is also LLDPE, which stands for linear low-density polyethylene, with much shorter branches than LDPE, as well as UHMW-PE, which stands for ultra-high molecular weight polyethylene.

A summary of their mechanical properties is given in Table 11.3 below:

Table 11.3: Material parameters of different types of polyethylene

Material	$\overline{M_n}$ (a.u.)	ρ (g/cm^3)	E (GPa)	T.S. (MPa)	ϵ_f	Izod (kJ/m)
LDPE	20-50k	0.92	0.25	10	4	no break
LLDPE	20-100k	0.93	0.3	20	5	>1
HDPE	200-500k	0.96	1.2	30	1.5	0.1
UHMW-PE	2-5M	0.975	0.55	40	3	no break

The parameters in Table 11.3 are mostly the usual ones for mechanical properties: E is the elastic modulus, T.S. the tensile strength, and ϵ_f the elongation at fracture. Izod refers to the so-called Izod impact strength from a test measuring toughness under impact conditions. The Izod test is similar to the Charpy test described in Chapter 9: Izod also uses a notched sample of fixed dimensions, but with a different geometry than Charpy.

The data in the table have been assembled from several industrial sources. It should be clear that the numbers for each type of polyethylene are representative averages. The numbers for $\overline{M_n}$ are to be taken as broad ranges without sharply defined boundaries. The variation in reported values of other parameters seems to be fairly large, especially for impact strength. Some sources report an Izod strength for LDPE and UHMW-PE of >1kJ/m.

Thus the numbers should be used only for rough correlations. What we do know is that they will depend on material details such as the degree of polymerization, the molecular weight distribution, and the processing of a material. For example, it is important whether a material has been cast into block form, extruded into a more complex shape, or drawn into a film or fiber.

Table 11.4. lists a few of the common uses of the various types of polyethylene. It is understood that the list could be twice as long.

Table 11.4: Common uses of polyethylene

Material	Common uses
LDPE	bags, food and textile packaging, electrical insulation
LLDPE	packaging, containers, pipes, tubing
HDPE	milk bottles, food packaging, containers
UHMW-PE	bottles, containers, gears, guides, dump trailer linings

Additional properties of interest for polyethylene are: good flexibility to temperatures as low as -73 °C, excellent corrosion resistance, excellent electrical insulation, low permeability for water.

Polypropylene, Poly(vinyl chloride), and Polystyrene

The next group of polymer materials we will discuss is the one originating from compounds with the formula $CH_2{=}CHR$, yielding the mer $CH_2{-}CHR$, where R is not an H atom but something else. Here R can be either a methyl group (CH_3) for polypropylene, or a Cl atom for poly(vinyl chloride), or a phenyl group (C_6H_5, essentially a benzene ring) for polystyrene. We will use the abbreviations PP, PVC, and PS for them.

In comparison to polyethylene, one can say that all four materials consist of the the same kind of polymer chain, but the latter three have a side group, something sticking out to the side. Side groups in a polymer chain will inhibit bond rotation and thus will render the chain effectively stiffer. In addition, polymer chains will not uncoil or slide passed each other as easily. These molecular characteristics are reflected in the mechanical properties of the materials in Table 11.5 below.

Again, the values for the mechanical parameters should be taken as broad averages, give or take 25 % for E and T.S., a factor 2 for ϵ_f, and even more for the Izod impact strength. These three materials are available commercially in different grades, differing in average molecular weight, but $\overline{M_n}$ or $\overline{M_W}$ are usually not specified.

The table has two entries for PVC. The rigid type refers to pure PVC, which is rather brittle and thus of somewhat limited use. It can be made more ductile, without substantial loss of strength, by blending it with a plasticizer compound. Other additives may be used as well with PVC, such as lubricants, fillers, or colorizing pigments.

Table 11.5: Material parameters of PP, PVC, and PS

Material	ρ (g/cm^3)	E (GPa)	T.S. (MPa)	ϵ_f	Izod (kJ/m)
PP	0.90	1.4	3.6	2.3	0.2
PVC (rigid)	1.45	3.2	4.6	0.6	0.1
PVC (plasticized)	1.25	-	1.7	3	var.
PS	1.05	3.4	40	0.02	0.02

The side groups also have a marked effect on the tendency of these compounds to crystallize. The general rule is that more regular the chain, the easier it is to form crystals. In practice it turns out that PVC does not crystallize at all, nor do the atactic forms of PP and PS. However the iso- and syndiotactic forms of PP and PS can crystallize. Table 11.6. summarizes a few main characteristics and common uses of PP, PVC, and PS.

Table 11.6: Main characteristics and common uses of PP, PVC, and PS

Material	Characteristics	Common uses
PP	chemically inert, excellent heat stability, sensitive to UV light	containers, bottles, carpets, rope, straps, auto fixtures
PVC (rigid)	low-cost, moderate heat stability	pipe, siding, window frames
PVC (plasticized)	low-cost, moderate heat stability	furniture, luggage, auto trim, appliance parts, flooring
PS	good heat stability, high optical transparency	appliance parts and housings, auto interior trim

As with metals and ceramics, a major issue for the application of polymers is the trade-off between strength and toughness or impact strength. We will mention two methods by which this trade-off can be engineered to some extent for polymers. The first example concerns the ***toughening of polystyrene*** by incorporating an elastomer during polymerization. The relevant data are presented in Table 7:

Table 11.7: Toughening polystyrene

Material	ρ	E	T.S.	ϵ_f	Izod
	(g/cm^3)	(GPa)	(MPa)		(kJ/m)
PS (gen. purpose)	1.05	3.4	40	0.02	0.02
PS (high impact)	1.04	2.1	20	0.45	0.14

You can see the large improvements in ductility and impact strength, with a reduction of the tensile strength by only a factor 2. The way this toughening is achieved is made clear by the following transmission electron micrograph:

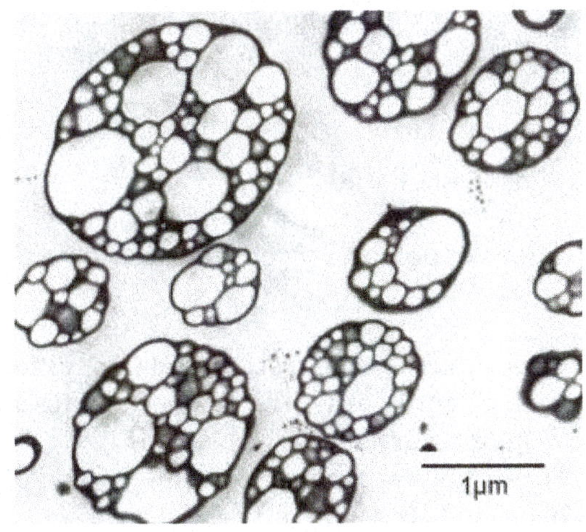

Figure 11.12: Microstructure of HIPS: Polystyrene toughened by inclusion of an elastomer component. The dark parts are elastomer, and the light parts are amorphous PS.

HPIS is in effect a composite, and the rubberized inclusions provide the toughening, presumably by deflecting and arresting cracks which start to grow in the amorphous PS matrix.

It is also worth mentioning that most of our discussion so far has been about **homopolymers**, i.e. materials consisting of molecules that contain only one kind of mer. However, many may also be used as part of copolymers, for the purpose of improving or fine-tuning material properties. A good example is polystyrene copolymerized with acrilonitrile to form ***styrene-acrylonitrile*** (SAN). Acrylonitrile is also of the form $CH_2{=}CHR$, but with $R = C{\equiv}N$, yielding the mer $CH_2{-}CHCN$. SAN is a random, thus amorphous, copolymer of styrene and acrylonitrile, giving a material with better chemical resistance, higher use temperature, and better toughness than polystyrene.

A compact representation of the range of properties over which polypropylene and some related polymers can be engineered is given in Fig. **??** below:

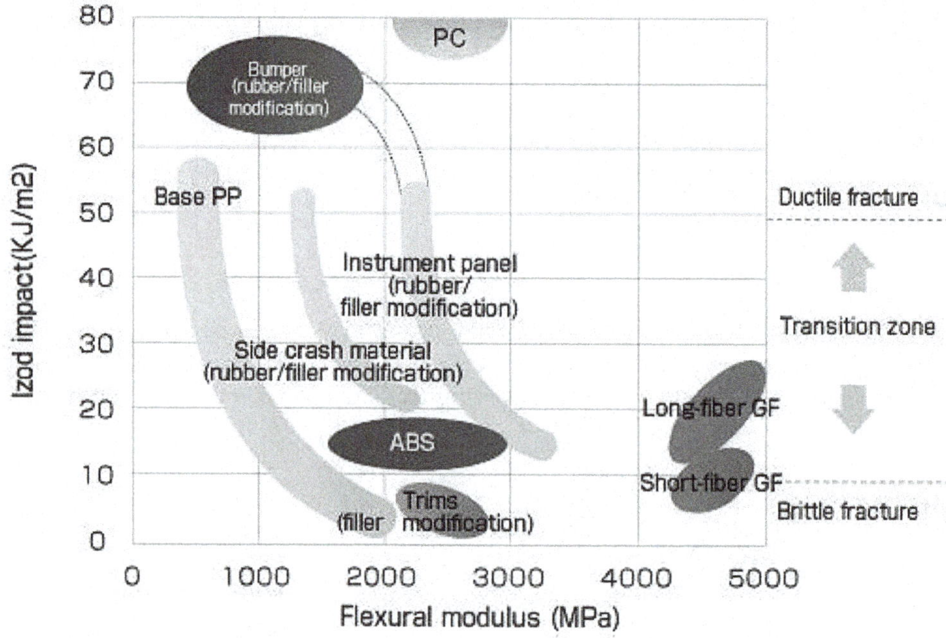

Figure 11.13: Automobile applications of polypropylene and a few other polymers.

The flexural modulus in Fig. **??** is similar to the elastic modulus, except it is obtained in a bending rather than a tensile test. The figure includes PP modified with filler and reinforced with glass fibers (GF), as well as two other polymers we have not discussed: PC for polycarbonate (see below), and ABS (a copolymer of acrylonitrile, butadiene, and styrene).

Other Polymers

The four major homopolymers we have examined to this point – polyethylene, polypropylene, poly(vinyl chloride), and polystyrene – constitute the most important thermoplastics in terms of amounts, or tonnage, produced worldwide. However, many other materials are of interest and in wide use. A few of those, which we use frequently, perhaps without realizing it, are listed in Table 11.8 below:

Table 11.8: Selected Thermoplastics

Material	Chemical Repeat Unit (Mer)
Polyester (Poly(ethylene terephthalate) PET)	$-\overset{O}{\overset{\|\|}{C}}-C_6H_4-\overset{O}{\overset{\|\|}{C}}O-CH_2CH_2-O-$
Polycarbonate	$-O-\overset{O}{\overset{\|\|}{C}}-O-C_6H_4-C(CH_3)_2-C_6H_4-$
Polyacrylic (Poly(methyl methacrylate) PMMA: Plexiglas, Lucite)	$-CH_2-C(CH_3)(C{=}O\,OCH_3)-$
Polytetrafluoroethylene (PTFE: Teflon)	$-CF_2-CF_2-$
Polysiloxane (Poly(dimethyl siloxane), PDMS: Silicone)	$-Si(R_1)(R_2)-O-$ $-Si(CH_3)_2-O-$

The polyester example is poly(ethylene terephthalate), which we used earlier in this chapter to illustrate polymerization by condensation (cf. Fig **??**). The example for polyacrylics is poly(methyl methacrylate), or PMMA for short, which you will probably know as Plexiglas®. (Note the German spelling of -glas for the english -glass). PTFE (polytetrafluoroethylene) is well-known as Teflon®. All of these as well as the carbonates are thermoplastics.

The last one is poly(dimethylsiloxane), the simplest example of the class of polysiloxanes, Si-based organic polymers where R_1 and R_2 stand for generic side groups. The interesting thing about these compounds is that their polymer backbone is not a chain of C atoms but a chain of –Si–O– . These are thermoset polymers. You may wish to reconsider Figs. 3.23 and 3.25 to see the similarity in structure of the –Si–O– and –C–C– chains. (For example, focus on the (0 0 1) plane and the [1 1 0] direction in it).

Table 11.9 displays a very brief summary of some of the main characteristics and uses of this set of polymers:

Table 11.9: Main characteristics and uses of polymers in Table 11.8

Material	Characteristics	Common uses
PET	chemically inert, moisture resistant	food packaging, film, fiber for cords
Polycarbonate	high toughness, high impact strength	safety shields, helmets, appliance and computer housings
PMMA	good strength, chemical resistance, highly transparent	shields, housings for lights, safety goggles
PTFE	high chemical resistance, temperature stability, low strength, high toughness	chemical piping, high temp. electrical insulation, nonstick coatings, sealing tape, gaskets
PDMS	highly hydrophobic, very low T_g	grease, lubricants, caulking, microphotonics, microfluidics

It should be understood that, as always with polymers, properties and uses will vary widely within a category of material, depending on the molecular weight and the detailed chemical structure. For example, silicones with low molecular weight are used as components of fluids, but with high molecular weight or cross-linked as polymeric solids.

Image sources

Title picture: reproduced with permission by Elsevier
Microstructure of Poly(methyl methacrylate) – Polyethylene-Polypropylene Copolymer Blended Polymer
L. Leibler, Prog. Polymer Sci. 30 (2005) 898-914
`http://www.sciencedirect.com/science/article/pii/S0079670005000742`

Fig. ??:
`www.hitachi-hitec-science.com/en/documents/technology/thermal_analysis/application_TA_068e.pdf`

Fig. ??: reproduced with permission by Elsevier
Microstructure of HIPS
L. Leibler, Prog. Polymer Sci. 30 (2005) 898-914
`http://www.sciencedirect.com/science/article/pii/S0079670005000742`

Fig. ??:
`http://www.primepolymer.co.jp/english/technology/material/pp/05.html`

References

Shackelford, Smith, and Callister and Rethwisch provide detailed coverage of polymer processing. Smith also discusses many more polymer materials.

For a polymer tutorial focusing on structure, you may wish to consult
`http://plc.cwru.edu/tutorial/enhanced/files/polymers/struct/Struct.htm`

For a polymer tutorial giving you much more on chemistry, go to
`http://chem.chem.rochester.edu/~chem421/classes.htm`
`http://www2.chemistry.msu.edu:80/faculty/reusch/VirtTxtJml/polymers.htm`

You will find almost anything about almost any polymer in the Sigma Aldrich Materials catalog
`http://www.sigmaaldrich.com/materials-science/material-science-products.html`

A few suppliers of polymer materials discussed in this book are:
`http://www.aschem.com.tr`
`http://www.eplastics.com`
`http://www.plasticsintl.com`

You may find the following Plastic Material Selection Guide useful:
`http://www.curbellplastics.com/technical-resources/pdf/plastic-material-selection.pdf`
It has ranked lists of many polymers according to a variety of properties.

Exercises

1. What is the degree of polymerization of the polyethylene in Fig. ?? ? How many mers does the most probable molecule contain ?

2. Calculate $\overline{M_n}$, $\overline{M_w}$ and the PDI for the materials shown in the figure.

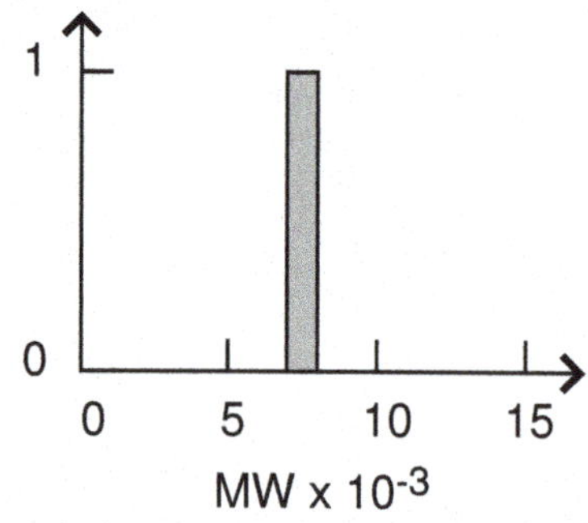

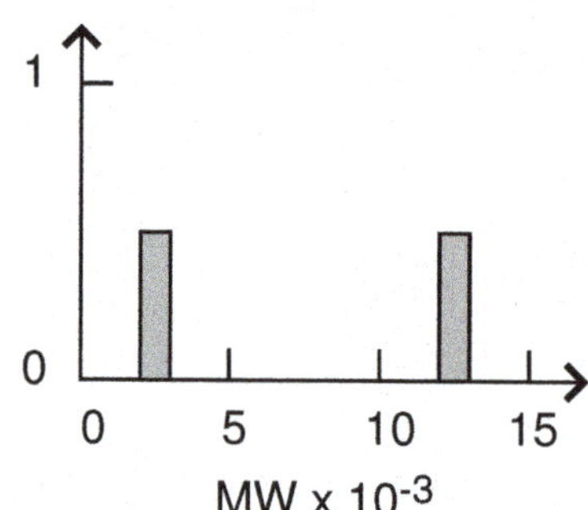

Discuss the numerical results in light of what you know about $\overline{M_n}$ and $\overline{M_w}$ and their relation to the shape of the molecular weight distribution.

3. Calculate $\overline{M_n}$, $\overline{M_w}$ and the PDI for the materials shown in the figure.

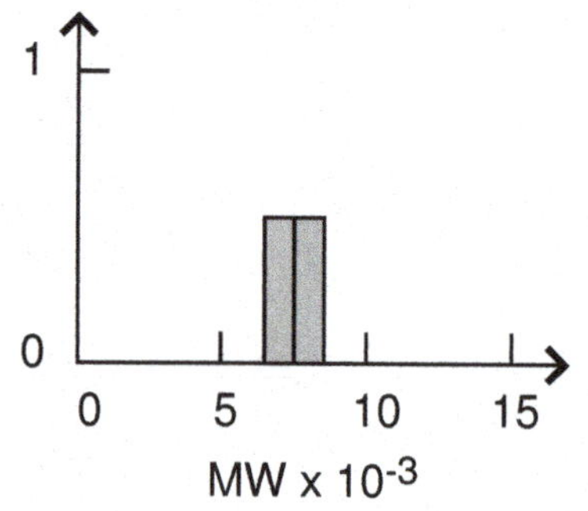

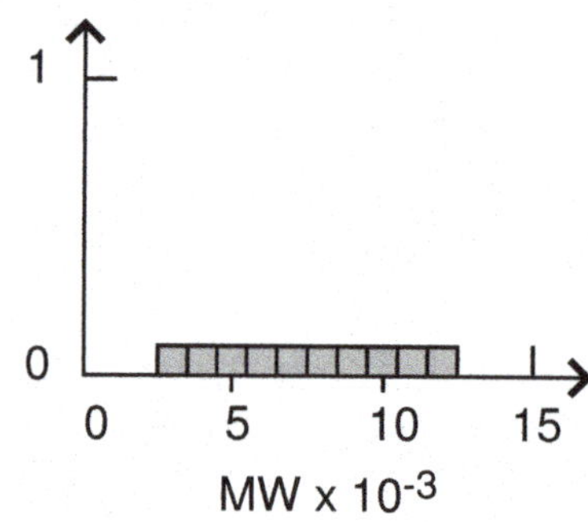

Discuss the numerical results in light of what you know about $\overline{M_n}$ and $\overline{M_w}$ and their relation to the shape of the molecular weight distribution.

4. Assume that LDEP is 20% crystalline by volume and has a density of 0.92 g/cm^3, and HDPE is 96% crystalline and has a density of 0.98 g/cm^3. What are the densities of completely amorphous PE and a perfect PE crystal, assuming that LDPE and HDPE are made up of the respective portions of amorphous and crystal material?
   Hint: Start by figuring out a way to express the mass of partially crystalline PE as a sum of an amorphous and a crystalline part.

5. Why is LDPE highly translucent, even transparent as a thin film, but HDPE much less so ?

6. How would you expect the degree of crystallinity to vary, on average, between the different types of polyethylene, and why ?

7. Especially in the area of polymers, some rather unusual units are still in wide use.
   a) Densities may be given in lb/ft^3. What is the conversion to g/cm^3 ?
   b) Izod impact strength is often given in ft lb/in. What is the conversion to J/m ?

8. How will T_g of siloxanes depend on the nature of the side groups R_1 and R_2 ?

9. From the data on Table 11.7, draw properly scaled stress-strain curves for general purpose PS and high impact PS.

10. Plot the data in the table for various samples of polyethylene fiber in such a manner that you can answer the questions below:

Sample Nr.	$\overline{M_n} \times 10^{-3}$	$\overline{M_w} \times 10^{-3}$	T.S. (GPa)
1	17	54	0.68
2	13	100	0.75
3	18	280	1.10
4	110	120	1.38
5	120	800	1.55
6	13	100	0.75
7	28	115	1.15
8	110	120	1.38

a) Is there a systematic dependence of the tensile strength T.S. on $\overline{M_w}$?
b) Is there a systematic dependence of the tensile strength T.S. on the PDI for a fixed $\overline{M_w}$?

Data from: Tensile Strength of Highly Oriented Polyethylene, P. Smith, P.J. Lemstra, and J.P.L. Pijpers, J. Polym. Sci., Polym. Phys. Ed. 20 (1982) 2229-2241
`http://alexandria.tue.nl/repository/freearticles/619807.pdf`

Problems

1. What is the ideal density of a crystal of polyethylene? Does the density of HDPE in Table 11.1 agree with this value ? If not, why not ?
2. Determine $\overline{M_n}$ and $\overline{M_w}$ for the following polyethylene sample:

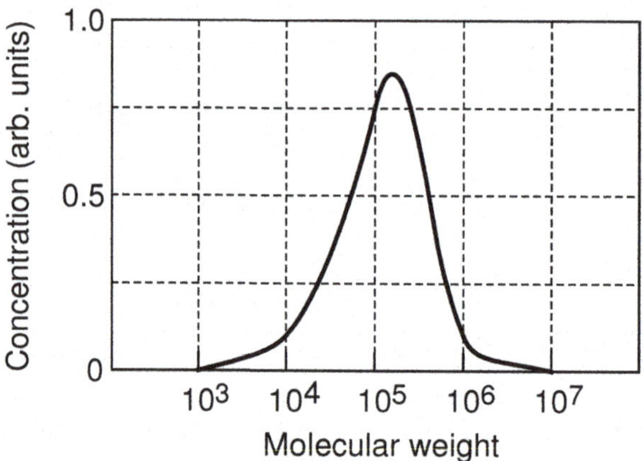

Note the semilog nature of the graph.

3. How would you expect polyethylene to interact with adhesives, and why? If you had to join two pieces of polyethylene, how would you do that ?
4. Assume you have the task of separating different types of plastics for recycling. Based on the data in Tables 11.3 and 11.5, can you think of a simple way to separate the materials listed in those tables ?
5. Find another example of a copolymer of styrene, i.e. styrene with something other than acrylonitrile, that improves material properties over styrene by itself. Describe what the improvements are and by what mechanism they are brought about.

6. Research the properties of PTFE. In particular, find out whether it is amorphous or partially crystalline, and if so, what its degree of crystallinity is. Also indicate its T_g and T_m and relate these parameters to those of polyethylene.

7. A variant of PTFE is poly(chlorotrifluoroethylene), also known as PCTFE. What are the similarities and differences of PTFE and PCTFE, and how can you explain those in terms of molecular chemical characteristics ?

8. An important use of PTFE is as a non-stick coating in pots and pans. How can PTFE itself be made to adhere to the metal pan such that the PTFE-metal bond is strong enough to survive fairly high temperatures over a period of time ?

9. What would be the point, if any, of making a PE-PP copolymer ?

10. Research the topic of polymer blends. Think of them as the polymer analog of metal alloys. What are the similarities and differences between the two kinds of alloys ? What are the special issues, if any, with polymer blends ?

Chapter 12
Electronic Properties

Overview

In a metal, electrical current is proportional to the applied electric field, due to abundant free electrons. This gives rise to two equivalent forms of Ohm's law. The electrical conductivity depends equally on the concentration and mobility of the electrons. In a semiconductor, free electrons are not naturally abundant because the electron energy states lie in two separate energy bands. The lower band normally has all its states doubly occupied by electrons, whereas in the upper band normally all states are empty. Electrons are free to conduct only if excited from the lower to the upper band. Such excitation is possible, but very limited, by thermal energy. It can be greatly enhanced by alloying the material with select impurities. Insulators do not conduct current since their energy bands are such that normally no free charge carriers are present. For the same reason, insulators are optically transparent and exhibit dielectric behavior.

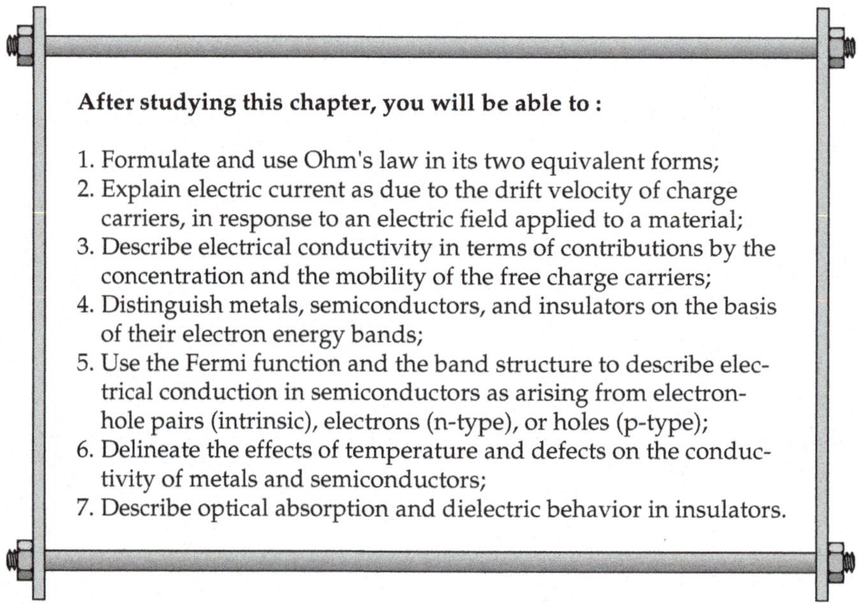

After studying this chapter, you will be able to :

1. Formulate and use Ohm's law in its two equivalent forms;
2. Explain electric current as due to the drift velocity of charge carriers, in response to an electric field applied to a material;
3. Describe electrical conductivity in terms of contributions by the concentration and the mobility of the free charge carriers;
4. Distinguish metals, semiconductors, and insulators on the basis of their electron energy bands;
5. Use the Fermi function and the band structure to describe electrical conduction in semiconductors as arising from electron-hole pairs (intrinsic), electrons (n-type), or holes (p-type);
6. Delineate the effects of temperature and defects on the conductivity of metals and semiconductors;
7. Describe optical absorption and dielectric behavior in insulators.

Cross-section through a Modern Electronic Integrated Circuit

The picture displays a scanning electron micrograph of an integrated electronic circuit, which was used by IBM some ten years ago to demonstrate their technology. The active element is a single transistor inside the magnifying glass. It measures about 200 nm across and has been fabricated in the bottom layer of single-crystal Si. The structure above it is a scheme of wires consisting of six layers of Cu (light gray, with granular texture), connecting different circuits. These Cu layers are separated from each other by insulating SiO_2 (dark gray). The features in white are from a layer of W between the bottom Si and the Cu higher up. The W layer is necessary to keep the Cu away from the Si, as the Cu atoms would diffuse readily into the Si and poison it, if Cu and Si were in direct contact.

12 Electronic Properties

One of the main themes of this book has been to explain macroscopic properties of solids on the basis of forces and bonds between atoms. These forces are ultimately due to the interactions of electrons and nuclei within atoms as well as between atoms. We will now revisit that subject and even go a little deeper, in the sense that electronic properties of solids depend on the behavior of electrons in more detail than we have seen to date. Our goal is to come up with an atomic-level description of the electrical conduction in solids and to use it for explaining a variety of electronic devices.

With respect to electrical conduction, solids can be divided into three classes: conductors, semiconductors, and insulators. Conductors are in fact good conductors of electrical current and comprise essentially the materials class of metals. Insulators do not conduct current at all, and consist of ceramics, inorganic glasses, and normal polymers. Semiconductors are intermediate between conductors and insulators, in the sense that they conduct much worse than metals, but much better than insulators. The ability to conduct spans about 20 orders of magnitude between the best metals and good insulators.

12.1 The Two Forms of Ohm's Law

A fundamental observation about electrical currents through a conductor is that for a given conductor, the current I is proportional to the applied voltage V: $I \propto V$.

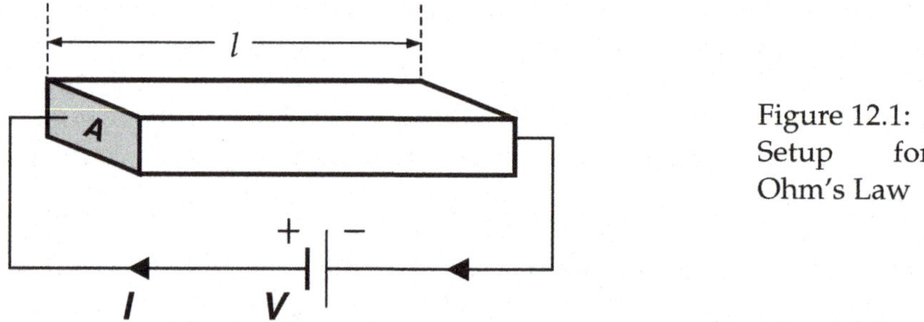

Figure 12.1: Setup for Ohm's Law

This proportionality relation can be written as an equation, in the first form of **Ohm's Law**:

$$V = I\,R \tag{12.1}$$

where R is called the **electrical resistance**, or just **resistance** for short, of the conductor. The SI units for V are Volt (1 Volt = 1 V = 1 Joule/Coulomb = 1 J/C) and for I are Ampere (1 Ampere = 1 A = 1 C/s). The units for R are Ohm, abbreviated Ω, and 1 Ω = 1 V/1 A .

Furthermore, I depends on the geometry of the conductor as follows: At constant V, $I \propto A$ and $I \propto l^{-1}$. This implies that $R \propto A$ and $I \propto l^{-1}$. Thus the proportionalities for R can be incorporated into a single equation as follows:

$$R = \rho \frac{l}{A} \tag{12.2}$$

The proportionality factor ρ in Eq. 12.2 is called the **electrical resistivity**, or resistivity for short. The units of ρ are Ωm.

It is important that you make a clear distinction between the resistance R and the resistivity ρ: R depends on the material of the conductor but also incorporates geometrical characteristics. On the other hand, ρ is a materials parameter independent of geometry.

There is an equivalent way of expression the normal form of Ohm's Law, Eq. 12.1, which can be obtained as follows. Let us replace R in Eq. 12.2 by V/I and solve for ρ. The result is:

$$\rho = \frac{V\,A}{I\,l} \tag{12.3}$$

If we rearrange the variables in Eq. 12.3 slightly, we arrive at

$$\rho = \frac{V}{l}\frac{A}{I} = \frac{V}{l}\frac{1}{\frac{I}{A}} = \frac{\mathcal{E}}{\mathcal{J}} \tag{12.4}$$

where $\mathcal{E}$ is the electric field driving the current across the conductor, and $\mathcal{J}$ is the current density, i.e. the current per unit area, with units A/cm^2 or A/m^2.

Now it is convenient to introduce a new variable σ called **electrical conductivity** by defining

$$\sigma = \frac{1}{\rho} \tag{12.5}$$

which allows us to write Eq. 12.4 as

$$\frac{1}{\sigma} = \frac{\mathcal{E}}{\mathcal{J}} \tag{12.6}$$

or equivalently

$$\mathcal{J} = \sigma\,\mathcal{E} \tag{12.7}$$

Eq. 12.7 is the ***second form of Ohm's Law***. It is important that you realize that Eqs. 12.1 and 12.7 express exactly the same thing. Also note that electrical conductivity σ is a materials parameter, just as is the electrical resistivity ρ. The units of σ are $\Omega^{-1}\text{m}^{-1}$. The unit of Ω^{-1} is sometimes called Siemens. For historical reasons, σ is often also listed in units of $\Omega^{-1}\text{cm}^{-1}$. The conversion between the two units is: $1\ \Omega^{-1}\text{cm}^{-1} = 100\ \Omega^{-1}\text{m}^{-1}$.

12.2 Atomic Picture of Electrical Conduction

We start with the fundamental assumption that electrical current through a solid is carried by electrons which are mobile enough so that under the influence of an electric field $\mathcal{E}$, they are able to move in the direction of the field. (Literally, the electrons move in the direction *opposite* to $\mathcal{E}$, since the direction of the electric field, by definition, is the direction in which positive charges would move under its influence. This is also how the direction of electric current is given.)

However, these mobile electrons are only free to move up to a point. If an electron experiences an electric field, it will be accelerated, not indefinitely, but only until it makes a collision with an atom. Such a collision instantly changes its velocity and direction, after which it is accelerated again, makes another collision with an atom, and so on.

Thus the motion on an individual electron inside a solid should be visualized as follows:

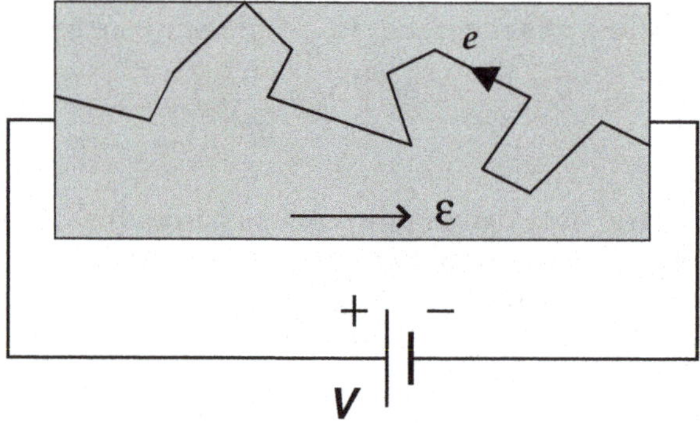

Figure 12.2: Directed random motion of an electron inside a solid under the influence of an externally applied electric field.

The motion of the electron would look somewhat random, in the sense that the change of direction due to a collision with an atom, and the times in-between such collisions, would vary statistically, but overall there would be a net component of motion in the direction opposite to the applied electric field.

Note also that when no electric field is present, that does not mean that the electrons are at rest ! They still move around on account of their thermal energy, but the thermal motion is completely random in all directions and does not produce an electric current.

Hence in the steady state set up by an applied electric field, the mobile electrons move with a certain velocity v_d along the electric field $\mathcal{E}$. Except in very unusual situations, it turns out that $v_d \propto \mathcal{E}$, and thus

$$v_d = \mu \mathcal{E} \tag{12.8}$$

The proportionality constant μ is called the **mobility** of the electrons, and the velocity v_d is referred to as **drift velocity**. From Eq. 12.8 it follows that he units of μ are m^2/Vs .

Now imagine electron current moving through a wire as in Fig. 12.3:

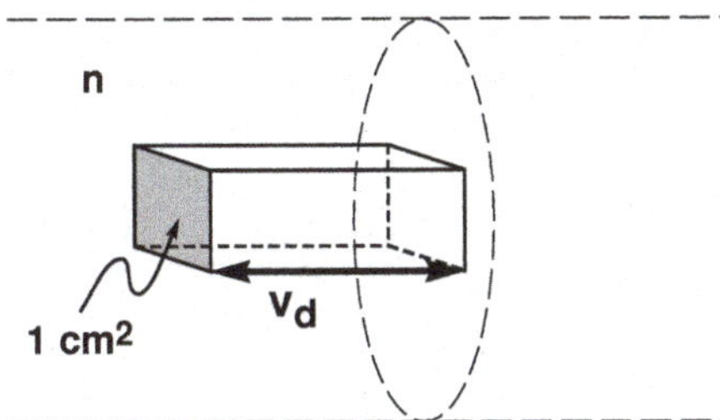

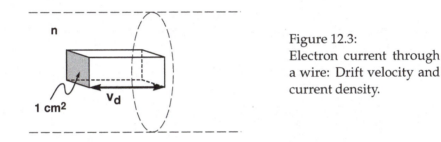

Figure 12.3: Electron current through a wire: Drift velocity and current density.

The concentration of mobile electrons is n, and they are moving with a directed velocity v_d. Therefore, the number of electrons moving through an area of 1 cm^2 per second, i.e. the current density $\mathcal{J}$, is given by

$$\mathcal{J} = e n v_d \tag{12.9}$$

where e is the electron charge.

If we combine Eqs. 12.7, 12.8, and 12.9, we can write

$$\sigma = \frac{\mathcal{J}}{\mathcal{E}} = \frac{e n \mu \mathcal{E}}{\mathcal{E}} = e n \mu \tag{12.10}$$

The beauty of this expression is that it tells us something important about the atomic mechanisms that determine the electrical conductivity: σ has two components, n and μ. Specifically, σ is proportional to 1) the density of electrons free to move, and 2) the mobility of these electrons.

A conclusion we can draw is that high σ requires high n ***and*** high μ. These two variables are quite independent. For example, semiconductors turn out to have much lower conductivity than metals because their n is much lower. Also, polymers exist with a certain amount of conductivity, which is generally limited not so much by n as by a low mobility μ.

Representative values of σ at room temperature (20°C) are given in Table 12.1. below for different types of materials:

Table 12.1: Conductivities and Resistivities for Different Materials

Material	σ (Ω^{-1}m^{-1}) at 20°C	ρ (Ωm) at 20°C
Ag	6.3×10^7	1.6×10^{-8}
Cu	6.0×10^7	1.7×10^{-8}
Al	3.8×10^7	2.6×10^{-8}
W	1.8×10^7	5.6×10^{-8}
Ni	1.4×10^7	7.1×10^{-8}
Ta	8.0×10^6	1.3×10^{-7}
Cr	7.8×10^6	1.3×10^{-7}
Constantan [1)]	2×10^6	5×10^{-7}
Nichrome [2)]	9×10^5	1.1×10^{-6}
Si (very pure)	5×10^{-4}	2×10^3
Si (high σ)	5×10^3	2×10^{-4}
SiO_2	$<10^{-12}$	$>10^{12}$
polyethylene	$<10^{-13}$	$>10^{13}$

[1)] usually 55Cu 45Ni; [2)] typically 80Ni 20Cr.

The data for metals in Table 12.1 are for well-annealed polycrystalline samples. The first thing to note about these data in Table 12.1 is that σ varies only by about an order of magnitude among the pure, elemental metals. Nichrome and Constantan are alloys. Constantan has the unique property of a conductivity which is practically independent of temperature. Nichrome, due to its low σ, is used for electric heating elements. Please note that in both cases, σ for the alloy is lower than for either of the individual pure metal components.

The entry for Si is representative of semiconductors in general. Even a type of Si with very high conductivity has a σ which is much lower than that of a metallic alloy, let alone a pure metal. On the other hand, σ for Si itself can vary by many orders of magnitude, for reasons you will get to know shortly.

The conductivities of ceramics and glasses such as SiO_2 and polymers are so low as to be zero for practical purposes. This is, of course, why those materials are classified as electrical insulators. In certain special cases, the conductivity of a ceramic can be increased markedly, by the presence of defects. Moreover, polymers have been synthesized with semiconducting characteristics which are due to special structural properties or the addition of impurities.

12.3 Electron States and Energy Bands in Solids

In order to appreciate the fundamental differences between conductivities of different types of solids, we need to reexamine atomic energy levels. Way back in Fig. 2.6 we gave a picture of these energy levels for Li and Li_2. We saw there that when two Li atoms come close together, their 2s electron states interact to create two new energy levels. By occupying the lower of these two levels the two 2s electrons can reduce their total energy and thus form a molecular bond.

Now we will extend the same picture to the interaction of a large number of Li atoms in Fig. 12.4 below:

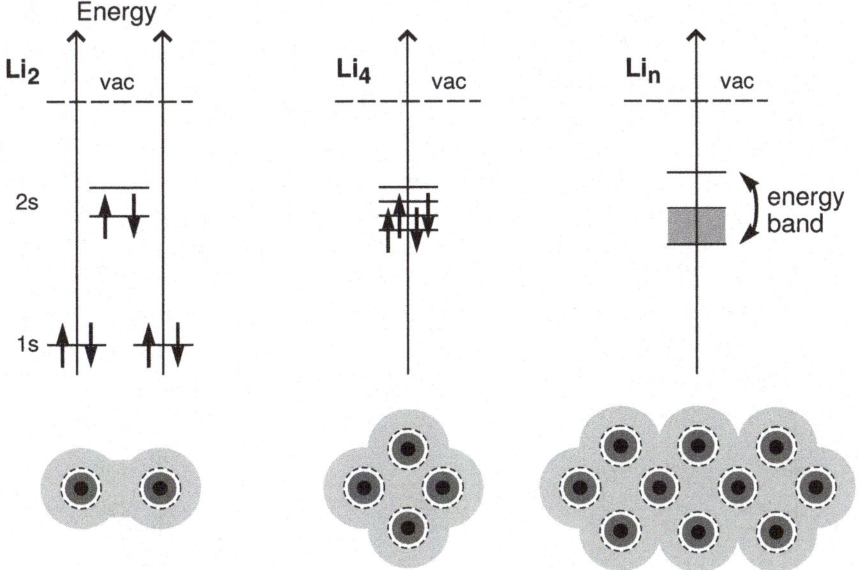

Figure 12.4: Schematic of the electron energy levels of Li_2, Li_4, and Li_n. The 1s states are not shown on top of the middle and right panels.

The left panel in Fig. 12.4 is the same as the right panel in Fig. 2.6, showing the Li_2 molecule. The procedure for going from 2 to a large number n of Li atoms should be transparent. For example, in the center panel, when 4 Li atoms are close, their 2s states create four new energy levels, and the lowest two levels are occupied by two electrons each. Once again, the four electrons together have been able to lower their total energy, and thus a bond between the four atoms has been formed. Think of this as a Li_4 cluster, the nucleus to a piece of Li metal.

As the process continues, with each additional Li atom another state is added. The key feature of an assembly with a large number n of Li atoms is that the difference between the highest and the lowest of these energy levels is limited; it does not keep increasing very much. This means that more and more levels are squeezed into a finite energy interval. Ultimately, we are dealing with Avogadro's number of levels within an energy interval of the order of typically a few **electron Volts** (**eV**). (When dealing with atomic energy levels, the unit of electron Volt, or eV, is in common usage. It refers to an energy of 1.6×10^{-19} Joule, which is the energy of an electron having gone through a potential difference of 1 V).

The consequence of having a very large number of energy levels in a small energy interval is that the individual levels are extremely close to each other in energy. Therefore, one says that these energy levels form an **energy band**, as indicated in the panel on the right of Fig. 12.4. An energy band refers to an energy range within which lie a very large number of states.

The crucial thing about the energy levels in this energy band is that only the lower half of them are occupied. Those are sufficient to accommodate all the electrons, since per the Pauli Principle two electrons can be in one level. But note: The upper half of the levels are still present in the energy band; they just do not have any electrons in them. In other words, they are empty.

The second consequence of this picture is that all these electrons, occupying the lower half of the energy band, are not only close in energy but also close in space. In fact they are spatially so close as to overlap to a considerable extent (see the bottom of the panel on the right in Fig. 12.4). Therefore we can say that, since these electrons are indistinguishable, they are shared universally. This universal sharing is the physical root cause why the electrons carrying the electric current in a metal have high mobility.

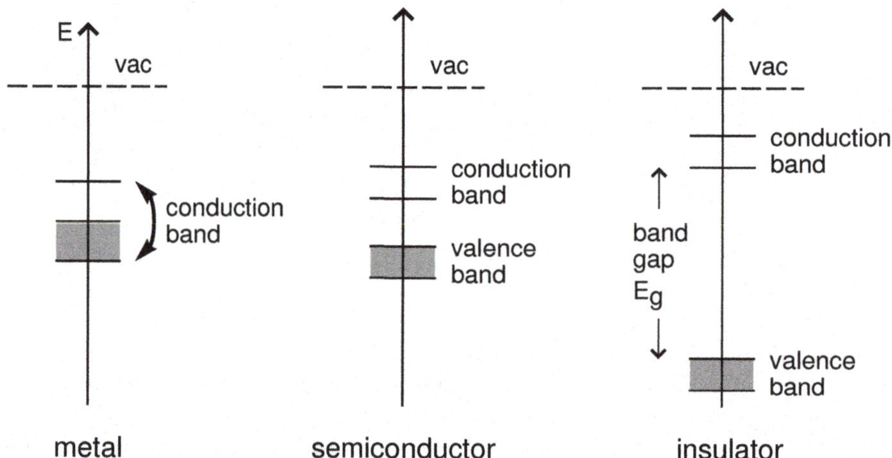

Figure 12.5: Layout of energy bands for different types of materials. The occupied portions of the bands are marked gray.

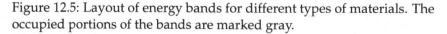

The formation of an energy band from atomic electron states when the atoms are very close is a general phenomenon, and is observed for all types of materials. The specific layout of the electron energy states in a band, or in two bands as the case may be, is what determines the nature of the material in question. The three basic possibilities are displayed in Fig. 12.5.

Note that the part of the states in a band which is occupied by, or filled with, electrons, is indicated in gray. The critical features of these energy bands can be summarized as follows:

- For a ***metal***, only one band is important. It is the one containing the electrons with the highest energies, and it is always only partially filled. We will call this band the **conduction band**. (Some authors call it the valence band).
- For a ***semiconductor***, two bands are important. It is as if the one band for the metal has been split into two. The upper band is also called conduction band, and normally ***none*** of its states are occupied by electrons. This does not mean that is is irrelevant, however, and its role will become apparent shortly. The lower band is called **valence band**, and normally ***all*** of its states are occupied by electrons. The energy difference between the lower edge of the conduction band and the upper edge of the valence band is the major parameter characterizing a semiconductor, and is called the **bandgap**, with the symbol E_g.
- For an ***insulator***, the layout of the bands is similar to the one for a semiconductor, with a conduction and a valence band. The main difference is that E_g is much larger for an insulator than for a semiconductor.

The detailed layout of the energy bands, including the position in energy of the bands, the widths of the bands, and the magnitude of E_g among other things, is referred to as the **band structure** of a material.

A few numerical examples for bandgaps of common semiconductors and insulators are listed below:

Table 12.2: Bandgaps of common electronic materials

Material	bandgap E_g (in eV)
metals	—
Ge	0.67
Si	1.1
GaAs	1.4
CdS	2.4
Diamond	5.5
SiO_2	9

The energy bands, and how they are occupied by electrons, have profound consequences for the electrical conduction properties of the materials. The crucial point is the following:

> When two electrons occupy a state together, as they certainly can, they are not free to participate in electrical conduction.
> In order for them to conduct current, one of the electrons has to be elevated into a state of higher energy, so that each of them occupies a state by itself.

This is illustrated schematically in Fig. 12.6. The figure shows a blown-up portion of the band structure of a metal, with the three topmost occupied energy levels and four slightly higher unoccupied levels.

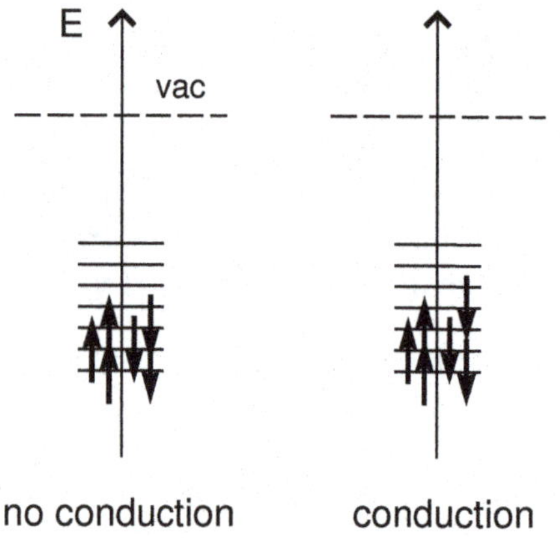

Figure 12.6:
Left panel: All energy levels doubly occupied.
Right panel: Two top most levels singly occupied.

In the panel on the left, all energy levels are doubly occupied. So there are no free electrons, and hence no conduction. In the panel on the right, one of the two electrons in the topmost occupied level has been elevated to the next level nearby. Now there are two electrons in singly occupied states, hence two electrons free to conduct.

Here is where the critical role of the band structure emerges:
If there are empty states available immediately adjacent to doubly occupied states, then the process described above for "freeing electrons" costs almost nothing in energy. This is precisely the case for metals. In fact, the thermal energy present in a metal at room temperature is enough to create large numbers of free electrons in the conduction band.

In a semiconductor, on the other hand, it takes an energy equal to the bandgap E_g for each electron to be elevated into a state where it is free to conduct. This is because empty states are available only in the conduction band (cf. Fig. 12.5). On the thermal energy scale, E_g represents a large amount of energy, and thus the process is exceedingly rare. In an insulator, the bandgap is so large that it is not possible to create free electrons.

12.4 Conduction in Metals: Influences on Electron Mobility

In Sections 12.1 and 12.2. I outlined the basics of electrical conduction in materials, in the forms of Ohm's Law. We take it as a given that in a metal the concentration of free electrons n is large enough to be favorable for efficient electrical conduction. Hence we will pick up from Eq. 12.10, and ask what factors will limit σ, which amounts to asking what factors limit the mobility μ.

The general answer is: Anything that reduces the drift velocity v_d, or even more fundamentally, anything that causes additional scattering for the electrons trying to follow the force of the electric field. Three factors are of practical importance in this regard.

It turns out that even though the conductivity is the more fundamental quantity, the practical effects show up most simply when we consider the resistivity. For metals around room temperature, one finds that to a good approximation one can write for the total resistivity ρ_{tot}:

$$\rho_{tot} = \rho_t + \rho_i + \rho_d \tag{12.11}$$

where ρ_t stands for the thermal contribution, ρ_i for the impurity contribution, and ρ_d for the deformation contribution. Eq. 12.11 is known as Matthiesen's Rule.

The thermal contribution is caused by the scattering of conduction electrons by the thermal motions of solid atoms around their ideal crystal positions. The impurity contribution comes from the scattering of electrons by impurity atoms in the solid. The deformation contribution is due to scattering from defects induced by mechanical deformation, such as dislocations and grain boundaries.

Eq. 12.11 expresses the fact that the best medium for electrons to move in is a perfect crystal. Any deviation from the periodicity of a crystalline material increases the scattering of electrons and thus impedes their movement.

12.4.1 Dependence of Resistivity on Temperature

Experiments show that over a not too large temperature range around room temperature T_0, ρ_t is a linear function of T:

$$\rho_t(T) = \rho_0[1 + \alpha(T - T_0)] \tag{12.12}$$

The parameter α is called **temperature coefficient of resistivity** and denotes the fractional change in $\rho_t(T)$ per degree C. Some representative data for α are listed in Table 12.3 below.

Table 12.3: Temperature coefficients of resistivity

Material	α (°C^{-1})
Ag	0.0061
Cu	0.0068
Al	0.0043
Ni	0.0069
Ta	0.0038
W	0.0045
Brass (60Cu 40Zn)	0.0015
Constantan	3×10^{-5}
Nichrome	0.0004

Note that Constantan is an example of an alloy developed specifically for its low temperature coefficient of resistivity. Such materials are used for temperature-independent standards of resistivity.

12.4.2 Dependence of Resistivity on Impurities

The fundamental observation about ρ_i as a function of the impurity concentration c_i is that a small amount of impurities always raises the resistivity of the pure metal. If you now consider brass as an example, you could start with pure Cu and add a little Zn, but you could also start with pure Zn and add a little Cu. The observation above applies either way.

This means that the curve $\rho_i(c_i)$ rises at both ends of the compositional range. It follows that if you look at $\rho_i(c_i)$ of an alloy over the entire compositional range, $\rho_i(c_i)$ must have a maximum at an intermediate c_i. The example of the resistivity of brass is shown below.

The black dots in Fig 12.7 are data points for annealed commercial alloys. The solid curve is an approximation to the data. The dashed lines approximate the rise in ρ_i for small additions of impurities to either pure material. The position of the maximum and the slope on the right side are of course just guesses, in the absence of alloy data for $c_i > 0.4$, but the figure should make it clear that there must be a maximum. ρ_i from Eq. 12.11 would be the difference between the ρ of the alloy and the pure metal.

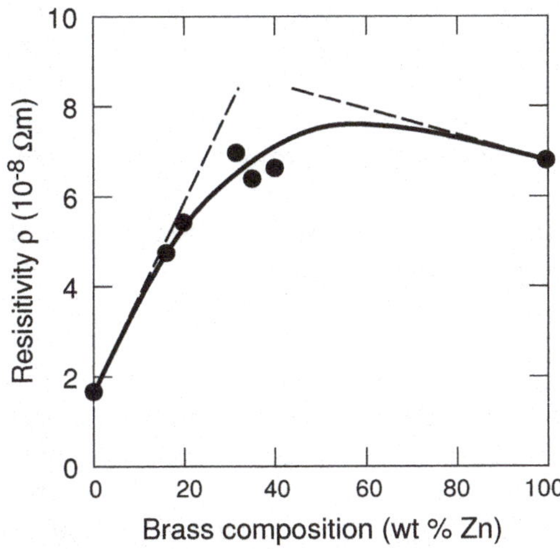

Figure 12.7: Room temperature electrical resistivity of annealed brass as a function of alloy composition.

12.4.3 Dependence of Resistivity on Deformation-Induced Defects

The defects in question here are dislocations and grain boundaries. Mechanical deformation, as you know, causes dislocations and breaks up the equilibrium grain structure. Both these kinds of defects act as obstacles to the motion of electrons.

A set of data is displayed in Table 12.4 below. Two samples of 70Cu-30Zn brass are listed in the table. $\Delta\rho$ is the change in resistivity of the original ρ caused by the deformation. $\Delta\rho$ can be identified with ρ_d in Eq. 12.11.

Table 12.4: Resistivities of annealed and deformed metals at 20°C

Metal	ρ annealed (Ωm $\times 10^8$)	Deformation $\ln(t/t_0)$	$\Delta\rho/\rho$
Cu	1.68	2.9	0.033
Ni	9.94	2.9	0.034
75Cu 25Ni	30.6	2.2	0.037
70Cu 30Zn	6.16	2.9	0.22
	6.17	3.5	0.23

The original samples were 2.5 mm thick and annealed, and were then cold rolled to a final thickness of 0.15-0.25 mm. This represents a very large amount of cold work of as much as about 2000 % or a factor 20 (Note that the deformation is given as $\ln(t/t_0)$ in Table 12.4).

The relative change in resistivity $\Delta\rho/\rho$ was also examined as a function of the amount of deformation. $\Delta\rho/\rho$ was observed to grow continuously with the amount of cold work for a given type of sample. The thinner the rolled sample, the higher was its resistivity.

Keep in mind that the original resistivity can be reclaimed by careful annealing, i.e. by restoring an equilibrium grain structure.

12.5 Semiconductors

When we examined the energy band structures of metals, semiconductors, and insulators in Fig. 12.5 above, I pointed out that in semiconductors, the possible energy states are split into two groups, called the conduction band and the valence band, separated by the bandgap E_g. I also said that the consequence of this was that normally, all states in the valence band were filled and no states in the conduction band were populated. This is in fact a simplification which only applies at a temperature of absolute zero.

The thermal energy present in the material at a finite temperature T is sufficient to excite a few electrons from the valence band up into the conduction band. These few electrons will then be free to move and respond to an electric field. That is why the conductivity of a semiconductor is very small, but not zero, at a finite temperature.

It should be evident from this picture that for semiconductors, the critical parameter in Eq. 12.10 defining conductivity is the concentration of electrons n in the conduction band. We should also expect n to be a strong function of T, and we need to figure out how to describe $n(T)$.

12.5.1 The Fermi-Dirac Distribution

If you recall our discussion of atomic structure and atomic energy states in Chapter 2, it was straightforward to answer the question as to which state overall an atom was in. You needed to know the atom's possible energy states, and there were only a few of those. Then you filled those states one-by-one, from the bottom up, with up to 2 electrons.

In a solid, the situation is complicated by the fact that we have to deal with huge numbers of states, bunched together in bands. Now, the question of how those states are filled becomes a statistical problem. If we assume that the energy bands of a material are known, the specific questions is:

What is the probability of any one of these allowed states to be populated by an electron?

The answer to this question is given by the so-called Fermi-Dirac distribution, sometimes also just called **Fermi function**. It arises from a consideration of how a very large number of particles can populate a very large number of energy states that are really close together, under the condition that you can put at most two particles in one state (the Pauli principle).

The Fermi function $f(E)$ denotes the probability that an allowed electron state in the solid with energy E is in fact populated by an electron. The statistical mechanics of electrons require that $f(E)$ have the following form:

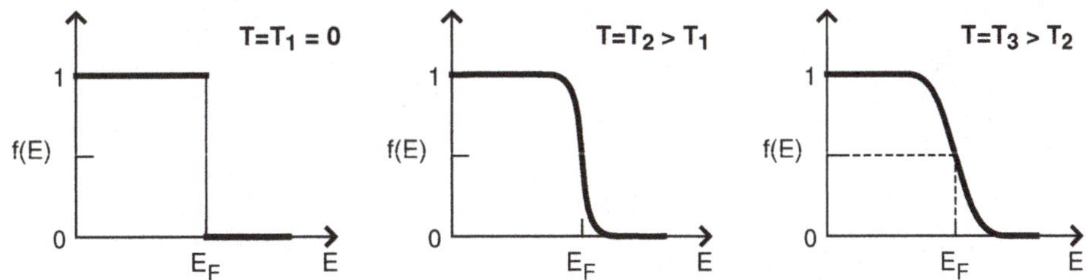

Figure 12.8: Fermi function $f(E)$ at different temperatures.

The function $f(E)$ depends on the temperature, but its shape is fairly intuitive. At $T = 0$ it is a step function at $E = E_F$ where E_F is called the **Fermi energy**. It illustrates what we know about electrons via the Pauli principle: Electrons will occupy the lowest possible energy states, but there can be only two in one state. Therefore, we need a lot of states to accommodate all these electrons, from $E = 0$ up to $E = E_F$.

At temperatures higher than absolute zero, the thermal energy in the solid will give a few electrons an extra push so that their energy is a little above E_F. These electrons come from states with energy just a little lower than E_F. Thus the gradual fall-off of $f(E)$ around E_F from a value of 1 down to 0. Moreover, the higher T, the more gradual this fall-off is.

The function $f(E)$ can be expressed mathematically as follows,

$$f(E) = \frac{1}{1 + \exp(\frac{E - E_F}{kT})} \tag{12.13}$$

where k is the Boltzmann constant and T is the absolute temperature (in K, or degrees Kelvin). Recall that $k = 1.38 \times 10^{-23}$ J/K $= 8.62 \times 10^{-5}$eV/K. (As an exercise in visualization, convince yourself that in the limit of $T \to 0$, $f(E)$ is indeed a step function).

The Fermi energy E_F is a parameter which is an individual characteristic of each material. It designates the energy at which, in that particular material, the probability for an electron state to be occupied is 0.5.

The complete picture of how electrons populate the available energy states can now be obtained by combining the band structure (allowed energy states) with the Fermi function (probability of states being occupied). The result for metals and semiconductors is indicated schematically in Fig. 12.9. The heavy solid curves represent the Fermi function $f(E)$ from above, with the x- and y-axes exchanged.

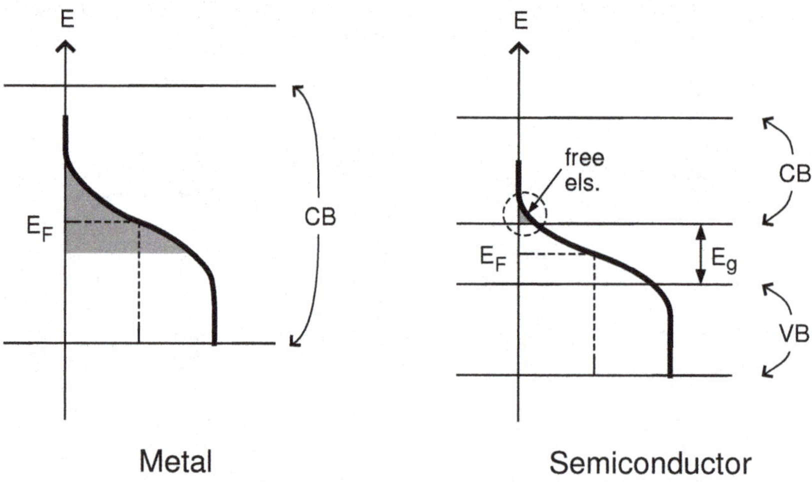

Figure 12.9: Energy bands and Fermi function for metals and semiconductors. CB stands for conduction band and VB for valence band.

For a metal, in the panel on the left, E_F falls somewhere in the middle of the conduction band. The number of states with free electrons, indicated schematically by the gray area, is quite large. For a semiconductor, in the panel on the right, E_F falls in the middle of the bandgap E_g, and the electrons free to conduct are only the ones under the $f(E)$ curve up in the conduction band, in the tiny gray area inside the small dotted circle. Thus, in the semiconductor the number of free electrons is very small, as it turns out smaller by several orders of magnitude compared to a metal. This is, in a nutshell, the reason for the large difference in electrical conductivities between metals and semiconductors.

12.5.2 Bands and Bonds in Semiconductors

There is another point of view with respect to electrons and the states they occupy, which can shed more light on the issue of free and not-free electrons. This point of view focuses on the chemical bonds between the semiconductor atoms.

Recall Chapter 3 where we introduced Si, the premier semiconductor material, as a crystal in which each Si atom forms four covalent bonds with its nearest neighbors in a tetrahedral cubic arrangement (see Fig. 3.23). We will use Si as the example for most of our discussion of semiconductor materials.

Fig. 12.10 displays a two-dimensional analog of a Si crystal. It is meant just to show the coordination of the Si atoms, i.e each Si atom having four nearest neighbors with which it forms covalent bonds. A covalent bond is indicated by a gray oval between two Si atoms and two electrons inside.

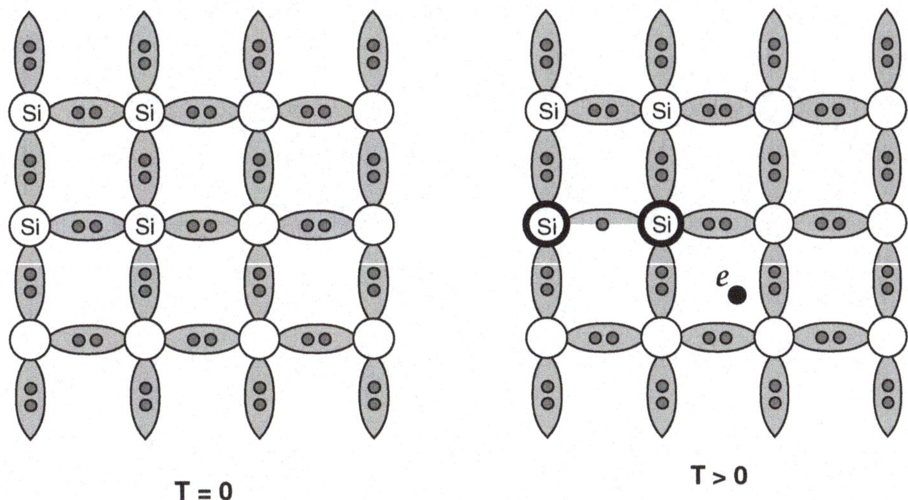

Figure 12.10: Schematic of four-fold coordination and covalent Si-Si bonds in a Si crystal. Note the broken Si-Si bond and free electron.

The left panel shows the Si crystal with all bonds intact. This corresponds to the T = 0 condition: All states in the valence band VB are filled, and all states in the conduction band CB are empty. The right panel shows a broken Si-Si bond: One electron is still between the original two Si atoms, and the other electron from this bond has moved away to the right. This second electron, marked black, is now floating around in the Si crystal as ***if it were free*** ! It is free in the sense that with respect to its energy, it makes no difference where the electron is, as long as it is not near the broken it came from. For the free electron, the environment of Si atoms looks the same everywhere except near the original bond.

From the point of view of energy bands, breaking the Si-Si bond and creating a free electron is equivalent to exciting an electron from the VB to the conduction band CB. This process may be brought about by the thermal energy in the crystal at a finite temperature although, as we shall see shortly, it is rather rare.

12.5.3 Intrinsic Semiconductors: Free Electrons and Holes

The creation of free electrons costs energy in a semiconductor, as I just argued, whereas in a metal it does not. But there is another difference between the two types of materials that follows from the picture in Fig. 12.10 above. This is illustrated more clearly in Fig. 12.11 below.

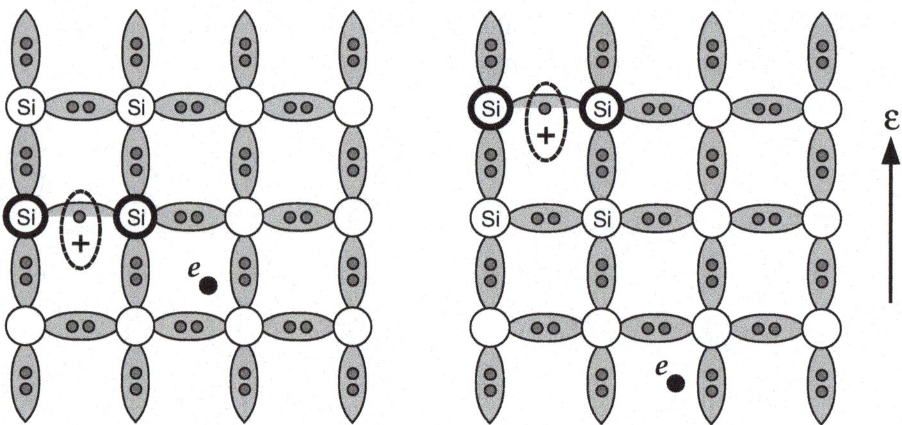

Figure 12.11: Semiconductor with a free electron and a free hole.

Note that the negatively charged free electron leaves behind a broken bond with a single electron. ***This broken-bond configuration represents a positive charge which itself is free to move.*** The broken bond is free in the same sense as the electron is free: The broken bond can be anywhere, between any two Si atoms, and its energy is still the same !

Another way to visualize the motion of the positively charged region is to think of the original broken bond as taking up an electron from another intact bond, thus creating a new broken bond. Again, in terms of an energy balance, there is no difference.

From the point of view of charge transport, this positively charged region acts as if it were a positive "particle" and therefore it has its own name: It is called a **hole**. This picture can be connected back to the energy bands by saying that a hole denotes an electron missing in the VB. (Recall that an intact Si-Si bond is equivalent to two electrons in the VB).

Our arguments so far can be summarized as follows: Let us define an **intrinsic semiconductor** as a very pure semiconductor, with no impurities or defects. Then we can say, again using Si as our example:

> In intrinsic Si, for every broken Si-Si bond you get:
> - one free electron in the conduction band, and
> - one free hole in the valence band.

The real consequence of having electrons and holes present in a semiconductor becomes apparent when we apply an electric field $\mathcal{E}$ (see Fig. 12.11). Since electrons and holes have opposite charge, physically they move in opposite directions under the influence of $\mathcal{E}$, but ***the electrical currents by electrons***

and holes add up ! A negative charge moving in one direction counts the same as a positive charge moving in the opposite direction.

This now requires that we modify Ohm's Law, Eq. 12.10, for semiconductors. Since we have to deal with two types of charge carriers, whose effects add up, we must set

$$\sigma = en\mu_e + ep\mu_h \tag{12.14}$$

where n and p are the electron and hole concentrations, respectively, e is the absolute value of the electron charge (i.e. $e = +1.6 \times 10^{-19}$ C, and the charge of a real electron is $-e$), and μ_e and μ_h are the electron and hole mobilities.

Moreover, in an intrinsic semiconductor free electrons and free holes are created in pairs, so that $n = p$. Therefore, we can write Eq. 12.14 as

$$\sigma = en(\mu_e + \mu_h) = ep(\mu_e + \mu_h) \tag{12.15}$$

In general, μ_e and μ_h are not equal in a semiconductor. Although both electrons and holes are free to move, usually $\mu_e > \mu_h$. This seems reasonable since an electron by itself should be able to move more easily than a hole (a partial covalent bond).

The electron concentration n in the CB was represented graphically in Fig. 12.9 above. Using Eq. 12.13, n can be expressed mathematically as

$$n = n_0\, f(E) = n_0 \frac{1}{1 + \exp(\frac{E-E_F}{kT})} \tag{12.16}$$

where n_0 is the concentration of energy states available to the electrons in the CB, and the Fermi function $f(E)$ denotes the probability that a state in the CB is occupied by an electron.

Eq. 12.16 looks cumbersome to use, but it can be simplified with some algebra. First, it can be shown that in an intrinsic semiconductor, the condition of $n = p$ requires that $E - E_F = E_g/2$. In other words, E_F must lie in the middle of the bandgap of the semiconductor. Second, it can be shown that around room temperature $\exp(\frac{E-E_F}{kT}$ is much larger than 1. So the rather complicated expression for $f(E)$ reduces to $\exp(-E_g/2kT)$, and to a very good approximation

$$n = n_0 \exp(-\frac{E_g}{2kT}) \tag{12.17}$$

Physically, $E_g/2$ in Eq. 12.17 represents an excitation energy. The factor 1/2 is due to the fact that by exciting an electron from the VB to the CB, thus spending an energy E_g, one actually creates two free charge carriers, an electron in the CB and and a hole in the VB. Hence the excitation energy per carrier is $E_g/2$. Note that you have seen this type of expression with an exponential probability factor before, for example in Chapter 4 on defects and Chapter 6 on diffusion.

For the purpose of this book, we will consider n_0 to be an experimentally determined parameter, which is also characteristic of a particular semiconductor, just as is E_g. The same applies to the mobilities μ_e and μ_h. However, they are all fundamental materials parameters. With knowledge of quantum theory, they can be calculated with excellent accuracy.

Numerical example: intrinsic Si

For Si at a temperature of 300 K we take $E_g = 1.1$ eV and $n_0 = 2.8 \times 10^{25}$ m^{-3} as given.

Also $\mu_e = 0.14$ m^2/Vs and $\mu_h = 0.05$ m^2/Vs. Then from Eqs. 12.15 and 12.17 we obtain:

$$
\begin{aligned}
E_g &= 1.1\,\text{eV} \\
n_0 &= 2.8 \times 10^{25}\,\text{m}^{-3} \\
n &= 2.8 \times 10^{25} \times \exp\left(-\frac{1.1 \times 1.6 \times 10^{-19}}{2 \times 1.38 \times 10^{-23} \times 300}\right) = 1.6 \times 10^{16}\,\text{m}^{-3} \\
\sigma &= 1.6 \times 10^{16} \times 1.6 \times 10^{-19} \times (0.14 + 0.05) = 5.0 \times 10^{-4}\,(\Omega\text{m})^{-1}
\end{aligned}
$$

As with all things electrical, units are tricky. You may find that you make the fewest mistakes if you use SI units throughout.

Table 12.5 displays a summary of numerical values for Si and Ge. Values for Cu are also listed, for a comparison with a typical metal. All the numbers are taken at a temperature of 300 K.

Table 12.5: Charge carrier concentrations, mobilities, and conductivities

Material	E_g (eV)	n (m^{-3})	μ_e (m^2/Vs)	μ_h (m^2/Vs)	σ (Ω^{-1}m^{-1})
Si	1.1	1.6×10^{16}	0.14	0.05	5×10^{-4}
Ge	0.67	2.6×10^{19}	0.38	0.18	2.3
Cu	—	$\sim 2 \times 10^{28}$	~ 0.02	—	6×10^{7}

The numbers in Table 12.5 make it clear that the much higher conductivity of metals relative to intrinsic semiconductors is due to much higher values of n for metals. On the other hand, the mobilities are comparable in the two types of materials. If anything, electron mobilities are lower in metals.

You will also notice that the temperature behavior of metals and semiconductors is totally different. For intrinsic semiconductor, σ increases rapidly, in fact exponentially, with rising temperature. For a metal, the conductivity decreases slightly with rising temperature.

Eq. 12.17 may be used to determine E_g of a semiconductor. If we take the natural log of Eq. 12.17, we obtain

$$\ln n = \ln n_0 - \frac{E_g}{2kT} \tag{12.18}$$

Therefore, if we measure the current I through a semiconductor as a function of T at constant voltage, as in Fig. 12.1, then Eq. 12.18 implies that ln I will be a linear function of $\frac{1}{T}$ with slope $-\frac{E_g}{2k}$.

12.5.4 Extrinsic Semiconductors: Doping

As we have seen, intrinsic semiconductors are inherently rather poor conductors. By themselves, they would not be very useful. The power of semiconductor materials comes mostly from the fact that by adding small amounts of the right kind of impurity, one can greatly increase the concentration of charge carriers. Furthermore, this can be done in two different, independent ways, so that either the concentration of free electrons or the concentration of free holes is greatly enhanced. In other words, one can modify a semiconductor so that electrical current is carried only by electrons, or only by holes.

To introduce these types of impurities is referred to as **doping** the pure material. The impurities themselves are called **dopants**. The method is general and works with all kinds of pure (i.e. undoped) semiconductors, but for a given material, only a handful of elements are suitable as dopants. I will explain the principle by focusing on the most important material, namely Si.

> The ***principle of doping***:
>
> In order for an impurity to be able to act as a dopant (in a manner to be described), the impurity atoms must have one valence electron more or less than the atoms they replace. This means that for Si, being a Group IVA material in the periodic table, the dopants must be either from Group IIIA or Group VA.

How doping works is best explained with a picture, Fig. 12.12. Our example is ***Si with P as the dopant***.

The P atoms, being from Group VA, have five valence electrons. When a P atom replaces a Si atom, only 4 of the 5 valence electrons make covalent bonds with the Si neighbors. The 5th electron is bound only very loosely to the P atom and is easily removed. Once it has moved away from its atom of origin, it is free !

The crucial thing in this case is that no strong Si-Si bonds need to be broken in order to create free electrons from the dopants. The energy required for shaking loose the 5th electron on the P atom is much smaller than the Si-Si

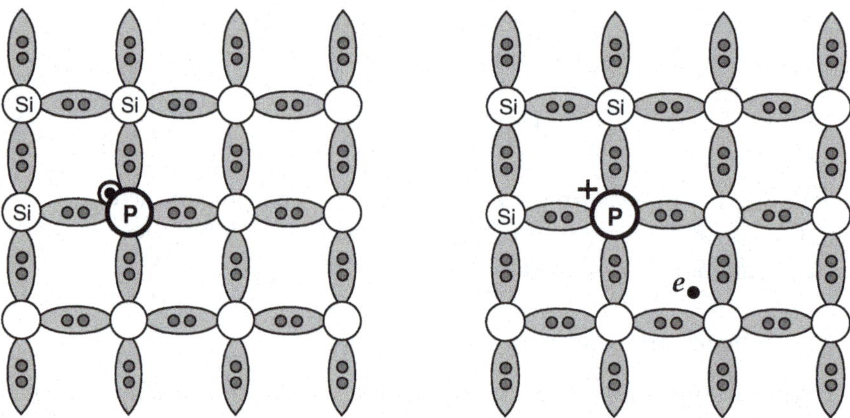

Figure 12.12: Si doped with P. Left panel: The 5th valence electron is still on the P atom. Right panel: The electron has been detached from P and made free.

bond energy. At the same time, please note that as an electron is freed from a P atom, ***no hole is created*** ! The positive charge left behind by the freed electron remains on the P atom, in effect making it an immobile P^+ ion. Hence, this type of doping preferentially enhances the concentration of free electrons.

Doping with a Group VA dopant makes the Si a so-called **n-type semiconductor**. The n in n-type refers to the extra negative charge carriers, i.e. the additional free electrons provided by the dopants. An n-type dopant is also called a **donor**, since it donates an electron to the conduction band.

In semiconductors it is often true that what applies to electrons also applies, in a slightly modified fashion, to holes. So it is with doping. Again let us consider an example, ***Si with B as the dopant***, a Group IIIA atom.

On the left in Fig. 12.13, the B atom is shown in its normal, neutral state. It has only three valence electrons and thus can make covalent bonds to only 3 of its 4 Si neighbors. However, it takes only very little net energy to remove an electron from a Si-Si bond and use it to complete the fourth B-Si bond. The broken Si-Si bond is the same as in Fig. 12.11, a hole free to move around the Si crystal.

This type of doping with a Group III dopant makes the Si a so-called **p-type semiconductor**. The p in p-type refers to the extra positive charge carriers, i.e. the additional free holes created by the dopants. A p-type dopant is also called an **acceptor**, since it takes on an electron from the valence band. Note that for each positive hole created, a negative charge remains on the B atom, in effect making it an immobile B^- ion.

Now we are in a position to connect our discussion of doping, breaking and reforming electron bonds, back to the picture of energy bands and electron energy states. In fact, some people argue that energy bands, and how they are

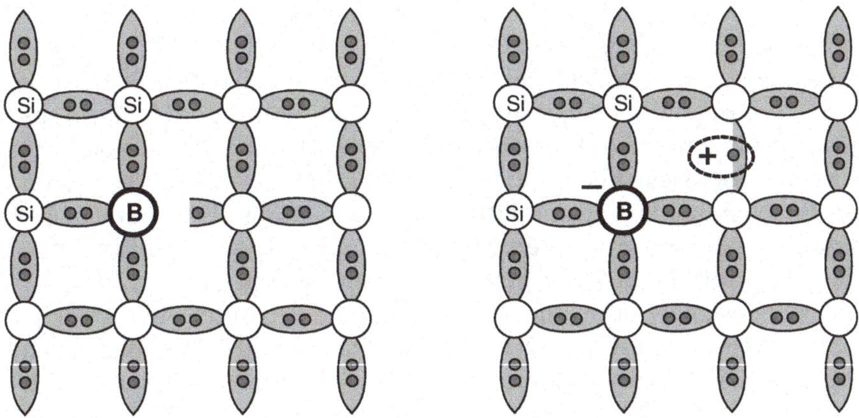

Figure 12.13: Si doped with B. Left: B atom with covalent bonds to three Si neighbors. Right: An electron from a Si-Si bond is used to make the fourth B-Si bond, and create a free hole.

populated by electrons, are the more fundamental description. However, both bonds and bands provide useful, but different, perspectives on the ultimate physical reality.

On the basis of what we know about what happens in doping, we can construct corresponding energy diagrams as follows:

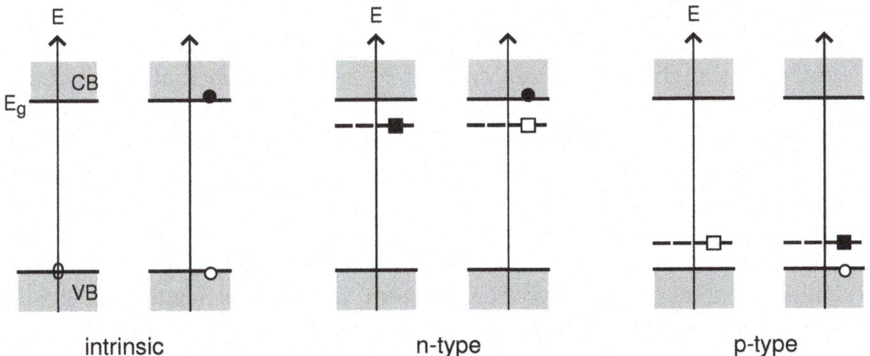

Figure 12.14: Energy band structures of intrinsic, n-type, and p-type semiconductors. The two graphs in each panel illustrate the non-excited state, and the state with free charge carriers either in the CB (n-type), or in the VB (p-type), or in both bands (intrinsic).

You will recognize that these bands are basically the same as those for the semiconductors in Fig. 12.5. The minor difference is that the bandgap region is magnified, and only the band edges are shown, not the entire bands. The zero of the energy scale is arbitrary. It is set conveniently at the top of the VB. This puts the bottom of the CB at $E = E_g$.

Free charge carriers are indicated by circles: electrons in the CB, and holes in the VB. The dashed lines denote extra energy states introduced by the dopants, let us say again either P or B. For n-type material, the fifth electron on P is held only loosely and excited easily to the CB. But as long as this electron is on the P, it is in an energy level just a little below the bottom of the CB. For p-type material, an initially empty energy level on B is located just a little above the top of the VB, but can easily be populated by an electron from the VB.

Squares designate immobile states. The open square in the middle panel denotes a P atom which has given off an electron to the CB. The filled square in the right panel stands for a B atom which has taken on an electron from the VB.

The ***point of doping*** a semiconductor material is really twofold:

1. Doping allows one to ***increase the conductivity of the base semiconductor material*** by many orders of magnitude. More specifically, by introducing dopants, one can ***engineer the conductivity of the semiconducting material*** quite precisely.
   We saw in the discussion of intrinsic material that around room temperature the intrinsic concentration of electrons n_i is typically of the order of $10^{16} m^{-3}$ in Si. By contrast, the purest Si material which can be produced contains impurities at a level of about $10^{19} m^{-3}$. (As an aside, this is incredibly pure compared to other engineering materials, say pure metals). Practical levels of doping range from about $10^{20} m^{-3}$ to $10^{25} m^{-3}$. They are often limited by how soluble a dopant is in Si.
   This means that under normal conditions and with typical doping levels, it is the charge carriers created by the dopants that dominate the material's conducting properties. The intrinsic concentrations are much smaller than the concentrations caused by doping. Therefore, intrinsic Si is not a practical material, but is used mostly as a hypothetical standard of comparison.

2. The fact that there exist two types of dopants enables the materials engineer to create two types of conductors, with either negative or positive charge carriers, i.e. with free electrons or free holes. Again this is very different from metals. It is this new option with semiconductors that makes it possible to ***construct all sorts of new kinds of electronic devices,*** by combining differently doped materials in clever ways, as we shall below.

It is much simpler to determine the conductivity σ for a doped material than for an intrinsic material. The reason is that the majority charge carrier, i.e. the one due to the dopant, is present in much higher concentration than the other carrier. Moreover, for Si at around room temperature, to a very good approximation all dopants are ionized. Thus the two types of doping work out as follows:

n-type:
Assume the donor concentration is N_d. Then Eq. 12.14 reduces to

$$n = N_d;\ p \approx 0: \quad \sigma = en\mu_e = eN_d\mu_e \tag{12.19}$$

p-type:
Assume the acceptor concentration is N_a. Then Eq. 12.14 reduces to

$$p = N_a;\ n \approx 0: \quad \sigma = ep\mu_h = eN_a\mu_h \tag{12.20}$$

The assumption that all dopants are ionized is a simplification. But the extra energy levels induced by the dopants (cf. Fig. 12.14) are usually close enough to the band edges so that the thermal energy in the solid at around room temperature is enough to ionize practically all dopants. In other words: For n-Si at room temperature, all dopants have given their extra electron to the conduction band; for p-Si, all dopants have filled their extra energy level with an electron from the valence band.

At very low temperature the above assumption breaks down. The thermal energy becomes too low to ionize all, or even just a good fraction, of the dopants. On the other hand, at high enough temperature, the contribution of intrinsic electrons may have to be included. The three ranges of temperature can be illustrated schematically as follows for n-type Si:

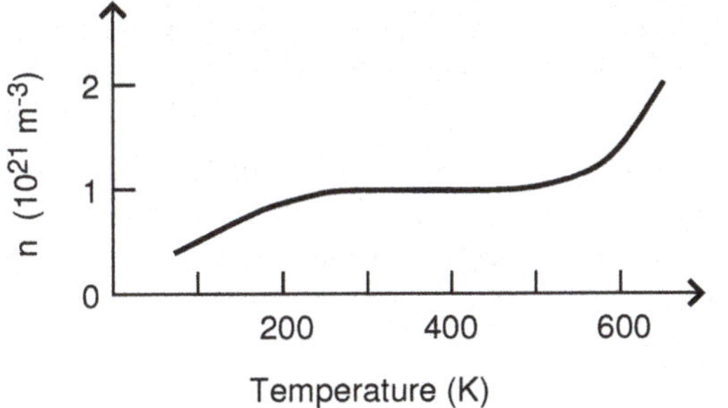

Figure 12.15: Schematic electron concentration vs. temperature for n-type Si with $N_d = 10^{21} m^{-3}$.

You will note that even at a rather low dopant concentration of $N_d = 10^{21} m^{-3}$, one must raise T to around 600 K for the intrinsic carrier concentration to become comparable to the extrinsic one from the doping. The figure also illustrates the completely different behavior of $n(T)$, thus $\sigma(T)$, of a semiconductor compared to a metal.

Our discussion of semiconductors so far has focused primarily on the concentrations of free charge carriers. However, the mobilities of electrons and holes are also affected by experimental variables, similar to the mobilities of conduction electrons in metal. For example, an increased temperature reduces these mobilities for the same reason as before with metals, namely because of increased scattering of charge carriers by vibrating lattice atoms. Also, at large dopant concentrations, there comes a point where scattering by the dopants reduces the mobility of the free charge carriers that were in fact created by these dopants.

12.6 Dielectrics, or Insulators

In order to round out our examination of electronic materials, we will discuss briefly the third type of such materials, namely insulators, also referred to as **dielectrics**. As you well know, these are materials which will not conduct electrical current when subject to normal laboratory conditions. By this I mean temperatures not close to the melting point and not too large electric fields. (These restrictions are a little fuzzy, but they can be made more quantitative if need be).

The classes of materials that exhibit this type of electrical behavior are the ceramics, inorganic glasses, and polymers. Again this is a rather broad statement that is supposed to capture the most important distinction, and there are exceptions to it. Certain ceramics may approach semiconductor-like conductivities due to a large amount of defects in them. Also, certain polymers have been synthesized with semiconducting properties.

The fact that insulators are materials which do not conduct electrical current does not mean that nothing electronic of interest happens to them. They are often used for their optical properties, since they are transparent. Optical properties are electronic in origin and derive from the band structure of a material. We will also see that dielectrics are useful for storing electrical charge.

Moreover, from a practical device point of view, insulators are just as important as the real conductors. Metals and semiconductors enable current to flow where it is supposed to, but insulators make sure it does not flow where it is not supposed to. Electronic devices make use of all types of materials and rely on their individual characteristics as well as their compatibility.

12.6.1 Optical Properties

Perhaps the most apparent property of insulators is that they are optically transparent when in single-crystal or amorphous form. On the other hand, metals and most semiconductors are opaque for visible light. As we will illustrate below, transparency is tied closely to electronic excitations, thus to the electronic band structure. But also note that a basically transparent material can be rendered opaque by a geometrical effect: If the material is polycrystalline, especially if the grains are small, then scattering of light by the grain boundaries can greatly reduce the amount of transmitted light.

Whether or not a material transmits light or, equivalently, does not absorb light, can be visualized readily on the basis of its band structure:

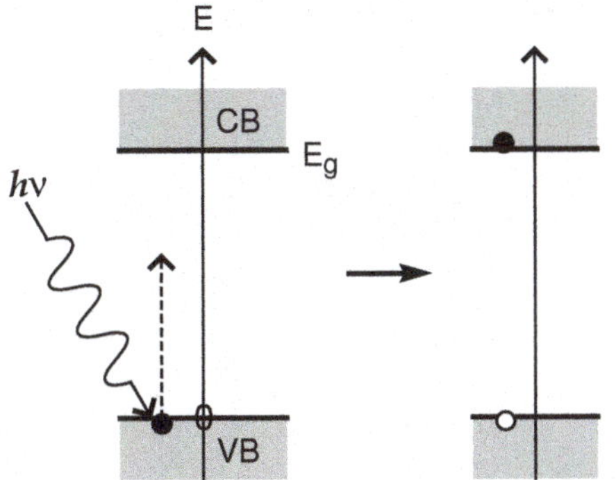

Figure 12.16:
Energy bands of an intrinsic semiconductors. In the right panel, a photon with enough energy has been absorbed to create an electron-hole pair.

In order to interpret the figure, you need to note some quantum properties of light. First, a light wave of frequency ν and wavelength λ consists of small energy packets called **photons**. The energy of a photon with frequency ν has an energy E_{ph} of

$$E_{ph} = h\nu \tag{12.21}$$

where h is **Planck's constant**, with a numerical value of 6.63×10^{-34} J s. In addition, you should recall that the frequency ν and wavelength λ of light are related to each other by the equation

$$\nu\lambda = c \tag{12.22}$$

where c is the speed of light (3×10^8 m/s).

When light is absorbed by a material, one photon gives its energy $h\nu$ to one electron at a time, thus promoting that electron from an energy state at the top of the VB to a higher state at the bottom of the CB. This process can only take place if the photon energy $h\nu$ is at least equal to the bandgap E_g. The only states available for an electron in the VB to be excited to are in the CB.

Fig. 12.16 shows on the left the VB and CB of an intrinsic semiconductor, with an impinging photon with energy energy $h\nu$ (cf. Fig. 12.14). Assuming $h\nu \geqslant E_g$, the photon is absorbed, creating an electron-hole pair, on the right of Fig. 12.16. If the photon energy $h\nu$ were less than E_g, no electron excitation could occur, and the photon would be transmitted through the semiconductor.

The threshold condition follows from Eqs. 12.21 and 12.22:

$$E_g = h\nu = \frac{hc}{\lambda}, \quad \text{thus} \quad \lambda = \frac{hc}{E_g} \tag{12.23}$$

Numerical values for this condition calculated from Eq. 12.23 are listed in Table 12.6:

Table 12.6: Conditions for the absorption of light

Material	E_g (eV)	Threshold λ for absorption
Ge	0.67	1.85 μm
Si	1.1	1.13 μm
GaAs	1.4	0.88 μm
metal	—	none

The numbers for the threshold λ indicate the wavelength of light above which the material is transparent.

You will note from Eq. 12.23 that the photon energy $h\nu$ increases of course with increasing ν but decreases with increasing λ, given that the speed of light c is constant.

The process of absorption of light, or radiation in general, is of course central to the issue of detecting radiation. The same type of mechanism forms the basis for the inverse of the process, namely the emission of light. If one can produce a non-equilibrium population of electrons and holes in a semiconductor by other means, then the electrons and holes may recombine to release the bandgap energy in the form of emitted light.

This general picture also explains why metals absorb essentially all kinds of light, i.e. all kinds of photons. Since a metal does not have an energy gap, there is no restriction on the photon energy for absorption. All photon wavelengths are absorbed (see Table 12.6).

12.6.2 Electrical Capacitance

Consider the following setup that should remind you somewhat of our discussions of Ohm's Law (cf. Fig. 12.1):

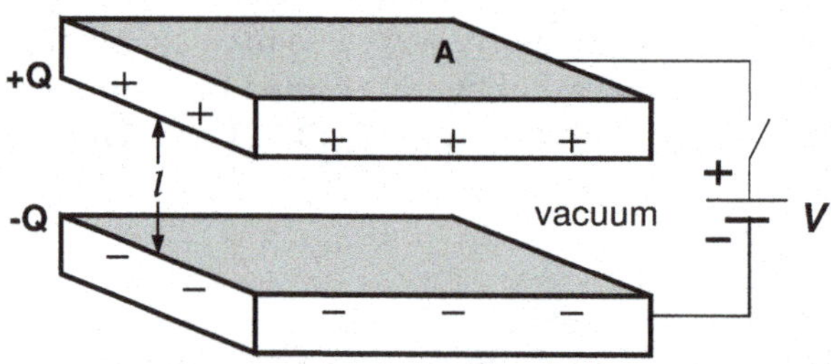

Figure 12.17: Setup for measuring electrical capacitance.

The main element of the circuit is a **capacitor**, consisting of two parallel metal plates with area A in vacuum (or, practically, in air), separated by a distance l. When the switch is closed and the voltage V is connected to the capacitor, the battery causes charges +Q and −Q to flow onto the plates. Thus the capacitor stores electrical charge.

One observes that the amount of charge Q is proportional to V. As usual, we can formulate this relation as an equation:

$$Q = CV \tag{12.24}$$

where C is called the **capacitance** of this capacitor.

The amount of charge stored, thus the value of C depends on the geometry of the plates. One finds that, keeping V fixed, $C \propto A$ and $C \propto l^{-1}$. Once again these relationships can be written as another equation:

$$C = \epsilon_0 \frac{A}{l} \tag{12.25}$$

where the proportionality factor ϵ_0 is called permittivity of the vacuum.

As regards units, again we will work within the SI system. The units of Q are of course Coulomb (I will abbreviate this here as Clb in order that there is no confusion with the capacitance C). Hence the units of C are 1 Clb/V, which has a name of its own, namely a **Farad** (1 F). This requires that the units of ϵ_0 be Clb V/m or F/m. The value of ϵ_0 turns out to be 8.85×10^{-12} F/m.

As I mentioned at the beginning of this section, the circuit in Fig. 12.17 is reminiscent of the one for Ohm's law in Fig. 12.1. The two circuits would be the same if, instead of vacuum (an insulator), a conducting medium connected the plates in Fig. 12.17. The correspondence becomes even more evident when we rewrite Eqs. 12.1 and 12.24 as follows:

$$R = \frac{V}{I} \quad \text{and} \quad \frac{1}{C} = \frac{V}{Q} \tag{12.26}$$

So R of a resistor is to flowing charge I (Clb/s) what 1/C of a capacitor is to static charge Q (Clb).

When the space between the two plates of a capacitor is filled with a dielectric medium, experiments show that at constant applied voltage V the charge Q' on the plates is higher than the charge Q before. In other words, the dielectric between the plates increases the capacitance C (Fig. 12.18). In this case C is given by the expression

$$C = \epsilon \frac{A}{l} = \frac{Q'}{V} \tag{12.27}$$

where ϵ is the permittivity of the dielectric material. Note the similarity of Eqs. 12.25 and 12.27.

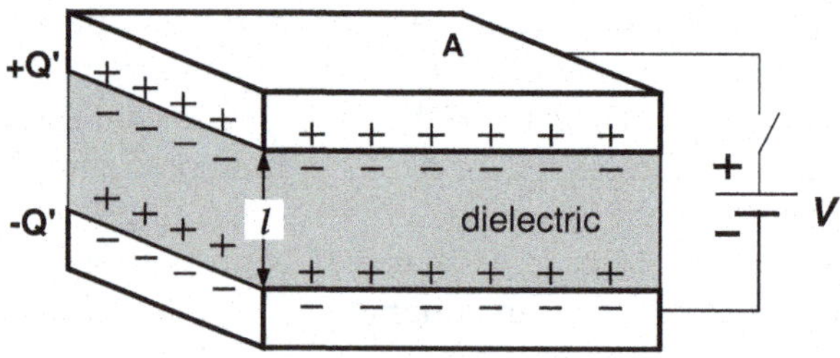

Figure 12.18: Capacitor with dielectric medium between the plates.

The constant ϵ can be written as

$$\epsilon = \epsilon_0\, \epsilon_r \quad \text{with} \quad \epsilon_r > 1 \tag{12.28}$$

The factor ϵ_r by which the capacitance has been increased due to the dielectric is called its **dielectric constant**. It is a fundamental property of the material between the capacitor plates.

The physical reason for the effect of the dielectric on the capacitance is suggested in Fig. 12.18. What really determines the charge on the capacitor plates is the electric field $\mathcal{E}$ between the plates, and for large plates close together $\mathcal{E} = V/l$. A dielectric material has certain positive and negative charges built in, either in the form of ions or just atoms. These charges move slightly from their normal positions in the presence of $\mathcal{E}$, such that the $+Q'$ charges on the upper plate face negative charges just inside the dielectric, and vice versa on the lower plate. This means that the dielectric becomes polarized, acquiring an electronic dipole moment. This **polarization** tends to oppose the effect of the externally applied potential. The consequence is that additional charges have to be supplied by the battery to the plates in order to maintain a fixed electric field $\mathcal{E}$.

A few numerical values of dielectric constants are listed in Table 12.7. These are static values: They are valid for a constant applied voltage. The values may be different for an applied ac voltage. In general, ϵ_r is a function of the frequency of the applied voltage (see below).

The lowest values of ϵ_r are observed for nonpolar gases. Common nonpolar liquids and polymers are a little higher. Many ceramics are higher yet by a factor of 2-3. The reason why quartz is listed with two values is that quartz is an anisotropic dielectric: Its ϵ_r depends on the crystallographic direction in which the electric filed is applied. Still much higher values are observed for liquids consisting of polar molecules, such as ethanol and water. $SrTiO_3$ (strontium titanate) and $BaTiO_3$ (barium titanate) are in a class by themselves. They belong to a group of materials called **ferroelectrics**, with extremely high ϵ_r. This is due to the fact that such materials have a permanent electric dipole moment built in, which is present even without an externally applied field.

Table 12.7: Static dielectric constants of various materials

Material	ϵ_r	Material	ϵ_r
Vacuum	1.0000	Dry air	1.0006
Methane	1.7	Chlorine	2.0
Gasoline	2.0	Dodecane	2.0
CCl_4	2.2	Olive oil	3.1
Ethanol	24	Water	80
PTFE	2.0	HDPE	2.3
PMMA	2.9	PVC	3.1
Quartz	4.69-5.06	fused SiO_2	3.8
Pyrex glass	5	Diamond	5.0
NaCl	5.9	MgO	9.6
$SrTiO_3$	300	$BaTiO_3$	1500

The listed ϵ_r values can be rationalized on the basis of the mechanism leading to electrical polarization in a given material. Recall that the internal effect of the electric field set up by the external voltage is to move charges slightly away from their normal equilibrium positions. The larger this charge separation, or polarization, the larger is the dielectric constant of the material.

There are three physical mechanisms which can give rise to polarization. Which one of the three is operative in a material depends on the material's internal structure.

Electronic polarization

This type of polarization involves a slight relative shift of the electron cloud around a nucleus or in a covalent bond between nuclei. Think of this mechanism as pulling the electrons slightly one way, thus making the charge distribution somewhat asymmetric.

Ionic polarization

This is polarization due to positive and negative ions being pulled slightly in opposite directions by the electric field.

Orientational polarization

In this case, polarization occurs from the forced alignment of molecules with permanent electric dipole moments along the applied electric field.

The observations outlined in Table 12.7 can be explained roughly as follows: For gases, liquids, and polymers not containing molecules with dipole moments, electrical polarization is primarily of the electronic type. Electron clouds respond readily to an applied electric field, due to the small electron mass, but are displaced very little, due the tight bonding to nearby nuclei. For ceramics and inorganic glasses, being substantially ionic solids, it is the positive and negative ions which respond to an applied field. The resulting ionic polarization is greater than the electronic polarization, but the response of the heavy ions is slower than the light electrons. In the case of liquids made up of polar molecules, polarization is orientational and large, but the alignment of the molecules to the applied field is rather slow.

These characteristics are summarized in Table 12.8:

Table 12.8: Characteristics of polarization mechanisms

Mechanism	Features
Electronic	small polarization, fast response
Ionic	medium polarization, medium response
Orientational	large polarization, slow response

The response time provides the clue as to how ϵ_r depends on the frequency of the applied field. If t_p is a characteristic time for the polarization to change, then the polarization cannot follow a change in the applied electric field that occurs in a time much shorter than t_p. This implies that the frequency t_p^{-1} for the applied field represents a limit with regard to the ability of that type of polarization being able to contribute to ϵ_r.

Thus we may conclude that the contribution from electronic polarization will be active up to very high frequencies of the order of 10^{15} Hz. On the other hand, ionic polarization will only be active over a limited frequency range, up to around 10^{12} Hz, and will drop to zero at very high frequencies. For orientational polarization, the active frequency range will be even more limited, up to around 10^{6} Hz.

The two materials $SrTiO_3$ and $BaTiO_3$ from Table 12.7. are in a separate class. Not only are they ferroelectric, as mentioned, with a very large ϵ_r due to a permanent internal polarization, but they are also **piezoelectric**. A piezoelectric material develops polarization, i.e. a potential difference between two sides, if it is stressed or deformed mechanically. Conversely, if a voltage is applied to such a dielectric, it expands or contracts as if it were under mechanical stress.

The piezoelectric effect was discovered by Jacques and Pierre Curie in 1880, studying quartz and rutile (TiO_2) crystals. It also depends on the temperature: It disappears above a certain critical temperature T_C, called Curie temperature. T_C is typically in the range of 120-300 °C. The effect, and its disappearance, are due to a change in crystal structure. Above T_C the crystal structure is cubic, but it distorts to the tetragonal structure (almost cubic, with one side slightly different) below T_C.

The piezoelectric effect is enormously useful for making all sorts of sensors, actuators, and transducers. The material used most often in applications is lead zirconate titanate (abbreviated by PZT), a mixture of $PbTiO_3$ and $PbZrO_3$ with the formula $Pb[Zr_xTi_{1-x}]O_3$, $0<x<1$. Doping of these PZT material with, for example, Ni, Bi, Sb, or Nb ions, makes it possible to adjust individual piezoelectric and dielectric parameters as required. In the engineering literature, these kinds of materials are also referred to a **piezoceramics**.

One ubiquitous application of a piezoceramic is the quartz tuning fork which forms the timing standard in wrist watches. Its mechanical resonance frequency is extremely stable, and its oscillation is maintained by a battery, using the piezoelectric properties of quartz. Piezoceramics are also in wide use in cell phones and other gadgets as acoustic transducers (microphones, speakers). In the science laboratory, the materials are being used as ultra-sensitive detectors of mass changes and in high-precision positioning devices.

12.7 Electronic Devices

Our discussion of electronic properties so far has been based on two concepts: Energy bands of electron states in a solid, and atom-atom chemical bonds in a solid. We saw that, in particular for semiconductors, the two points of view are complimentary and illuminate different aspects of the creation and movement of free charge carriers.

For the application of electronic properties to devices, we will focus on the energy bands. As we will see, the characteristics of electronic devices emerge naturally, and can be best understood, on that basis. We will be particularly interested in semiconductor devices, but they require an integrated use of metals and insulators as well.

It is convenient to classify individual electronic devices according to the number and type of materials used. The general question for all devices is what useful function they can perform, either by themselves, or as sensor responding to an external stimulus, or as actuators initiating an external response.

12.7.1 Single-Material Devices

Single material devices are somewhat limited in their range of applications, but examples of such devices do exist. I will list a few examples briefly, in no particular order.

Thermometers

We found that in both metals and semiconductors the electrical conductivity σ, or equivalently, the resistivity ρ depends on temperature T.

Metals may be useful as temperature sensors for two reasons. First, as we saw earlier, there is generally a certain range of temperature in which ρ is a linear function of T. This makes it straightforward, with a simple calibration, to use the value of ρ as a measure of T. At very high temperature, the resistivity of metals may also be useful simply because other materials or measuring devices cannot be operated under those conditions.

Semiconductors are used as thermometers in certain low-temperature applications, in a range where σ is a strong function of T because the dopants are only partially ionized. For example, for the material shown in Fig. 12.15, that would be below 150 °C or so. The advantage of this kind of thermometer would be high sensitivity at low temperature (in contrast to a metal), but neither ρ nor σ would be a linear function of T, and thus calibration would be more complicated.

Photoconductors

An individual piece of the right kind of semiconductor may be used as a light detector by exploiting the effect of photoconductivity. We mentioned earlier in our discussion of optical properties that photons can be absorbed by a semiconductor if the photon energy is equal to or greater than the bandgap E_g.

If that condition is fulfilled, then every photon absorbed creates an electron-hole pair. With a suitably designed device configuration, these charge carriers can be separated and collected, and an electrical current can be measured which is directly proportional to the photon flux impinging on the material. A problem with such devices is that the photon-generated electrical current may be rather small and hard to measure accurately.

12.7.2 Two-Material Devices: Semiconductor Diodes

More interesting devices can be made from combining two different semiconducting materials, specifically n-type and p-type of the same base material.

In the simplest case, an n-type and p-type piece of Si are "stuck together" to form what is called a **p-n junction**. For the device to work properly, it has to be fabricated all in one single crystal. The interface between the n- and p-region is just a transition in the type of doping. If two pieces of material were stuck

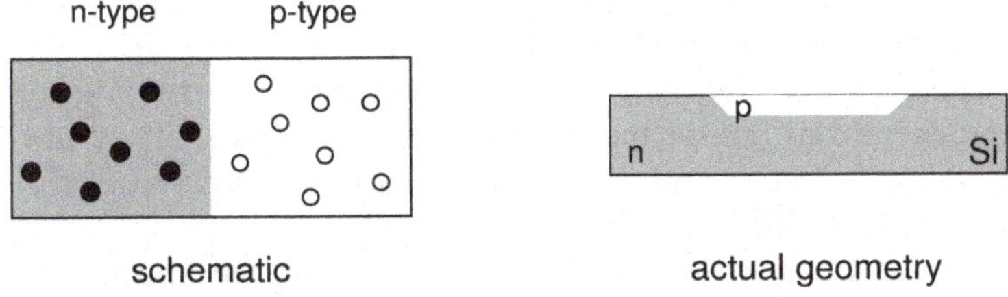

Figure 12.19: Schematic and actual layout of a pn-junction

together literally, structural defects at the interface would prevent the device from working properly.

Electrically speaking, this is not just a resistor but an asymmetrical device: The current through it depends on how the voltage is applied. Using the schematic representation from above, we can distinguish the following two setups:

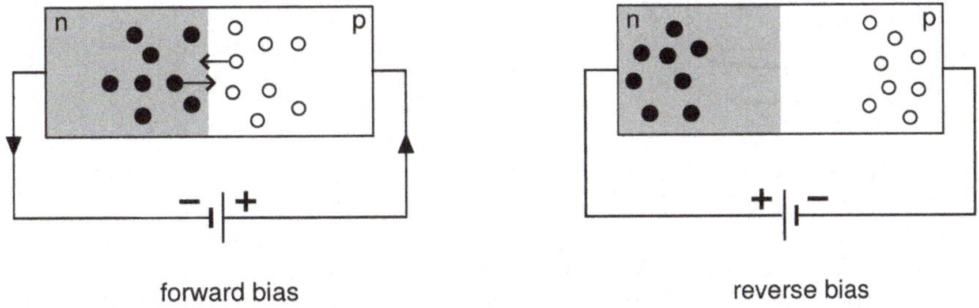

Figure 12.20: p-n junction in forward or reverse biased mode.

When the + from the battery goes to the p side and the − to the n side, the junction is said to be **forward biased**. This drives electrons on the n-side and holes on the p-side towards and across the junction. Now, in equilibrium, on the n-side electrons are the majority carriers and no holes are present. When holes from the p-side are injected to the n-side, a non-equilibrium situation arises. The consequence is that holes which make it to the n-side will recombine with electrons, giving off an energy E_g. On the p-side the exact opposite happens: Electrons which make it to the p-side will recombine with holes there. From the point of view of the circuit, this means that a continuous current is flowing. Electrons on the n-side are replenished by electrons moving from the battery into the material, and holes on the p-side are replenished by electrons moving out to the battery.

When the polarity of the battery is reversed, so that − goes to the p side and + to the n side, the junction is said to be **reverse biased**. In that case, the charge carriers are pulled away from the junction on either side. Hence no current flows across the junction, and therefore no current flows in the external circuit.

This means that the p-n junction is a unidirectional electrical conductor, or in device terms a semiconductor **diode**. In can conduct current only in one direction. Diodes are useful as devices for rectifying an alternating current, by suppressing current in one of the two possible directions. They are applied widely in practical circuits and can be designed in various physical sizes and for a wide range of power levels.

It turns out that this either/or property is actually an approximation. In reverse bias, there is a small residual current flowing across the junction, so that the I(V) (current vs. applied voltage) characteristic of the diode looks as illustrated below. In the figure, a positive V means that the + end of the battery is connected to the p side.

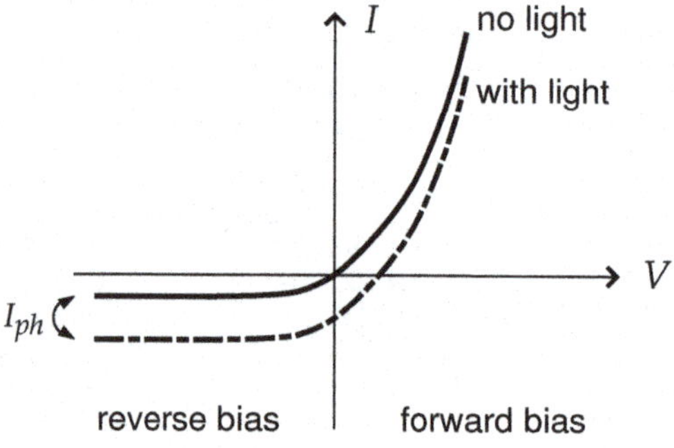

Figure 12.21: Current-vs-voltage characteristic of a diode, without or with light being absorbed.

The residual, voltage-independent current flowing under reverse bias is caused by diffusion of the charge carriers to the junction. The electric field inside the material in reverse bias creates a large gradient in the carrier concentrations, and as you know, a concentration gradient gives rise to a diffusional flux.

Semiconductor diodes can be made into efficient light detectors called **photodiodes**, in principle simply by providing a small window through which light can shine onto the p-n junction.

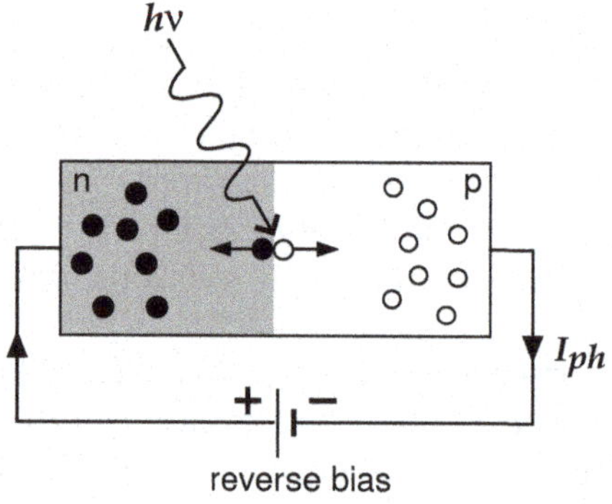

Figure 12.22: Semiconductor diode as a light detector. Absorption of photons in the junction region causes additional current to flow in the reverse direction.

When photons with energy $h\nu > E_g$ impinge onto the junction, they can be absorbed in the junction region. For every absorbed photon, an extra electron-hole pair is created which registers as extra current in the external circuit. Note that the photodiode is reverse-biased and the photo-induced current is in the reverse direction (cf. Figs. 12.22 and 12.21). This assures that the difference in current without and with photons, I_{ph}, is substantial, and much larger than in the case of the simple photoconductor discussed earlier.

12.7.3 Three-Material Devices: The Bipolar Transistor

Even more interesting devices can be built by combining p-type and n-type material to form two junctions. This can be done in two ways. Shown schematically in Fig. 12.23 is one way, a so-called n-p-n transistor. Its mirror image of a p-n-p transistor can be obtained by exchanging n- and p-type materials and simultaneously reversing the battery voltages.

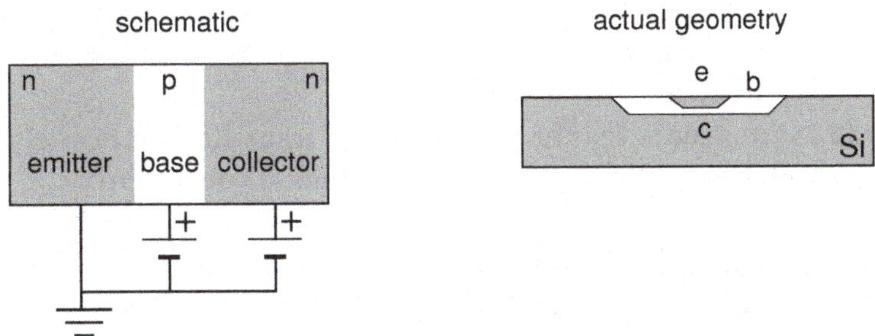

Figure 12.23: Layout of an n-p-n transistor.

An n-p-n transistor has three regions – emitter, base, and collector – and two junctions (Fig. 12.23). The emitter-base junction is forward biased, which causes injection of electron current into the base region. The base-collector junction is reverse-biased and would appear not to let any current go across. However, the base region is extremely thin (right panel in Fig. 12.23), so that most electrons entering the base end up diffusing into the base-collector junction, and from there are swept into the collector region.

Thus almost the entire electron current flows right through the transistor from emitter to collector, and only a small fraction leaves the transistor at the base. In device terms this means that a small base current turns into a large collector current. Hence the device acts as a current amplifier.

The exact same mechanism operates in a p-n-p transistor. One only has to switch all the signs: type of material, charge carrier, electrical current, and battery polarity.

12.7.4 The MOSFET Transistor

The **MOSFET transistor** is another three-terminal device, in which one terminal controls the current between the other two. But in its layout and operation it differs substantially from the n-p-n transistor. The MOSFET is the fundamental building block of all digital electronic circuits. MOSFET stands for Metal Oxide Semiconductor Field Effect Transistor.

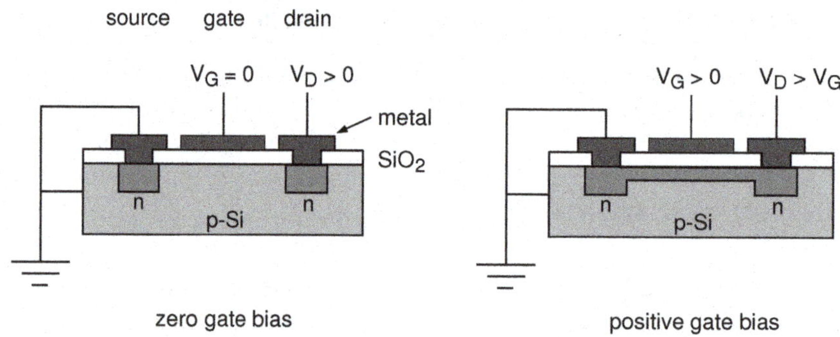

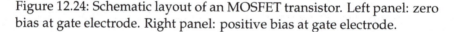

Figure 12.24: Schematic layout of an MOSFET transistor. Left panel: zero bias at gate electrode. Right panel: positive bias at gate electrode.

The device in our example is made in p-type Si as the base material. It has three terminals, called source, gate, and drain. The source and drain are two regions of n-type Si inlaid into the p-Si. The gate electrode is insulated from the central p-region by a thin layer of SiO_2. The metal pads are actually wires for making contact to other devices.

The panel on the left shows the transistor with zero bias V_G at the gate. This implies that the transistor is in a non-conducting state between source and drain because the junction between the central p-region and the n-type drain is reverse biased.

The key thing for the operation of the device is what happens when a positive voltage V_G is applied to the gate. Because of the SiO_2 insulating layer between the gate electrode and the p-Si underneath, the gate acts as a capacitor. In the p-Si the majority charge carriers are the holes. As V_G becomes more and more positive, the electric field pushes these holes down. At the same time, the few electrons still present in the p-Si are being pulled up to the SiO_2/p-Si interface. If V_G is large enough, a point is reached where the electron concentration at the top of the p-Si under the SiO_2 becomes large enough so as to turn a thin layer of p-Si effectively into n-type material. When this condition of inversion is reached, a continuous conducting path of n-Si exists from source to drain, and the device has become conductive.

So in the simplest case, the MOS transistor acts as a switch: It can be turned on and off by the voltage gate V_G. This is the basic operation of the device in a digital circuit. However, it is also possible to operate it in an analog mode.

The type of transistor described above is called an n-MOS transistor, in order to highlight the fact that the charge carriers have a negative charge (they are electrons). One can also make p-MOS transistors by reversing all the relevant polarities (dopants, charge carriers, applied voltages). Moreover, it is possible to have both n-MOS and p-MOS transistors on the same circuit. In that case, one speaks of CMOS technology, meaning complementary MOS.

Fig. 12.25 illustrates the process flow in manufacturing integrated circuits containing MOS transistors.

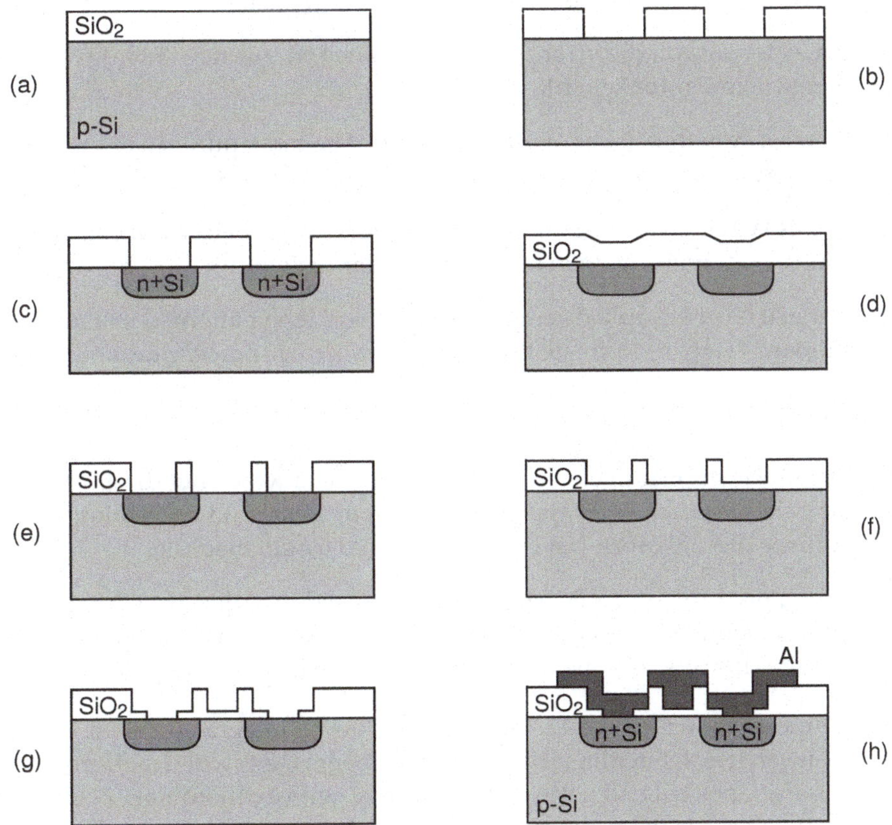

Figure 12.25: Simplified step-by-step process for manufacturing an integrated circuit containing n-MOS transistors.

The figure is schematic and simplified: Certain individual steps are lumped together. The key stages are the following:

(a) The process sequence starts with a p-Si wafer, on top of which a layer of SiO_2 is grown. This is a fairly simple process. It only requires that the wafer be exposed to an oxidizing ambient such as air at a high temperature for a certain period of time, so that oxygen can react with the Si.

(b) In areas reserved for the source and drain regions, the SiO_2 is etched away. (These areas are defined by a photographic process requiring several steps itself).

(c) The wafer is then exposed to a gas containing n-type dopants at high temperature. This causes dopants to diffuse into the Si surface and form the source and drain regions. (n+ Si means strongly n-doped Si).

(d) A second layer of SiO_2 is grown over the entire wafer.

(e) Areas for source, drain, and gate are defined in this new SiO_2 layer in a manner similar to Step (b).

(f) A third, very thin, layer of SiO_2 is grown. The main purpose of this layer is to serve as the insulator between the gate electrode and the Si substrate.

(g) In this last SiO_2 layer, small holes are opened up for electrical contacts to source and drain, again in a manner similar to Step (b).

(h) A metal layer is applied across the wafer, and then patterned in a manner similar to Step (b) to create the connections to the source, gate, and drain.

The power of the technology comes from two special features:

- The process is massively parallel. On a single Si wafer, hundreds of complex circuits, each with approximate size of 1 cm^2 and containing many millions of transistors, can be manufactured simultaneously.
- The same types of process steps can be utilized to realize the CMOS technology mentioned above, with n-MOS and p-MOS transistors on the same circuit. This provides the circuit designer with a great deal of flexibility.

The process shown in Fig. 12.25 illustrates how an integrated circuit is built layer-by-layer. It is a thin film technology: One begins with a Si wafer, and then materials are deposited and patterned one by one, with extraordinary precision, since every layer has to be aligned perfectly with the ones underneath. In the current technology, the smallest feature on an integrated circuit, the gate of an MOS transistor, is less than 50 nm wide.

Fig. 12.25 also makes it clear that integrated circuit technology involves the use of all types of materials – semiconductors, inorganic glasses, metals, sometimes even polymers – in a highly integrated fashion. A critical aspect is the compatibility of materials at their interfaces. The bond between two materials in contact should be strong, yet the two materials should not affect each other in a negative way.

Materials and processing development has been crucial in enabling advances throughout the history of microelectronics manufacturing. I will mention here just two examples in CMOS technology: The material used to connect circuits with each other, and the gate insulator material.

As regards the metal for circuit interconnections, during the early years the material of choice was aluminum. It was cheap, fairly straightforward to process, and had no compatibility issues. During the 1990s it became apparent that future advances required more efficient circuit interconnections, and the only potential substitute for Al was Cu. However, copper brought with it a host of issues: For example, Cu was poison for Si and had to be kept way from it. Cu also adhered very poorly to SiO_2. In addition, Cu required a complete change in the way patterned wires were fabricated. All these issues were resolved in a short period of time , and now Cu is the material for connecting circuits on a chip.

As transistors and circuit elements on a chip continued to shrink to ever smaller dimensions, it became clear in the 2000s that the insulating SiO_2 layer between the gate electrode and the Si was becoming so thin, of the order of 10 nm or less, that it was difficult to control its properties. Hence there was a need for replacing this SiO_2 with a different, thicker gate insulator, which in turn required a material with a higher dielectric constant. After extensive development, HfO_2 was found to be a suitable replacement for SiO_2: ϵ_r is 3.9 for SiO_2 and 25 for HfO_2.

Image sources

Title picture: IBM press release (circa 2000)
Scanning electron micrograph of a cross-section of an integrated circuit.

References

Shackelford, Smith, and Callister and Rethwisch, all cover electronic properties at a level similar to this book.

For technical data on engineering metals and alloys, including electrical properties, you may wish to consult
`http://www.espimetals.com/index.php/technical-data/`
`http://www.hpmetals.com/Metals-Alloys/Comparative-Data.aspx`

Specifically for spec sheets, go to
`http://www.hpmetals.com/spec-sheets/index.aspx`

For general resistivity data see
`http://www.engineeringtoolbox.com/resistivity-conductivity-d_418.html`

An exhaustive discussion of resistivities of metals from cryogenic temperatures to near the melting point can be found in:
`http://www.nist.gov/data/PDFfiles/jpcrd258.pdf`
for Hf, Ta, Mo, Zn, W, and
`http:\\www.nist.gov/data/PDFfiles/jpcrd155.pdf`
for Cu, Au, Pd, Ag.

For original work on the effect of mechanical deformation on the resistivities of metals, consult:
T. Broom, Aust. J. Sci. Res. A5 (1952) 128-134
`http://articles.adsabs.harvard.edu/full/gif/1952AuSRA...5..128B/0000128.000.html`

For information on insulators and dielectric constants, check out:
`http://hyperphysics.phy-astr.gsu.edu/hbase/tables/diel.html`
`http://my.execpc.com/~endlr/dielectric_const_.html`

You can find additional information on piezoelectric materials and various applications at
`http://www.piceramic.com/piezo_effect.php`
`http://www.morgantechnicalceramics.com/products-materials/about-ceramics/electro-ceramics/lead-zirconate-titanate-pzt/`

For a different perspective on Si properties and technology, have a look at
`http://www.tf.uni-kiel.de/matwis/amat/semi_en/kap_3/backbone/r3_1_1.html`

Exercises

1. The Ta wire of an electric heater is 0.6 mm in diameter and 30 cm long. What is its resistance ?

2. The conductor below is typical of a "wire" found in an integrated circuit.

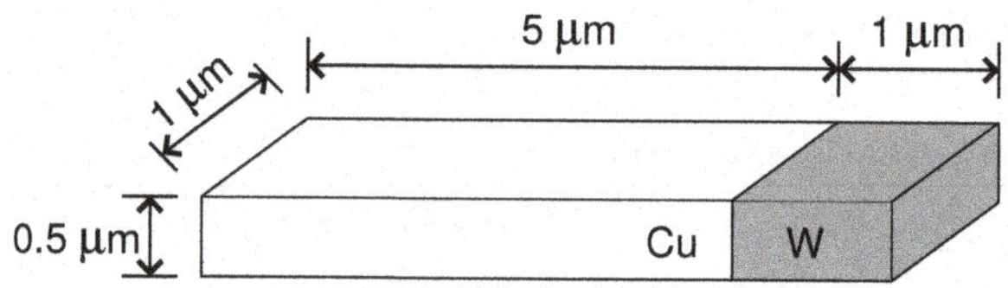

a) What will be the total resistance along the length of the conductor at room temperature ?
b) What fraction of this total resistance is due to the W part ?

3. You have designed an electrical resistance heater using Ta for the heating element. Your design was based on resistivity data at 20 °C. By how much will your resistance be in error if the heater is operating at 100 °C ?

4. From the data in the table below, determine how good the assumption of a constant α is for Ta in the temperature range of 0 - 200 °C.

T (K)	250	273	300	350	400	450	500
ρ ($10^{-8}\Omega$m)	11.1	12.2	13.5	15.8	18.2	20.6	22.9

5. At what temperature would the resistivity of Al be the same as the resistivity of Cu at room temperature (293 K) ?

6. Find another alloy with a temperature coefficient of resistivity of the same order as Constantan. Give the relevant parameters of this alloy and compare it to Constantan.

7. Do the data for Constantan and Nichrome in Table 12.1 agree with the type of behavior displayed in Fig. 12.7 ?

8. Data such as those in Fig. 12.7 can often be approximated quite well with an expression of the form

$$\rho_i = c\,\rho_B + (1-c)\rho_A + \gamma c(1-c)$$

where A and B are the two metal components, c is the fraction of B in the material, and γ is a suitable constant.

1) Come up with a qualitative argument that this expression is reasonable.
2) Determine a value for γ which provides a good description of the data in Fig. 12.7.

9. Determine the range of the energy E over which the Fermi function $f(E)$, Eq. 12.13, goes from a value of 0.95 to 0.05.

10. Measurements of current as a function of temperature, at constant voltage, for an unknown semiconductor material have yielded the following results:

T (°C)	0	20	40	60	80
I (A)	1.8×10^{-5}	4.6×10^{-4}	2.0×10^{-3}	9.8×10^{-2}	9.0×10^{-1}

What is the bandgap of this material ?

11. What is the probability for an electron to be excited into the conduction band for:
    a) intrinsic Si at 300 K, b) intrinsic Si at 500 K,
    c) intrinsic Ge at 300 K, d) intrinsic GaAs at 300 K ?

12. What kinds of dopants, if any, are the following atoms in Si:
    a) In; b) As; c) Ge; d) Zn ?

13. If Si is doped with P atoms at a concentration of $10^{20}m^{-3}$, what impurity level is this equivalent to in parts per million, or parts per billion ?

14. Consider n-type Si. At what temperature does the intrinsic concentration of free electrons become equal to the extrinsic concentration from the dopants for the following dopant concentrations:
    a) $N_d = 10^{21}m^{-3}$,
    b) $N_d = 10^{22}m^{-3}$?

15. a) Calculate the electrical conductivity for n-type Si doped with As at a concentration of $2 \times 10^{22}m^{-3}$ and compare it to intrinsic Si.
    b) Calculate the electrical conductivity for p-type Si doped with Sb at a concentration of $2 \times 10^{22}m^{-3}$ and compare it to intrinsic Si.

16. Devise an argument as to why ϵ_r is considerably lower for PTFE than for PVC.

17. Explain the statement:
    The position of the Fermi function $f(E)$ is given by the Fermi energy E_F and the shape of $f(E)$ by the temperature T.

Problems

1. What is the resistance of a filament in an incandescent light bulb ?

2. Why is copper used for power cables, but brass for power plugs ?

3. Your task is to develop a Cu-Ni alloy with the proper composition for use as an electrical resistor. You have the following data:

$$\rho(\text{Cu}) = 1.7 \times 10^{-8}\,\Omega\text{m}$$
$$\rho(\text{Ni}) = 7.1 \times 10^{-8}\,\Omega\text{m}$$
$$\rho(\text{90Cu-10Ni}) = 1.4 \times 10^{-7}\,\Omega\text{m}$$

Is there a way of coming up with a reliable estimate of ρ(50Cu-50Ni) just from the data above, without making the alloy and measuring it ?
Tip: Have a look at Exercise 8.

4. In Table 12.5, what is the cause of the much higher value of n for Ge compared to Si ?

5. Assume that in an integrated circuit employing intrinsic Si, the smallest Si "wires" conducting current are 1 μm wide, 0.5 μm thick, and 4 μm long. Current along the length of these wires is driven by a voltage of 0.2 V. What will be the current through the Si wires at room temperature ? On the basis of your result, do you think intrinsic Si is a feasible material for use in integrated circuits ?

6. If you formed a dilute alloy of Ge in Si, what would be the electronic and mechanical consequences ?

7. What would the bandgap of a semiconductor have to be so that no light from the visible range of wavelengths is absorbed ?

8. What is ϵ_r of water at 95 °C ? How does it compare to the value at room temperature ? What is the reason for the difference between the two values ?

9. From the purely electronic material properties point of view, Ge would have certain advantages over Si, yet essentially all integrated circuit technology involves Si and not Ge.
   a) What are the material advantages of Ge over Si ?
   b) What is the main practical stumbling block in making integrated circuits in Ge ?

10. The range of wavelengths over which Si photodiodes are sensitive is typically said to extend from 1100 nm to around 250 nm. Explain why this range is what it is.
    Tip: The reasons for the two limits are not the same.

Chapter 13
The Future of Materials

Overview

In this final chapter I will present a personal selection of a few case studies of fascinating materials science. They all concern subjects of current interest in research and engineering development and are likely to provide fertile ground for continued studies into the future. What I will cover is cutting-edge, and you know enough materials science now so that you will be able to appreciate this readily. More specifically, we will discuss atomic-level microscopy techniques for materials science, size effects in materials at the nanometer-scale, new carbon-based materials, advanced metals, optical fibers, and engineered materials for advanced microelectronics. In each case, we will highlight materials structure and properties, with related advanced processing methods.

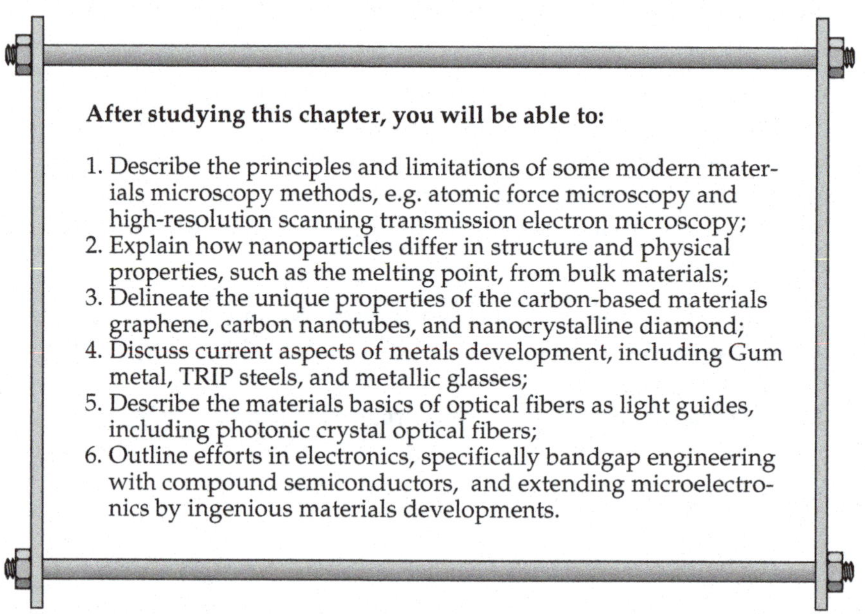

After studying this chapter, you will be able to:

1. Describe the principles and limitations of some modern materials microscopy methods, e.g. atomic force microscopy and high-resolution scanning transmission electron microscopy;
2. Explain how nanoparticles differ in structure and physical properties, such as the melting point, from bulk materials;
3. Delineate the unique properties of the carbon-based materials graphene, carbon nanotubes, and nanocrystalline diamond;
4. Discuss current aspects of metals development, including Gum metal, TRIP steels, and metallic glasses;
5. Describe the materials basics of optical fibers as light guides, including photonic crystal optical fibers;
6. Outline efforts in electronics, specifically bandgap engineering with compound semiconductors, and extending microelectronics by ingenious materials developments.

On-chip Multi-layer Wiring Scheme in a Modern Electronic Integrated Circuit

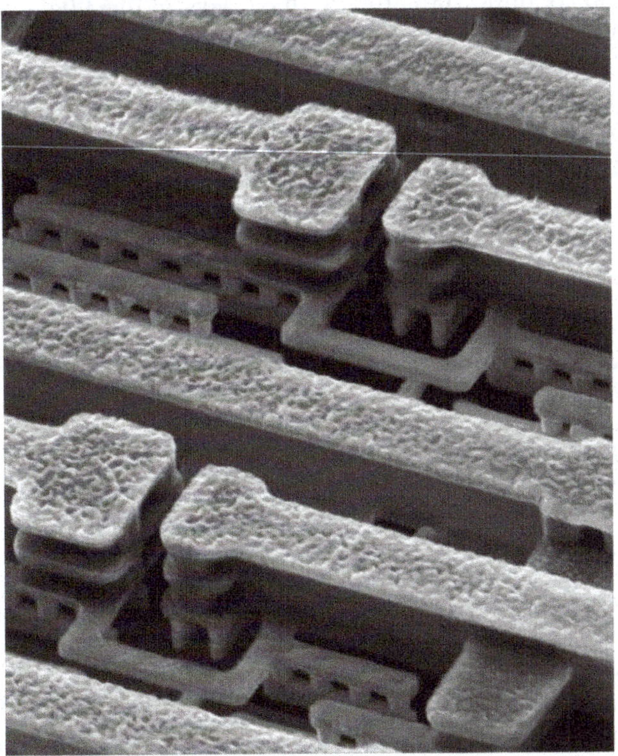

The picture displays a scanning electron micrograph of the scheme of copper wires interconnecting individual circuits elements in a modern integrated electronic circuit. The picture was used by IBM some ten years ago to demonstrate their technology. (For a cross-section through the circuit, see the title figure of Chapter 12).

On the actual circuit, these wires are embedded in SiO_2 insulation. For the purpose of illustration, the SiO_2 has been etched away. Note how the wires become larger as one goes to Cu layers higher up. The structure is formed one Cu layer at a time, with its insulation, such that each conductor layer is aligned perfectly with the layers underneath.

13 The Future of Materials

If you have stuck with the book to this point, your work is practically done, at least until you continue your study of materials. In this final chapter, I want to present a kind of personal outlook on the future of materials. I am certain that its future is bright and unforeseen. When I got into this business, a lot of material things we take for granted today were barely on the horizon. There is no reason why this should be any different now.

What makes the present time particularly exciting for materials is that I have no doubt materials science will continue to be the enabling science for the solutions of a vast range of engineering problems. I base this general outlook on two factors:

1. Analytical and synthetic techniques are now at a stage where it is becoming possible to examine, synthesize, and manipulate materials at the atomic scale;

2. Modeling and simulation techniques have evolved to a point where they have enough predictive power to provide crucial input into the design of materials, not just design with materials.

My goal is to present, in no particular order, some case studies of current and ongoing research and development in materials science and engineering that I find fascinating. I hope that you will agree with this assessment here and there. The purpose of this book would be truly fulfilled if it motivated you to pursue continual studies of some of this stuff we call materials. I am not an expert in all the areas I am going to cover, but I can appreciate what they are about. You have learned enough materials science that you are now in a position to do the same.

13.1 Materials Microscopy at the Atomic Scale

Examining a material with a microscope to discern its structure and properties has a long history, beginning with the light microscope. At the same time, the quest has always been to improve the technique to reveal ever finer details of a sample.

For most microscopic techniques one can say that the resolution, i.e. the minimum size of an object it can resolve, is a function of the radiation used to probe the sample. (See below for an exception that does not use radiation). The inherent physics of the process puts a limit on the resolution: It is generally of the order of the wavelength of the radiation involved. Therefore, for example in a light microscope, one can expect to resolve at the very best μm-sized objects.

In order to achieve higher resolution, one has basically two options: 1) use light with a shorter wavelength, or 2) use electrons. We will not say much about the first option, except to point out that we already know one such technique, namely X-ray diffraction. It uses wavelengths of the order of a nm and does reveal atomic structure, namely the separation between atom planes, but it is not a microscopic technique. Special light sources for wavelengths much smaller than visible light have also been developed.

Here we will focus on a brief overview of some electron-based techniques which have found wide application in materials science and engineering. We will begin with two techniques which launched the study of materials at the atomic scale.

13.1.1 Field Ion and Field Electron Emission Microscopy

The principles of the two techniques are illustrated schematically in Fig. 13.1. In both cases, the sample consists of a very sharp metal tip, and the entire setup is contained in an ultra-high vacuum chamber.

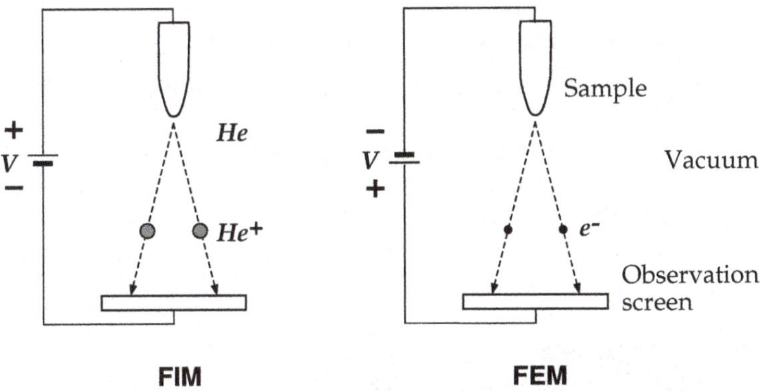

Figure 13.1: Schematic of the field emission (FEM) and field ion (FIM) microscopes. Sample and observation screen not drawn to scale.

The sample is shown greatly magnified in Fig. 13.2. Its radius of curvature is in the nm range, so that it exposes tiny pieces of individual crystal planes indicated by the dashed lines. At the very end of the tip you can see a (1 0 0) plane, and next to it on either side a (1 1 0) and a (2 1 0) plane.

Of the two techniques, the field emission microscope was invented first, by Erwin W. Müller, in 1936. The metal tip sample is cooled to a cryogenic temperature (~25K), and a large negative voltage (~5kV) is applied to it relative to the screen (cf. Fig . 13.1). This makes the electric field at the surface of the tip high enough that electrons are being pulled out of the sample. The locally emitted electron current depends on the surface atom configuration: It varies for dif-

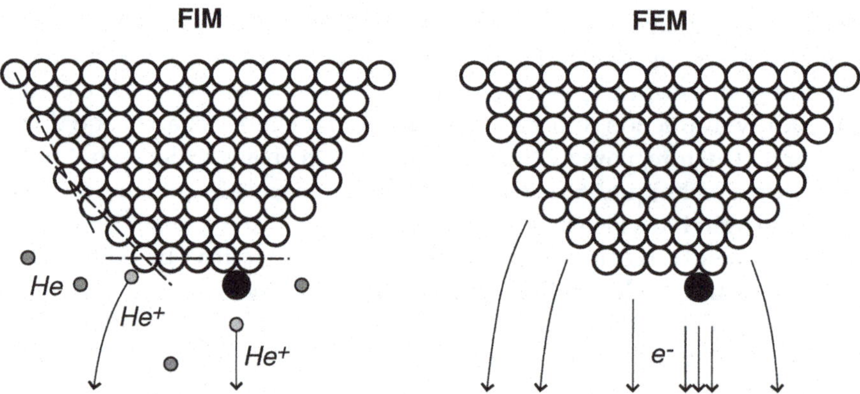

Figure 13.2: Magnified view of FEM and FIM metal samples. They have been etched into the shape of an atomically sharp tip. A foreign atom is indicated by the black circle.

ferent exposed crystal planes. The current distribution of electrons leaving the sample spreads out, and is collected and made visible at the phosphorescent observation screen. The atom configuration on the tip surface is thus rendered indirectly as a magnified image on the screen, via the spatial variations in emitted electron current density.

An example of a field emission image of a W tip is shown in Fig. 13.3. The central dark spot of this BCC metal corresponds to the (110) plane. (100) and (112) crystal planes are also identified. All these low-index planes are low-efficiency electron emitters, whereas the bright areas in-between show much higher emitted electron current.

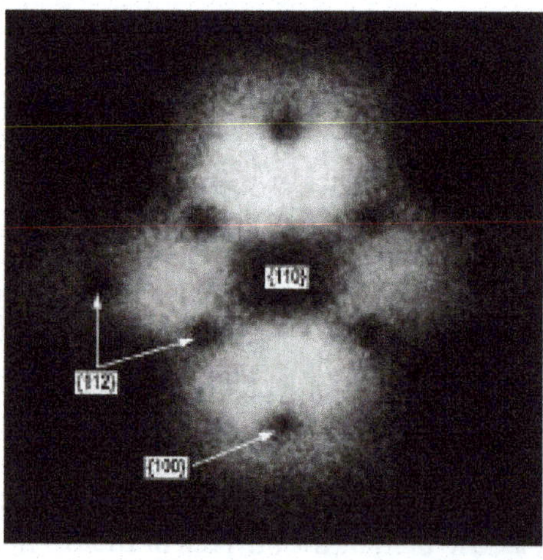

Figure 13.3: Field emission image of a W tip.

The field emission technique gave direct evidence of the presence of different crystal planes on a metal tip, but it did not make individual atoms visible.

This was achieved with the field ion microscope, also invented by Müller in 1951 (see Figs. 13.1 and 13.2). The FIM resembles the FEM in that it also makes use of a sharp metal tip as a sample, but the imaging is done by He^+ ions, not electrons.

In the FIM, the sample is also cooled to a cryogenic temperature and held at a high voltage (of the opposite polarity compared to the FEM), but the experimental chamber has been filled with He gas (after very careful evacuation). Again, the electric field at the tip surface is very strong, so that He atoms temporarily adsorbed on the tip become ionized. The He^+ is then accelerated away from the tip to the observation screen, where its impact is recorded as a bright spot. The trajectory of an ion is a function of its point of origin. In general, the electric field at the tip is strongest locally right above a surface atom. Hence that is the location where a He atom is most readily ionized. This provides a mechanism by which the surface structure of the tip at the atom level can be imaged.

A spectacular example of the resolution achievable in FIM was given in the title picture of Chapter 3 on p. 31. Literally individual atoms can be seen. The ones visible most clearly are along the edges of the little pieces of exposed crystal planes. The symmetry of the structure is revealed clearly.

It is also possible in some cases to observe foreign atoms on an FIM tip (see the black atoms shown in Fig. 13.2) since the efficiency of ionizing He atoms varies with the type of atom that the He is adsorbed on. Even the motion of atoms by surface diffusion has been observed, comparing FIM snapshots at different times.

13.1.2 Scanning Electron and Scanning Transmission Electron Microscopy

At present, the workhorses for examining the structure of solids are electron microscopes. They are the electron analog of light microscopes. There are two major categories: scanning electron microscopes, and transmission electron microscopes, and the transmission electron microscopes themselves come in different types. Fig. 13.4 shows schematics of a standard scanning electron microscope (SEM) and a scanning transmission electron microscope (STEM).

There are some similarities, but also important differences between the two techniques. In both cases, the source is an electron emitter. In an STEM this is always, and in an SEM often, an FEM tip at a high negative potential, which provides a very sharp, very bright source of electrons. These electrons are accelerated and then focused by a system of magnetic lenses onto the sample. This beam of electrons is scanned across the sample, and the detector measures the current of returned electrons (in SEM) or transmitted electrons (in STEM) as a function of the beam position on the sample.

The main difference between the two techniques lies in the detected electrons. In SEM, the sample is thick, and what is detected is secondary electrons re-emitted from the sample which are generated by the absorption of the pri-

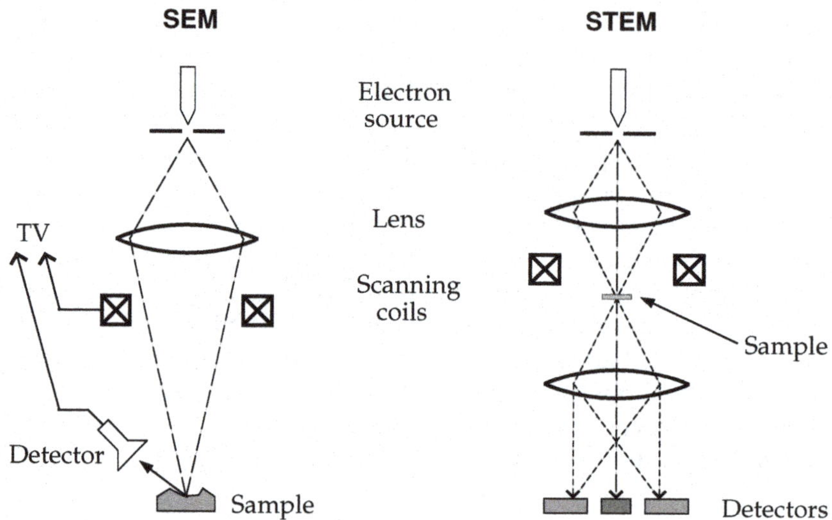

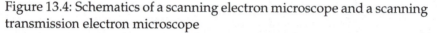

Figure 13.4: Schematics of a scanning electron microscope and a scanning transmission electron microscope

mary electrons in the incident beam. In STEM, the sample is extremely thin so that it transmits a large fraction of the electrons in the incident beam, and what is detected is transmitted electrons.

The properties of the detected electrons determine what can be seen and resolved by the two techniques. In SEM, the incoming electron beam is very well-focused, but as these electrons are stopped in the sample, they spread out somewhat, so that the detected secondary electrons come from a considerably larger area (typically 50-100 nm) than the one covered by the electron beam. It is this re-emission area that determines the resolution, in conjunction with the fact that variations in the sample topography give rise to differences in the yield of emitted electrons.

In STEM, the resolution is determined directly by the electron beam diameter (typically <1nm; as low as 0.2 nm). Variations in the transmitted beam intensity caused by atoms in the sample are recorded on position-sensitive detectors (cf. Fig. 13.4). This is why the sample needs to be very thin, or it may consist of small particles deposited on a thin carbon grid. In addition, a transmission electron microscope can usually be operated in two modes: either recording atomic-level features directly, or recording diffraction patterns of small areas on the sample.

A concise summary of the main characteristics of the two microscopy methods is displayed in Table 13.1. As always, the numbers in the table are broadly representative and may vary somewhat from one instrument to another.

Table 13.1: Main features of SEM and STEM

	SEM	STEM
Electron beam energy	5-15 keV	50-200 keV
Electrons detected	secondary	transmitted
Beam size at sample	to 5 nm	0.2-1 nm
Practical resolution	50 nm	0.2-1 nm

The practical resolution refers to the scale on which structural features can be distinguished. In SEM, a better resolution is sometimes quoted, but it usually means the smallest distance resolvable between two particles under optimal conditions. The numbers in the table make it clear that SEM is best suited for examining what we have called the microstructure of a material, whereas in STEM one may be able to observe features on the atomic scale. It goes without saying that this is only possible under strictly controlled experimental conditions (temperature, vibrations, etc.) and with optimally prepared samples.

13.1.3 Atomic Force Microscopy

The final example of a method to visualize material features at the atomic scale is the atomic force microscope (AFM). This is a type of instrument that does not rely on photon or electron radiation as a probe. Rather, it is a scanning mechanical probe with an extremely fine stylus mapping out the atomic geometry of a surface. The physical principle is illustrated in Fig. 13.5:

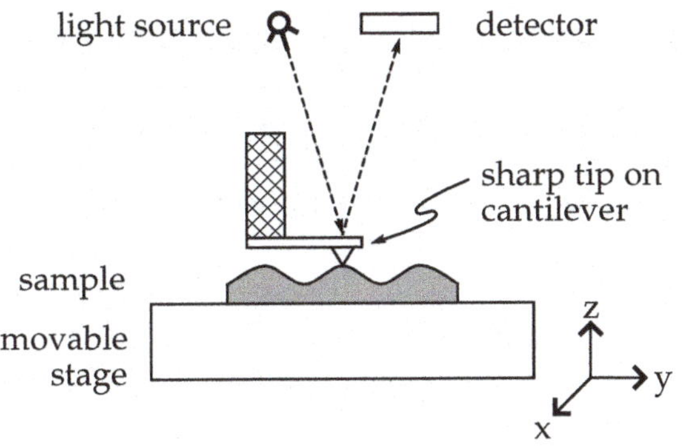

Figure 13.5: Schematic of an atomic force microscope. Parts not to scale.

The probe is a very fine, sharp tip on a small, flexible cantilever. The stage with the sample is brought very close to the tip so that there is observable interaction between the sample surface and the tip. The stage is then raster-scanned in the xy-plane, and the deflection of the tip in the z-direction is recorded as a function of its xy-position relative to the sample.

An AFM can be operated in different modes, depending on what the nature of the interaction is between the tip and the surface.

- In the so-called ***contact mode***, the tip touches the surface so that a measurable deflection of the tip results. A surface profile can be obtained in one of two ways:
  1) The tip deflection can be recorded as a function of the (x,y) position of the stage, with the stage being scanned horizontally at constant z, or
  2) The tip deflection is held constant by a feedback mechanism on the z-position of the stage, and the z-position of the stage is recorded vs (x,y).

- In the ***non-contact mode***, the tip does not touch the sample but is oscillating at its resonant frequency with a certain amplitude, at a distance close to the surface but subject to nearly zero force. As the stage is being scanned, a feedback loop between the z-position and the tip keeps the oscillation amplitude constant while the varying z-position is recorded.

- There is yet a third mode called ***tapping mode***, which is similar to the non-contact mode in that the tip is oscillating, but the tip is close enough to the surface so that the force on the tip reduces its oscillation amplitude by 50% or so. Again a feedback loop on the z-position of the moving stage keeps the tip amplitude constant as the z-position is recorded.

An AFM tip with cantilever and support is a monolithic structure, most often fabricated out of Si or Si_3N_4 using microelectronics techniques (see Fig. 13.6 below). Typical dimensions and physical parameters of cantilevers and tips are listed in Table 13.2 below:

Table 13.2: Typical specifications of AFM cantilevers and tips

	Contact mode	Tapping mode
Cantilever (l × w × t)	450 × 50 ×2 μm	125 × 30 ×4 μm
Tip height	17 μm	17 μm
Radius of curvature	<10 nm	<10 nm
Apex angle	45°	45°
Force constant	0.1 N/m	40 N/m
Resonant frequency	10 kHz	300 kHz

Parameters such as the force constant and the resonant frequency can be designed by the proper choice of the cantilever dimensions. It should be pointed out that in contact mode, the force exerted by the tip on the sample may be large enough to dislodge atoms and thus damage the sample. This is why the tapping mode is often the preferred method of examination.

A typical tapping mode Si cantilever with tip is shown below:

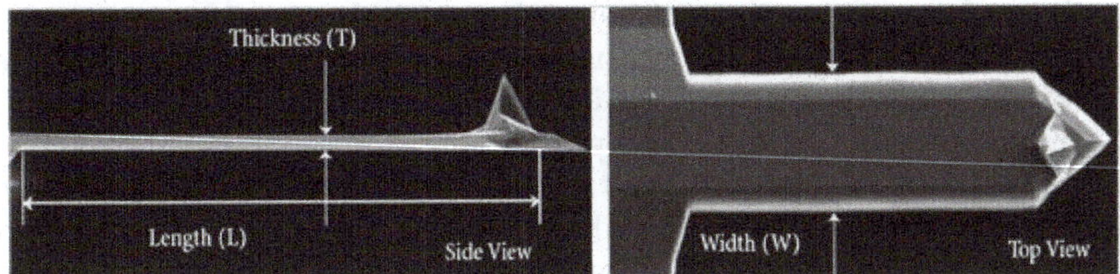

Figure 13.6: Typical tapping mode Si cantilever with tip.

As with all microscopies, the resolution is of interest for AFM, too. This can be illustrated with a simple geometrical argument:

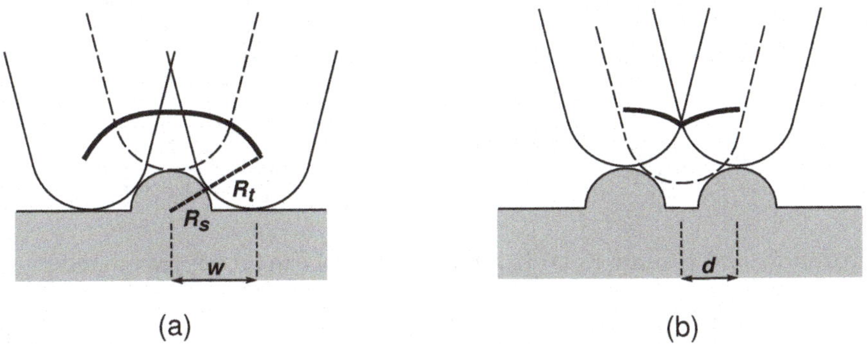

Figure 13.7: Geometries for AFM resolution. a) an individual feature; b) two closely-spaced features.

We assume that the AFM tip has a radius of curvature of R_t. In the panel on the left of Fig. 13.7, we are trying to detect an individual feature, a hemispherical particle with radius R_s. The heavy solid line indicates the trajectory of the center of the tip as it traces the particle. The excursion in the vertical direction is equal to R_s as the tip moves across. Thus the resolution of the AFM in the z-direction must be better than R_s in order that the *presence* of the particle is detected. The *apparent half width* of the particle w is

$$w = \sqrt{(R_s + R_t)^2 - R_t^2} = \sqrt{R_s^2 + 2R_sR_t} \tag{13.1}$$

This can be evaluated easily. When the goal is atomic-level resolution, usually $R_t \gg R_s$, in which case Eq. 13.1 can be approximated by

$$w \approx \sqrt{2R_sR_t} \tag{13.2}$$

The point to note here is that detection of an atomic-level feature can be achieved with a tip with dimensions much larger than atomic, as long as the sensitivity of the AFM in the z-direction is atomic. However, for a proper description of the shape of the feature, including its width, the measured tip trajectory has to be further analyzed taking into account the shape of the tip.

Now consider Fig. 13.7b, for the purpose of determining when two hemispherical particles separated by a distance 2d can be resolved. Again the heavy solid line indicates the trajectory of the center of the tip, as it moves from exactly above one atom to exactly above the other atom. Hence one must be able to detect the small dip in the heavy solid line.

Let us say that when the tip is exactly on top of a particle, its center is at a distance $z_1 = R_s + R_t$ above the reference plane. When the tip is exactly in the middle between the two particles, it is at a distance z_2 above the plane, where

$$z_2 = \sqrt{(R_s + R_t)^2 - d^2} \tag{13.3}$$

and the change in position of the tip, $\Delta z = z_1 - z_2$ is equal to

$$\Delta z = R_s + R_t - \sqrt{(R_s + R_t)^2 - d^2} = (R_s + R_t)\left[1 - \sqrt{1 - \frac{d^2}{(R_s + R_t)^2}}\right] \tag{13.4}$$

As above, we assume $R_t \gg R_s$, so that Eq. 13.4 can be approximated by

$$\Delta z \approx \frac{1}{2}\frac{d^2}{R_s + R_t} \tag{13.5}$$

As a numerical example, let us consider two hemispherical particles with R_s = 2 nm and a space of 2 nm between them, i.e. d = 3 nm. Let us also assume that R_t = 10 nm. Then Eq. 13.5 yields Δz = 0.15 nm. Note that in this case the separation of the two particles is given accurately by the trace of tip, but Δz is much less than the particle radius. Of course, the sharper the tip, i.e. the smaller R_t, the better the resolution.

A large variety of AFM tips are available commercially. These include not only tips with varying physical dimensions, but also e.g. extra sharp tips, tips coated with a thin diamond layer for extra wear resistance, and films coated with metal for magnetic interaction measurements.

The great advantage of AFM over the electron microscopies is that AFM can deal easily with all kinds of samples: insulating materials, biological samples, particles in an aqueous environment, etc. The electron microscopies generally require an evacuated instrument, and some types of samples may be damaged by the electron beam. On the other hand, with AFM one has to make sure that no mechanical damage is inflicted on the sample.

Figure 13.8:
AFM image of the basal plane of graphite.

We have encountered an example of AFM earlier, in Fig. 7.15, which showed the very detailed structure of a growing polymer spherulite. Another example, in Fig. 13.8, displays the surface of graphite.

13.2 Nanoscale Materials - Size Effects

Given that materials consist of assemblies of atoms, one could argue that all material properties are determined at the ***nanoscale***, i.e. at nm dimensions. What we have emphasized up to this point is what we have called the microstructure of a material. This means, roughly, structure that appears on the μm scale, for example grains and grain boundaries in normal polycrystalline engineering materials.

Here we will examine briefly the question whether materials have different properties when their internal structures are much smaller, for example materials containing nm-size particles or nm-size grains. We will refer to such materials as **nanocrystalline**. The general answer is that many material properties change at that scale, sometimes in beneficial ways. Again, we will look at a few interesting examples.

13.2.1 Surface vs. Bulk Atoms in Small Particles

You will recall that in most of our prior discussions, we made the assumption that the outer surface of a material, strictly speaking a 2D defect, did not have a significant effect on the material properties. On the other hand, one instance where we did deal with surfaces was in connection with the effects of grain boundaries as internal surfaces.

The basis for this assumption was that the number of atoms in the bulk was much larger than the number of atoms on the surface. For small particles this assumption is not valid anymore, and this is important because surface atoms generally differ in their properties and reactivities from bulk atoms. In other words: For small particles, the material properties become size-dependent. This fact has long been recognized by chemists and chemical engineers working with catalyst particles.

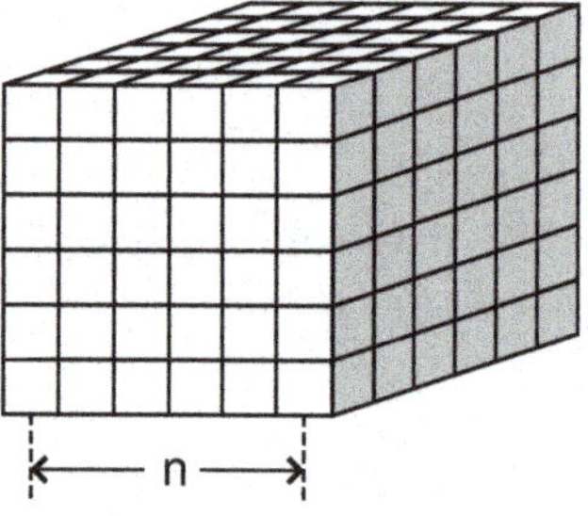

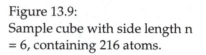

Figure 13.9:
Sample cube with side length n = 6, containing 216 atoms.

As a simple illustration of this size effect, let us determine the fraction of atoms residing on the surface of a small particle. Let us assume the particles have simple cubic crystal structure, with a cube length given by the parameter n. In the example of Fig. 13.9, there are six atoms along the side of the little cube, and thus $n = 6$. For an arbitrary n, the total number of atoms in the particle is obviously n^3. The number of bulk atoms, i.e. atoms not on the surface, is $(n-2)^3$. Hence, the fraction f_s of particles on the surface is

$$f_s = \frac{n^3 - (n-2)^3}{n^3} \tag{13.6}$$

Equation 13.6 can be evaluated as follows:

$$f_s = \frac{6n^2 - 12n + 8}{n^3} = \frac{6}{n} - \frac{12}{n^2} + \frac{8}{n^3} \tag{13.7}$$

As long as n is not too small (more exactly, if $n \gg 2$), Eq. 13.7 can be approximated by just the first term on the right-hand side:

$$f_s \approx \frac{6}{n} \tag{13.8}$$

This says that the fraction of atoms on the surface of a particle increases roughly as the inverse of the linear dimension of the particle.

The fractions of surface and bulk atoms as per Eq. 13.7 are plotted in Fig. 13.10 below. For example, for a particle with 1000 atoms, about half of the atoms are on the surface, and the approximation of Eq. 13.8 is still quite good.

13.2.2 The Melting Temperature of Small Particles

One of the earliest observations of a size-dependent materials property has been the melting temperature T_m of small particles. Data for T_m of small Au

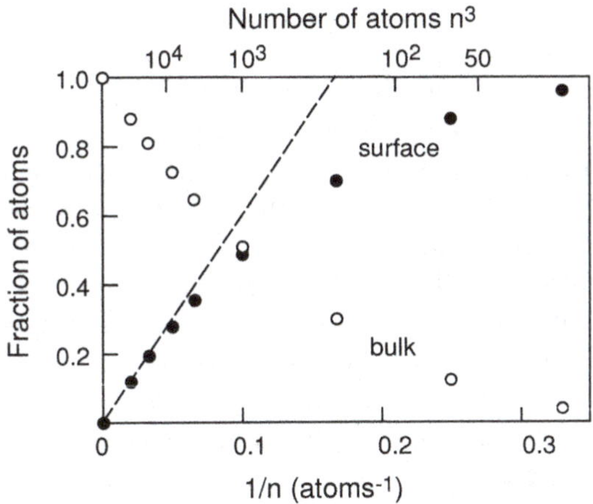

Figure 13.10: Surface and bulk atom fractions vs. 1/n for SC cube-shaped particles. Top x-axis shows total number of atoms. Dashed line: Eq. 13.8.

particles of various sizes are displayed in Fig. 13.11. These data are plotted as the ratio of the absolute actual melting temperature to its bulk value, T_m/T_∞, vs. the inverse of the particle radius r.

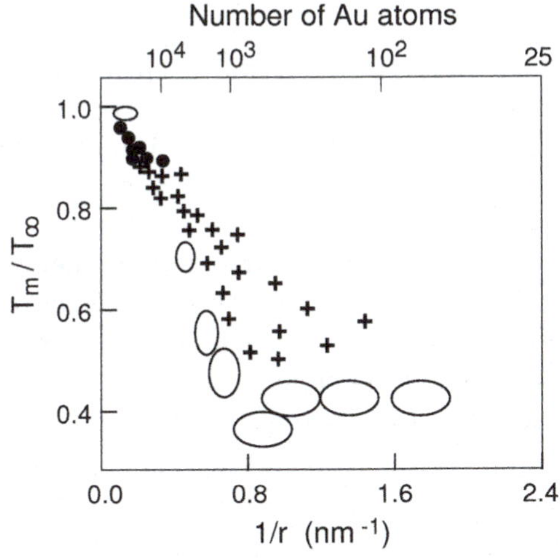

Figure 13.11: Experimental data for the size dependence of the melting temperature of small Au particles.

The data are from three different sources (see references). The black circles are from evaporation rates of small particles. The crosses are from X-ray diffraction measurements of small particles as they turn from crystalline to amorphous upon melting. The ellipses represent field emission microscopy data, where the electron emission current from a W tip changes abruptly as adsorbed Au particles melt. The size of the ellipses is a measure of the experimental error bars. The data sets of circles and crosses overlap at the largest particle sizes, i.e. for values of r^{-1} approaching 0.

The data are in good agreement with each other and clearly demonstrate the dependence of the melting temperature on the particle size. Moreover, the data suggest that the ratio T_m/T_∞ is well approximated by a linear function of r^{-1}. This is exactly the same behavior we noted above for the bulk atom fraction f_b of a particle as a function of n^{-1}. Given that r and n are just two different parameters characteristic of the linear dimension of a particle, it seems reasonable to conclude that the data can be explained as being due to the increasing influence of the surface atoms on the melting of a particle. Since surface atoms are held more loosely in the crystal lattice than bulk atoms, the more surface atoms there are, the easier it is, in an average sense, to cause disorder among all the atoms, i.e. to melt the particle.

The reality may be more complicated, however, and using a single number for the melting point is itself an approximation. There is good evidence from modeling studies that the melting of nanoparticles occurs in two stages. First, a surface layer of width equal to 2-3 lattice spacings becomes disordered, and then the entire particle melts.

The particle size effect of melting is not just of academic interest. Nanoparticles have been investigated as "low melting glue" at a material interface and for the purpose of facilitating sintering at lower temperatures.

13.2.3 The Shape of Small Particles

Although we have focused here on particle size, it should not be surprising that the shape of small particles also is an issue of interest. We should expect shape to matter when surface atoms are important. One way in which this shows up is that very small particles may have shapes that differ from what one would expect on the basis of macroscopic arguments.

From our discussion of crystal structures we know that the energetically most favored planes are the close-packed ones. Therefore, we expect a thermodynamically equilibrated particle to be bounded by close-packed planes. For example, for an FCC material this would mean that a particle assumes an octahedral shape, the surface being formed by (1 1 1)-type planes (Fig. 13.12b).

However, this argument once again neglects the special character of the edge and corner atoms, both high-energy locations. Note, for example, that the number of nearest neighbors for a corner atom in an octahedron is only 4, not 12. Hence it is not obvious what is the best configuration for a particle consisting of only a small number of atoms: a cube, an octahedron, something in-between (Fig. 13.12c), or something else entirely?

We can illuminate the issue further by considering a specific example, namely finding the best geometry for a particle consisting of 13 atoms. We start with three mini-layers of close-packed atoms, a 7-atom hexagon and two 3-atom triangles, along the lines of our discussion on building crystals in Chapter 3 (see

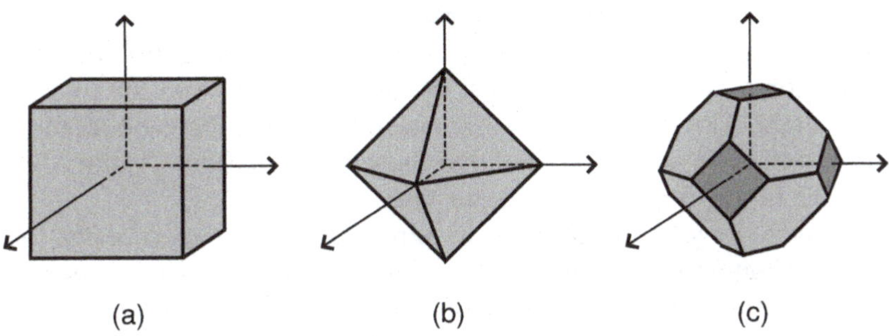

Figure 13.12: Possible small FCC particles. a) cube; b) octahedron; c) truncated octahedron.

Figs. 3.5 and 3.6 and Fig. 13.13 below). The question we want to answer is: What is the best 13-atom cluster we can build?

Figure 13.13: Building a 13-atom cluster

Two evident possibilities of stacking 3+7+3 atoms are displayed in Fig. 13.14:

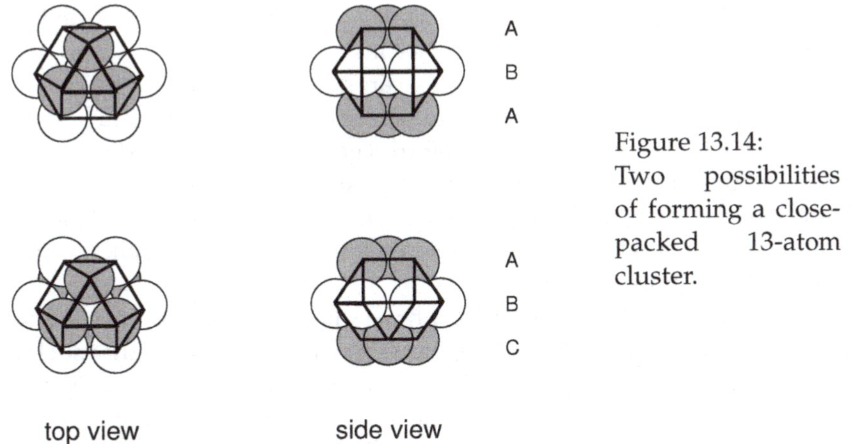

Figure 13.14: Two possibilities of forming a close-packed 13-atom cluster.

The two clusters differ in the sequence in which the three layers are stacked: ABA as for bulk HCP, or ABC as for bulk FCC (cf. Fig. 3.6). A good way to visualize the difference is to focus on the polygon spanned by the centers of the atoms. Both polygons consist of 6 isosceles triangles and 6 squares, but arranged differently. This is perhaps most apparent in the side views.

We noted in Chapter 3 that the difference between bulk HCP and FCC structures is slight and depends on subtle differences in the interaction potential between the atoms. So we would expect the difference between the two clusters also to be slight. However, for this discussion let us just focus on atoms which eventually will form an FCC crystal. For those kinds of atoms, is the ABC cluster in Fig. 13.14 the most stable? The answer is: No !

It has been shown by modeling, in agreement with experiments, that the most stable 13-atom cluster, based on an atom-atom interaction potential for FCC crystals, is a cluster in the shape of an icosahedron, Fig. 13.15.

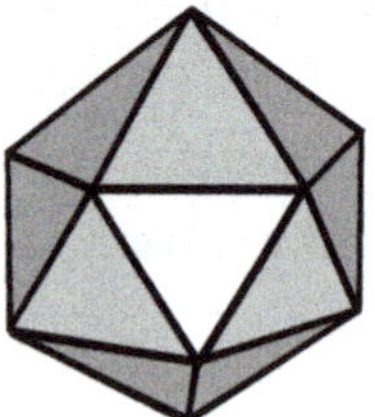

Figure 13.15:
Icosahedral 13-atom cluster. The atoms are located at the vertices and in the center.

You can think of the icosahedron as consisting of 20 isosceles triangles, each made up of three atoms. Alternatively, you can visualize it as a polyhedron with two staggered five-atom rings, and with an atom on top, in the center and at the bottom. The 13-atom icosahedron is the first of a series of so-called magic clusters. The next two contain 55 and 147 atoms. They are all characterized by a closed icosahedral shell. This makes them unusually stable compared to clusters containing a few atoms more or less. The 55-atom icosahedron has 20 triangular faces with six atoms in each face instead of three.

Results for clusters of FCC materials can be summarized as follows: For small clusters, the most stable structure is based on an icosahedral geometry. For intermediate-sized clusters, the preferred geometry is decahedral, and normal FCC crystals are favored above a clusters size of around 2×10^4 atoms.

For clusters of BCC and HCP materials, the detailed results will be different, but again the smaller clusters have a geometry which deviates from the expected bulk crystal structure.

Also keep in mind that everything I have said so far about particle shapes involves thermally equilibrated particles. In practice this means that such particles have to grow slowly from the melt, at a temperature close to the melting point. However, even under those conditions gradients of temperature and composition may cause unusual shapes to arise, as we saw in Chapter 7.

Moreover, research is very active in a broad area referred to as directed self-assembly. In most general terms this means growing ordered materials under non-equilibrium conditions, for example growing metal or ceramic particles from solution near room temperature. This may lead to unexpected or entirely new structures. An example is shown below:

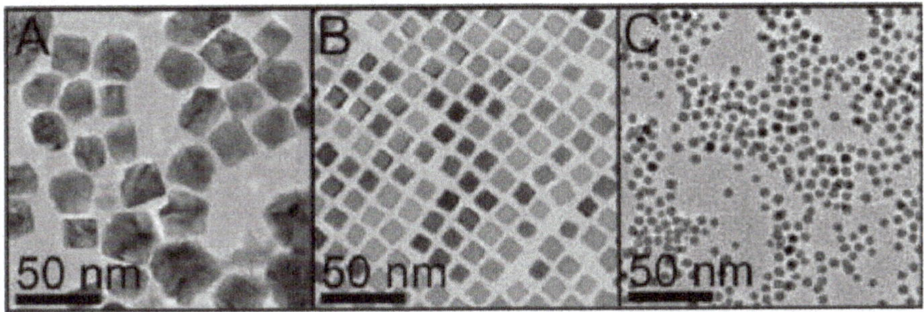

Figure 13.16: Ni particles grown from aqueous solution using different amounts of a surfactant.

In Fig. 13.16, the amount of surfactant in the solution determines the shape and size distribution of the particles. The shape may be irregular (Fig. 13.16a), or cubic with fairly constant size (Fig. 13.16b), or spherical (Fig. 13.16c). It is evident that the particle shape depends systematically on details of the properties of the solution. This suggests that under properly controlled conditions, growth of nanoparticles with desired shapes and sizes may be achieved.

13.3 New Carbon-Based Materials

In Section 3.6 on covalent crystals we introduced diamond as crystalline carbon (Fig. 3.23) and several of the solid carbon allotropes (Fig. 3.24). For convenience, the figure with the allotropes is reproduced below.

Some of these different forms of carbon have received a great deal of attention recently. We will discuss why this is so, focusing in particular on graphene, carbon nanotubes, and crystalline diamond films.

13.3.1 Graphene

Graphene consists of a single ordered hexagonal layer of C atoms (Fig. 13.17a). Think of it as an atomic layer taken off of a sample of graphite. This is literally how graphene was produced initially: In 2004 Geim and Novoselov discovered that by rubbing graphite against a smooth surface, a few layers could be scraped off and then thinned until only one layer was left. They also found that a single layer of graphene deposited on a thin layer of SiO_2 on a Si wafer could be detected in an optical microscope, thanks to a slight change in color under illumination with white light.

Graphene has extraordinary electronic properties. In its hexagonal plane, it is an excellent metallic conductor (it has no bandgap). The charge carrier mobility at room temperature is as high as 20 m^2/Vs, compared e.g. to 0.14 m^2/Vs for Si. Graphene can also sustain very large current densities. In addition, the thermal conductivity of graphene is very high, higher even than that

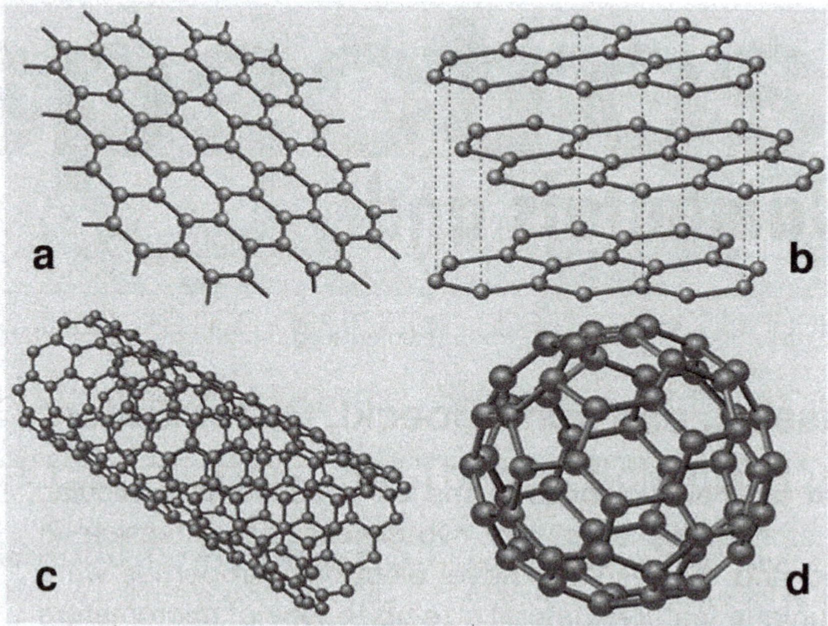

Figure 13.17: Allotropes of carbon. a) graphene; b) graphite; c) carbon nanotube; d) C_{60} molecule (fullerene).

of diamond, at room temperature.

These unusual properties of graphene come mostly from the fact that C atoms have four valence electrons, but they need only three of those to form the hexagonal layer, binding to three nearest neighbors. The fourth electron of each C atom sticks out normally from the basal plane. These extra electrons from neighboring C atoms interact with each other to become delocalized and essentially free to move laterally, in the plane of the C atoms. So from the point of view of electronic properties, along its plane graphene acts like a highly conductive 2D metal.

The method of mechanical exfoliation mentioned above makes it straightforward to produce individual graphene samples for study. However, for the purpose of electronic applications, it will be necessary to be able to produce reliably large, predictably oriented layers on a suitable substrate. A great deal of effort is underway to improve the synthesis of graphene, and some success has been achieved in growing it on SiC and a few metals.

13.3.2 Carbon Nanotubes

Another closely related new form of carbon is carbon nanotubes (Fig. 13.17c). Think of a carbon nanotube as a piece of graphene, rolled up and with edges joined into a seamless cylinder.

The discovery of carbon nanotubes is often attribute to S. Iijima, who in 1991 described thin needles growing on the carbon cathode in an arc discharge. However, other investigators had described hollow carbon fibers in earlier work. Iijima described each needle as containing a number of coaxial tubes. These tubes consisted of carbon atoms in hexagonal coordination and arranged in a helical fashion. Individual tubes had varying diameters and lengths. Two years later, the observation of single-wall carbon nanotubes was reported. In these works, the growth of the nanotubes had been catalyzed by Fe or Co particles.

The principle in all these experiments was evaporation of carbon atoms into a gaseous ambient, followed by condensation and aggregation on a cool surface. Other byproducts were fullerenes and, of course, graphitic carbon.

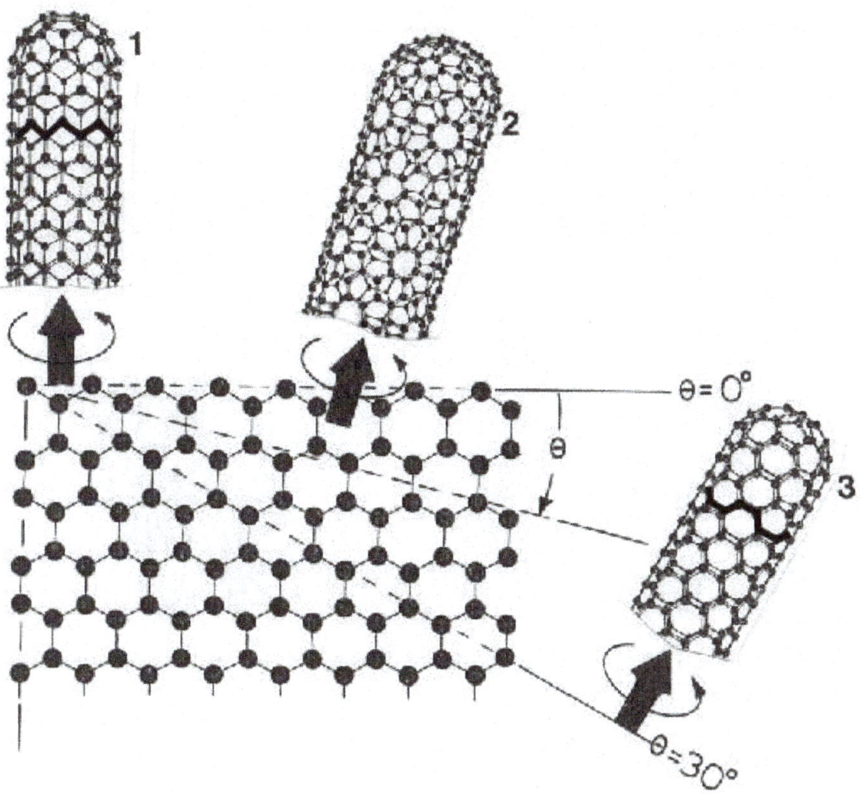

Figure 13.18: Different carbon nanotube structures, depending on the orientation of the axis about which the hexagonal sheet is rolled up.

Given how carbon nanotubes can be constructed from graphene sheets, we should expect that carbon nanotubes will also have interesting electronic properties. However, their structure is more complex. Since tubes have a curved surface, their properties depend on the radius of curvature, or equivalently on their diameter. In addition, the pitch describing how a sheet is rolled up into a tube is another important parameter. This amounts to defining the direction of the axis in the plane of hexagons about which a sheet is rolled up.

Distinct configurations are obtained with rotation axes between $\theta = 0$ and $\theta = 30°$. The configuration with $\theta = 0$ is referred to as zigzag, and the configuration with $\theta = 30°$as armchair. (Note the heavy black lines around the circumference of the tubes clarifying these names). Angles between 0 and 30°are possible. The ends of carbon nanotubes are typically terminated with a cap resembling half a fullerene molecule.

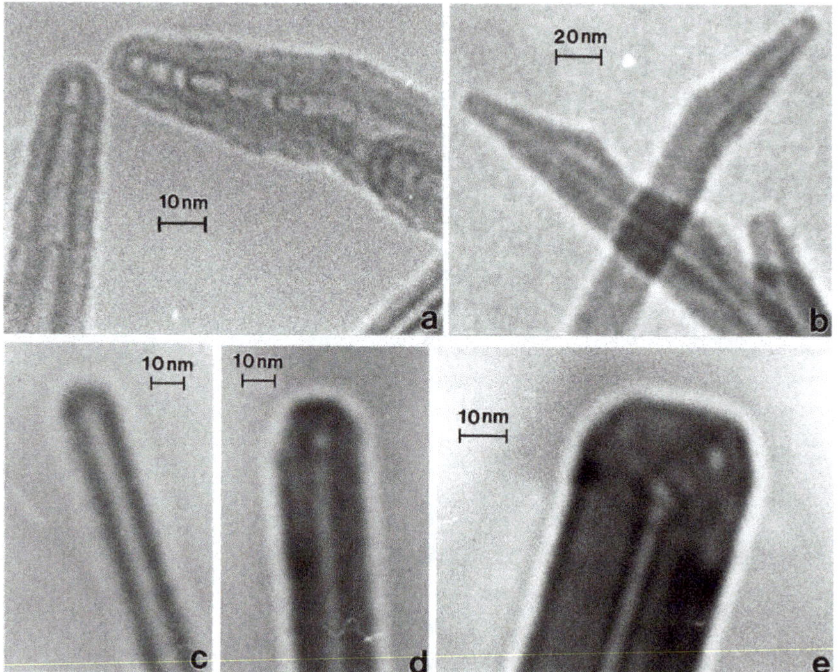

Figure 13.19: TEM images of multi-wall carbon nanotubes with varying diameters.

Multi-wall carbon nanotubes and fullerenes are also among the constituents of carbon soot, the carbonaceous particulate matter occurring in incomplete combustion processes as simple as the gas flame on a kitchen stove. TEM images of various multi-wall carbon nanotubes are displayed in Fig. 13.19. Note how each single-wall tube in Fig. 13.19 is capped individually. Caps may be conical with a hemispherical cap at the end, or more irregular in shape, depending on the diameter of a tube.

For well-defined electronic properties one needs to deal with single-wall nanotubes. Such tubes may be either metallic or semiconducting, depending on the structure and the diameter. In particular, tubes with the armchair structure (see Fig. 13.18) are metallic, and tubes with other roll-up axes are semiconducting. The bandgap decreases with increasing diameter and can be anything between about 1.5 and 0.1 eV.

FET transistors and even simple circuits have been built with single-wall nanotubes. The tube functions as the channel of the transistor. Metal contacts are applied for the source and drain. The gate electrode is formed on top of an insulating layer, which in turn covers the nanotube. The transistor is said to be of an ambipolar nature: The nanotube is essentially intrinsic, and whether electrons or holes flow in the channel depends on the type of metal used for the source and drain contacts.

Carbon nanotubes also exhibit exceptional mechanical properties, i.e. a very high elastic modulus and very high strength. The elastic modulus of multi-wall nanotubes was determined in bending tests with an AFM and was found to be around 10^3 GPa. An additional result of such experiments was that the concentric individual tubes can slide by each other, This implies that stress is transferred incompletely between different concentric tubes. The elastic modulus of a single-wall nanotube was estimated to be much larger than 10^3 GPa. Also, the tensile strength of multi-wall nanotubes was found to be in the range of 10-50 GPa.

Given the unusual properties of carbon nanotubes, it is not surprising that numerous applications have been proposed. But it is also apparent that in order to achieve predictable and repeatable electronic or mechanical material properties, the carbon nanotubes should preferably be single-wall and of uniform diameter. This points to a fundamental issue in nanotube synthesis: preparing practically useful quantities with tightly controlled characteristics. Since this is rather difficult, considerable efforts are ongoing not only in synthesis, but also to develop methods for selecting nanotubes by size and for separating them from catalyst particles.

For electronics built with nanotubes, it will be necessary either to grow single-wall carbon nanotubes directly over sizable areas on a suitable substrate, or to develop methods to deposit them as highly ordered films onto such substrates. In the area of mechanical applications, nanotubes would represent the ideal reinforcing agent in a composite polymer material, but there the issues are to insure that the nanotubes are dispersed properly and that their surfaces can be modified chemically such that they interact strongly with the matrix.

There are several other properties of interest with carbon nanotubes which we will not go into in detail here. These include a very high thermal conductivity, interesting optical properties, etc. It may be possible to take advantage of the high thermal conductivity using multi-wall nanotubes. A pertinent example, where the nanotubes were grown where needed, over a macroscopic area, is displayed in Fig. 13.20. The substrate was a Ti-coated Si wafer with a 2.5 nm layer of Co as catalyst. For some samples, an additional, very thin layer of 0.5 or 0.8 nm Mo between the wafer and the Co layer was used.

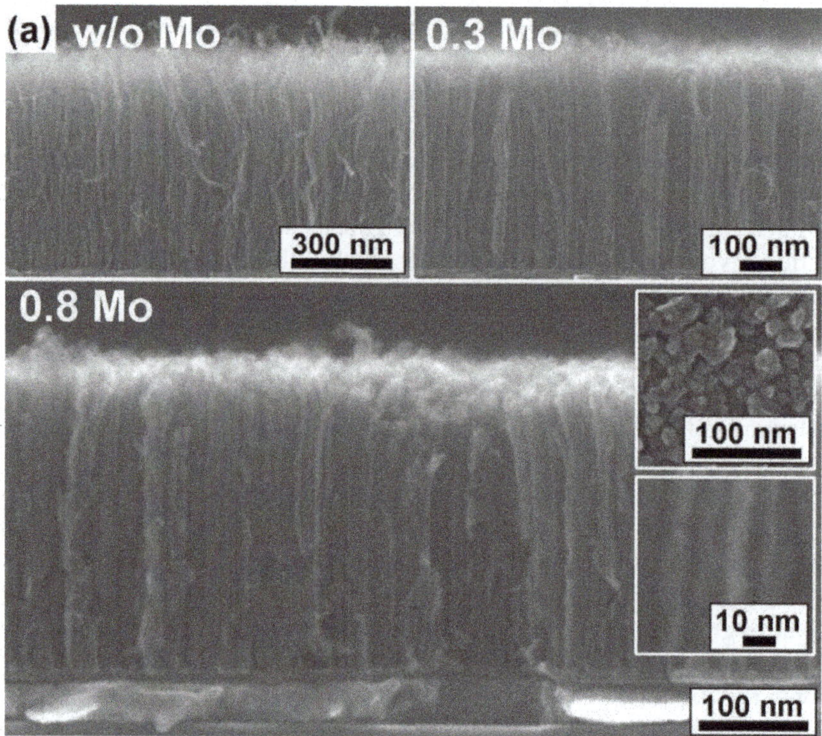

Figure 13.20: SEM micrographs of a Si wafer with a "carbon nanotube forest" layer grown on it. The samples differ in the thickness of the initial Mo layer (in nm) employed. The upper insert depicts a top view.

The "carbon nanotube forest" layer grows vertically and is said to be exceptionally dense, due to the presence of the Mo. Structures like this are envisioned as high-efficiency heat sinks in integrated electronic circuit applications.

13.3.3 Nanocrystalline Diamond Films

The formation of bulk diamond requires high temperature and pressure. It is thus of great interest that it is possible to grow polycrystalline diamond films under much milder conditions. The growth of such films provides an excellent example of how the interplay between nucleation and growth per se determines the outcome of the overall process.

A rough distinction is made between microcrystalline, nanocrystalline, and ultrananocrystalline diamond, depending on the average grain size. An example is displayed in Fig. 13.21.

The average grain size in the microcrystalline diamond was slightly larger than 1 μm , and the grains were columnar (long in the direction normal to the film). The grain size was about 300 nm in nanocrystalline diamond, and about 7

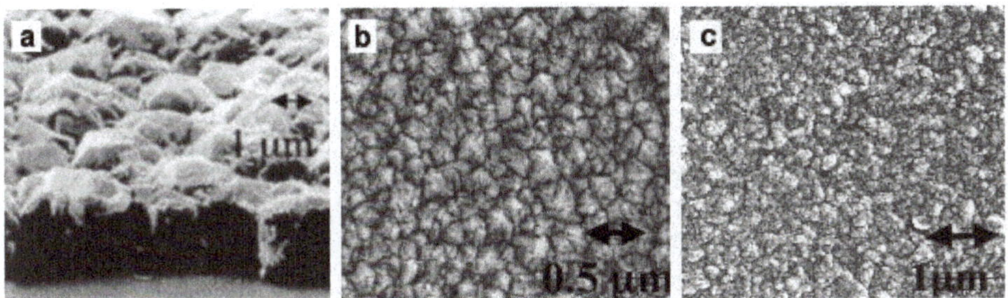

Figure 13.21: SEM micrographs of the microstructure of a) microcrystalline diamond; b) nanocrystalline diamond; and c) ultrananocrystalline diamond.

nm in ultrananocrystalline diamond, and in both those types of films the grains were equiaxed (about equal in size in all three directions). The surface roughness was typically about 100 nm for microcrystalline diamond and 15-25 nm for nanocrystalline diamond.

The films were grown on a Si wafer, and the resulting grain size was a function of the detailed growth conditions. Growth was achieved by a method known generally as chemical vapor deposition (CVD). This means that the wafer is exposed to a gas mixture under such conditions that molecules present in the gas phase react on the wafer to form the new material, in this case a diamond film.

In order to grow diamond films, a discharge with a mixture of about 2% CH_4 in H_2 was used, in conjunction with a wafer temperature of about 450-700°C. For the ultrananocrystalline films, the gas mixture was about 2% CH_4 + 1% H_2 in Ar, together with a wafer temperature of 400-650°C. Under these conditions typical growth rates of a few hundred nm per hour were achieved.

The other crucial ingredient in these processes is a diamond seed layer. That is, for the growing film to come out crystalline, there have to be small diamond nuclei present on the wafer to begin with. Without such nuclei, the film would turn out amorphous. The diamond nuclei can be put onto the wafer either by a mechanical polishing step with diamond powder, or by deposition from a solution containing diamond powder.

If after the deposition the Si wafer is etched away, free-standing diamond films, or diamond sheets, can be realized. These are extraordinarily strong and resilient, approaching the mechanical property values of bulk diamond for the elastic modulus and the fracture strength. Diamond sheets can sustain large elastic deformations and can adapt easily to curved surfaces (Fig. 13.22).

Development continues to improve the growth conditions of these films. Nanocrystalline and ultrananocrystalline diamond films are finding application in electromechanical micro-sensors and devices, and in integrated optics

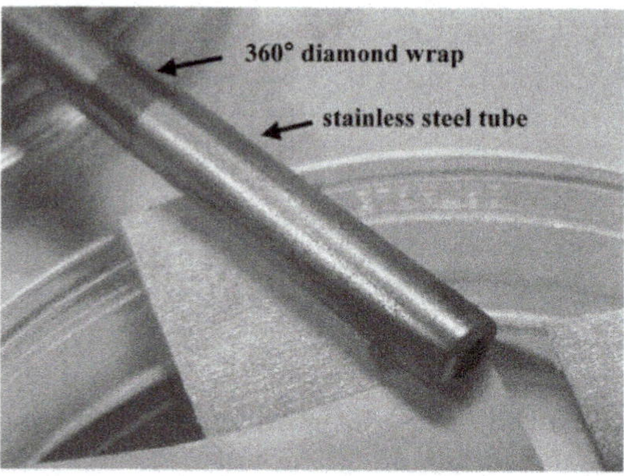

Figure 13.22: A 1.6 μm thick diamond sheet wrapped around a 0.25 inch steel pin.

and optomechanical devices. Monolithic diamond AFM tips and ultrasmooth diamond coatings are commercial products.

13.4 Advanced Metals

Even though metals have been studied by materials scientists and engineers for a long time, research and development into advanced metals continues. Below are some examples.

13.4.1 Advanced Steels

The driving force for advanced high strength steels (AHSS) is primarily the auto industry. Materials with an improved combination of strength and ductility allow it to meet today's requirements for improved crash performance, while at the same time enabling a reduction of vehicle weight for improved fuel efficiency and maintaining formability of the material for manufacturing. Advanced steels are also finding applications in construction.

Current developments can be visualized and put in perspective with help of the schematic Fig. 13.23, which once again illustrates the trade-off between ductility and strength.

The conventional steels are the ones you know about already, including high-strength low-alloy steels (HSLA) and austenitic stainless steels. DP and CP refer to dual-phase and complex phase steels, respectively. DP steels consist of a ferritic matrix with a second phase in the form of islands of martensite. CP steels are complex-phase steels consisting of a ferrite/bainite matrix with small amounts of martensite, austenite and pearlite. The newest variety is called TRIP steels (for transformation-induced plasticity). TRIP steels retain some austenite in a primary matrix of ferrite, but also include small amounts of martensite and bainite. When this type of material is deformed plastically, the austenite phase is transformed gradually into martensite, which leads to enhanced strain hardening.

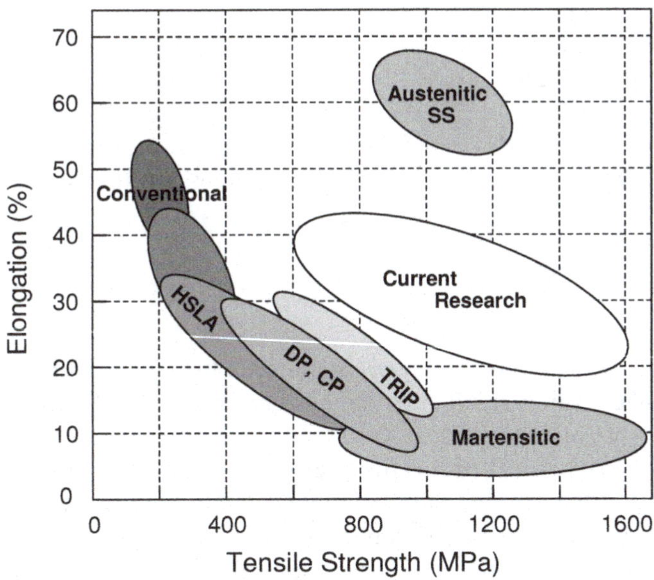

Figure 13.23: Illustration of various types of steel in terms of tensile strength and corresponding ductility.

13.4.2 Gum Metal

Gum metal is my example of a metal alloy designed and optimized for specific properties on the basis of first-principle calculations. Given that the material of interest was a multi-component alloy, the many experimental parameters would have made it impractical to evaluate a sufficient number of material compositions by experimental trial and error. Based on the modeling, the material was synthesized, and it confirmed the predicted properties. "Gum metal" is a registered trademark of Toyota Central R&D Labs and was initially developed there. The results of this research were first published in 2003.

Gum metal is a cubic Ti alloy with ultra-low elastic modulus, ultra-high T.S., very large elasticity, and very high plastic deformation at room temperature which occurs without work-hardening. Hence, plastic deformation must take place via a dislocation-free mechanism. In addition, with a suitable heat treatment after cold work, some elasticity can be sacrificed for even greater strength. Yield strengths close to 2000 MPa have been achieved, which is comparable to the strongest steels.

The Toyota researchers sought to optimize the mechanical properties of Ti alloys by varying the composition of several constituents. Optimal properties were achieved for alloys of the approximate composition $Ti_{0.75}X_{0.25}$ where X was another transition metal (Fig. 13.24).

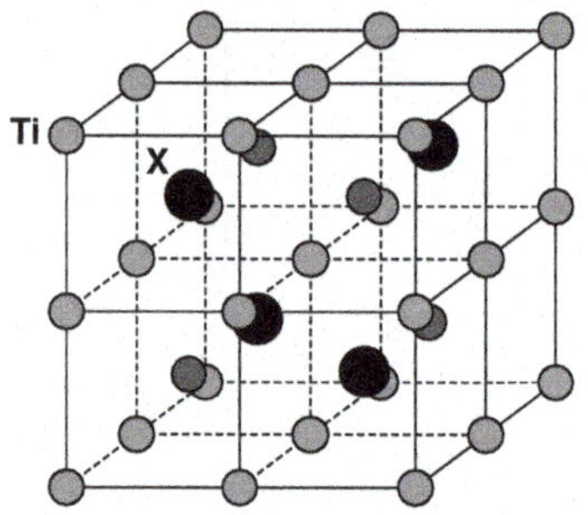

Figure 13.24:
Unit cell of gum metal. Dark gray Ti atoms and black X atoms are in the centers of the small BCC cubes.

In Fig. 13.24 the Ti atoms are in prinicple on a BCC grid, but the atoms in the centers of every other small cube have been replaced by X atoms, whether you go in the x-, y-, or z-direction.

The optimal mechanical properties were a result of the alloy's electronic structure, specifically of three "magic numbers" related to the effective number of valence electrons and the atom-atom bond order. The alloy properties could be further improved by adding small amounts of additional elements including oxygen. A practical example is 74Ti-23Nb-0.7Ta-2Zr-0.1O (mol%).

13.4.3 Metallic Glasses

In our discussion of metals, I made the point that it is almost impossible not to end up with a crystalline material. Specifically, this means that it is very difficult to cool molten metal fast enough not to form crystals, especially a pure metal or a simple alloy.

Therefore, if one is interested in amorphous metals, one has two options: 1) develop an experimental method to cool the molten metal fast enough or, 2) figure out a different way to prevent the molten metal from crystallizing. Both these approaches have been implemented, and it is indeed possible to produce amorphous, i.e. glassy, metals.

What is at issue can be illustrated with a crystallization curve similar to the one we discussed in Chapter 10 on ceramics (cf. Fig. 10.4).

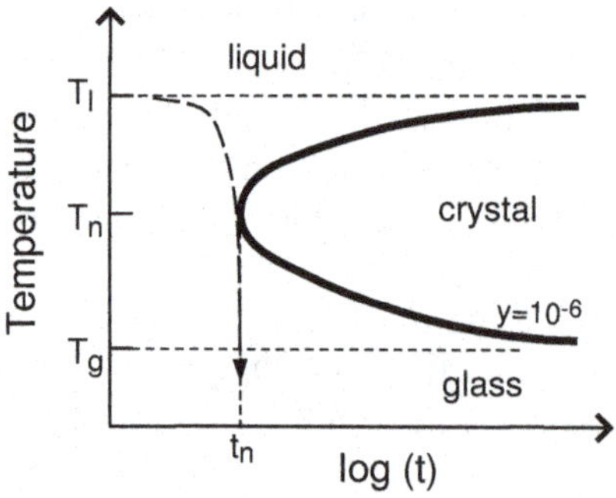

Figure 13.25:
Generic crystallization (TTT) curve for a metal alloy. The critical continuous cooling curve is indicated by the dashed line.

The time-transformation curve in Fig. 13.25 indicates how long it takes for crystallization to begin as a function of temperature. In order that crystallization is avoided entirely, the cooling rate must be larger than the critical cooling rate $r_c = (T_l - T_n)/t_n$, where T_l is the liquidus temperature and T_n the temperature of the nose for the given alloy composition. T_g is the glass transition temperature, below which the amorphous material has become rigid, i.e. essentially solid.

Research on metallic glasses began in the early 1960s with simple binary and ternary alloys. It became evident quickly that very large cooling rates, typically $r_c > 10^5$ K/s, were needed to produce amorphous metals. Such cooling rates were achievable by a process called splat-cooling. This basically involved dropping a small drop of molten material onto a cold surface. As the molten drop spread into a thin splat, it cooled down extremely quickly.

Early experiments involved binary alloys consisting of a transition metal and a non-metal (B, N, C, Si, P). These showed, for example, that alloys of Au, Pt, and Pd yielded ductile glasses, whereas Fe- and Ni-based glasses were brittle. It was also clear that the very high cooling rates made these materials rather impractical, although it was demonstrated that thin metallic glass wires and ribbons could be produced. In addition, there were indications that it may be easier to prevent crystallization in ternary alloys, an example being $Pd_{78}Si_{16}Cu_6$ vs. $Pd_{80}Si_20$.

The search for more practical materials has been ongoing, and still is. Generally, a multi-component alloy is necessary. Many of the investigated alloys have four or five components. The fundamental question is, of course, what properties of the alloy constituents make it likely that the material will form a glass. You are in a position to appreciate at least in broad outline the important parameters.

First, we can go back one more time to our approach to forming metals by packing spheres. A useful way of visualizing an amorphous alloy is to start with a random packing of spheres of the main constituent and then try to fill the holes in-between these main atoms with other, smaller atoms. It also helps if atoms interact more strongly with the other type rather than with their own kind. Both these characteristics stabilize the amorphous structure. Second, compositions near a eutectic point with a deep eutectic, i.e. with a large temperature difference between the liquidus and the eutectic lines, are favorable for glass formation. A deep eutectic indicates a stable liquid phase.

Since one has to deal with a multi-component system, it is clear that in terms of potentially useful compositions, the range is enormous. Hence, modeling approaches based on thermodynamic and kinetic considerations are being developed so that more compositions can be evaluated quickly, at least in an approximate fashion. Fig. 13.26 gives an example of a set of theoretical TTT curves. The figure is meant to illustrate how sensitive these curves are to small variations in composition. The numerical details, for example the exact conditions for the nose, depend on the model assumptions.

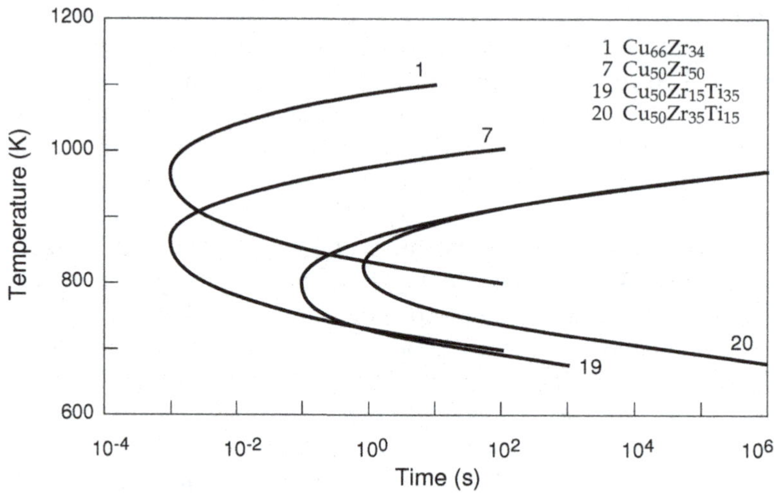

Figure 13.26: Calculated TTT crystallization curves for some CuZr alloys.

An experimental TTT curve for a $Zr_{58.5}Nb_{2.8}Cu_{15.6}Ni_{12.8}Al_{10.3}$ alloy is displayed in Fig. 13.27.

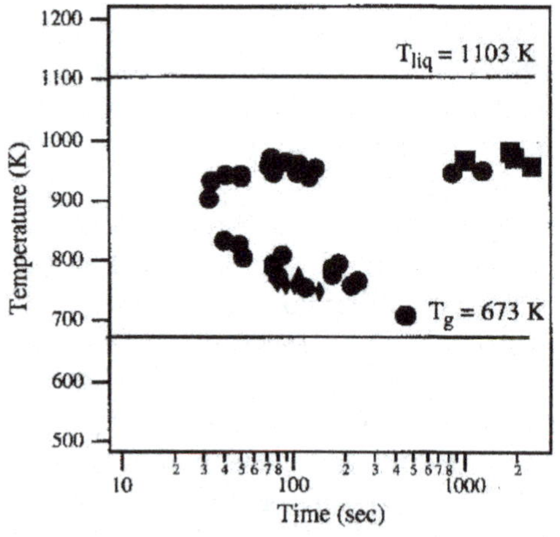

Figure 13.27: Experimental TTT curve for a $Zr_{58.5}Nb_{2.8}Cu_{15.6}Ni_{12.8}Al_{10.3}$ alloy.

The mechanical properties of a metallic glass are of special interest compared to polycrystalline metals. From the point of view of structure, the metallic glass is the same type of material as a silicate glass, namely amorphous. But the bonding between the atoms is totally different: metallic in the metallic glass vs. covalent in the silicate glass. Hence, the metallic glass is structurally amorphous, but not a network. The consequence is that a metallic glass is not brittle, but has a very high ductility. Furthermore, its plastic deformation is not by dislocation motion because, being a glass, it does not form dislocations. The mechanism giving rise to plastic deformation in a metallic glass is complicated but more similar to homogeneous shear. In terms of the elastic modulus and the tensile strength, a metallic glass shows behavior that is typical of a metal.

Recent developments in the application of metallic glasses have taken advantage of TTT curves such as the one in Fig. 13.27 implementing a process called thermoplastic forming. This involves plastic deformation of a metallic glass at a temperature above T_g but below the crystallization curve. Fig. 13.27 suggests that for the right kind of alloy, there will be sufficient time for the material in the viscous state to be formed into a desired shape while remaining amorphous. The deformations achievable under these conditions are, of course, much larger than at room temperature. There are companies offering this type of process as a commercial service.

13.5 Photonic Crystal Optical Fibers

The next case study I wish to present concerns optical fibers, and in particular recent developments in fibers involving a so-called photonic crystal. This will be more a study in advanced materials processing, but the achieved structures are equivalent to new man-made materials. Optical fibers have, of course, become a mainstay of modern communications.

The key feature of optical fibers is that they are able to guide light over long distances. The principle of optical fibers can be illustrated as follows:

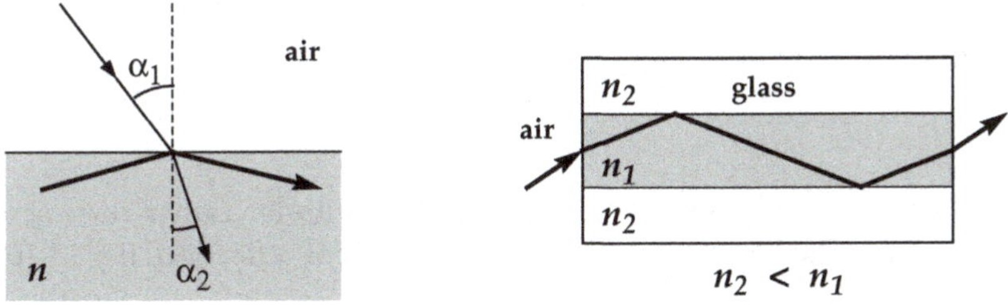

Figure 13.28: Refraction of light at a material interface, and guiding of light inside a fiber by internal reflection.

In the panel on the left, we consider light waves impinging from the air on a piece of planar glass. The optical properties of the glass are indicated by the so-called **refractive index** n, sometimes also called **index of refraction**. The refractive index is the factor by which the propagation velocity of the light wave is reduced in the glass compared to air (or vacuum). The refractive index n is related to the dielectric constant ϵ_r (see Eq. 12.28) by

$$n = \sqrt{\epsilon_r} \tag{13.9}$$

which is true if the material is an ideal dielectric, i.e. perfectly transparent or, equivalently, non-absorbing.

Experience shows that if the light wave is incident from the air at an angle α_1, it continues in the glass at a different angle α_2, where $\alpha_2 < \alpha_1$. The angles on either side of the interface can be shown to be related by

$$n = \frac{\sin \alpha_1}{\sin \alpha_2} \tag{13.10}$$

Here you should realize that the path of light waves is reversible: It could be from air into the glass, or from glass out into the air. Eq. 13.10 applies either way. Clearly, the angle α_1 can be at most 90°. Hence, it the light wave originates in the glass, there will be a critical angle $\alpha_{2,c}$ above which the light wave cannot exit the glass into the air. This is indicated by the path with the heavy line in the left panel of Fig. 13.28. At that point we have what is called total internal reflection.

One can take advantage of this effect in optical fibers. The situation is illustrated in the right panel of Fig. 13.28 which displays a lengthwise cross-section of an optical fiber. The fiber is basically a long, thin glass cylinder with two concentric regions: An inner part called the **core**, with a refractive index n_1, surrounded by an outer part called the **cladding**, with a refractive index n_2. The key feature of the fiber is that n_2 is a few percent less than n_1. This leads to total internal reflection and thus guiding of the light along the fiber.

Note that in the case of a fiber, which involves an interface between two dielectrics rather than just air and one dielectric, Eq. 13.10 must be modified slightly to read

$$\frac{n_2}{n_1} = \frac{\sin \alpha_1}{\sin \alpha_2} \tag{13.11}$$

Even though light propagation is now inside a cylinder rather than across a planar interface, the basic condition for total internal reflection, Eq. 13.10, still applies.

Optical fibers are produced in a variety of different geometries, leading to different optical properties. A basic distinction is made between multi-mode and single-mode fibers. Two implementations of a multi-mode fiber are illustrated in Fig. 13.29:

The designation of multi-mode refers to a type of fiber in which different light wave modes are confined and can propagate. Without becoming too technical, you can think of different modes as light with different wavelengths. Step index and graded index refer to an abrupt or gradual change in the refractive index as one goes from the inner core region to the cladding. The outside coating is a layer of polymer to protect the silica fiber inside.

The alternative to multi-mode is single-mode fibers. These are similar in geometry, except that the core is much smaller in diameter: typically several μm rather than roughly 50-500 μm for multi-mode fibers. Single-mode fibers are

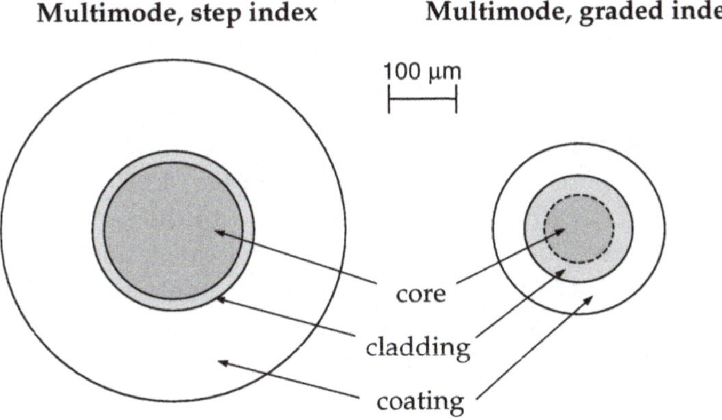

Figure 13.29: Typical cross-sections of step index and graded index multi-mode fibers. Fiber parts drawn roughly to scale.

generally step index, and the size and refractive index of the core and cladding are designed such that only a well-defined wavelength of light can propagate through it. The main advantage of single-mode fibers is that light pulses can propagate in them over longer distances without deterioration.

Representative geometrical parameters for different fiber types are summarized in Table 13.3. below:

Table 13.3: Typical parameters of optical fibers
All dimensions in μm

Fiber type	Core diam.	Cladding diam.	Coating diam.
MM step index	200	230	500
	400	430	730
	1000	1035	1400
MM graded index	50	125	245
	62.5	125	245
	100	140	250
SM step index	5.4 – 3.4	125	245

MM refers to multi-mode fibers and SM to single-mode fibers. The range of core diameters given for single-mode fibers is an abbreviation of the following:

5.4 μm for guiding light with $\lambda = 780$ nm, 4.6 μm for $\lambda = 680$ nm, 4.3 μm for $\lambda = 633$ nm, and 3.4 μm for $\lambda = 488/514$ nm. These are frequently used wavelengths in the visible range.

The method by which these fibers are fabricated is very simple in principle, but requires a great deal of control with regard to the starting material and the processing. A schematic process is displayed in Fig. 13.30. One starts with a so-called preform, a cylindrical piece of glass with the same compositional makeup as the desired fiber. The preform is then heated, and a fine thread of glass is pulled from it, coated, and picked up on a spool. It should be obvious that this sounds much simpler than it really is. In practice, the process requires a very carefully prepared, flawless preform, and then exquisite mechanical and thermal control with the entire procedure.

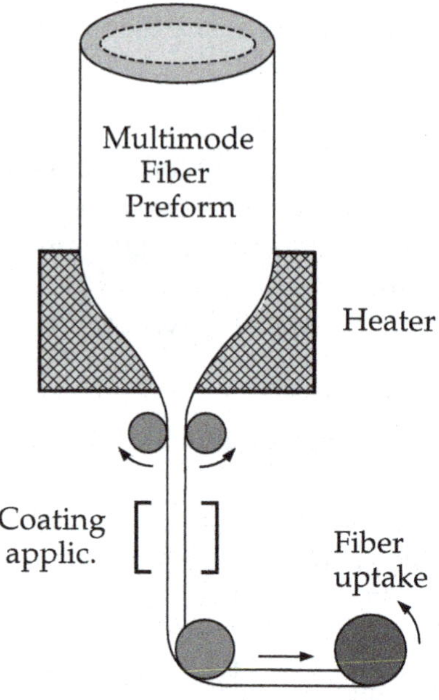

Figure 13.30: Schematic of the optical fiber drawing process, not drawn to scale. The entire apparatus is mounted on a tower.

Preforms are typically about 20 cm in diameter, about two meters in length, and can weigh up to around 250 kg. Some 100's of km of fiber can be drawn from one preform. Fabrication of the preform is an art in itself. The first stage of preform fabrication consists of producing a cylinder of extremely high purity silica. Most often, a rod of material for the core with slightly increased refractive index is produced by chemical vapor deposition. The preform is then finished as a rod in a tube.

Whereas multi-mode and single-mode fibers have been available commercially and in practical use for some time, a new generation of fibers is being developed with an entirely new structure, with additional design flexibility regarding optical properties, and potentially with improved performance. The principle of these new fibers is illustrated in Fig. 13.31.

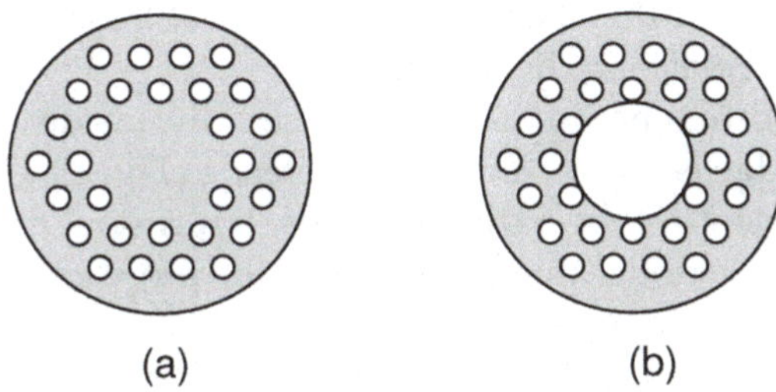

Figure 13.31: Cross-sections of photonic crystal optical fibers. Polymer coating not shown.

They are called photonic crystal fibers because of their main feature, a periodic array of axial cylindrical holes. In the fiber of Fig. 13.31a, there is a solid central core guiding the light, whereas in the fiber of Fig. 13.31b, the light is guided by the hollow space in the center.

You can think of the fiber of Fig. 13.31a as somewhat similar to the traditional fibers, in that it implements a cladding of lower average refractive index using a set of holes, but in addition the refractive index varies periodically along the circumference.

The fiber of Fig. 13.31b is a different story altogether, since light is guided in the central hollow, i.e. in a space with ***lower*** refractive index than the surrounding region. The design of this type of fiber is referred to as an optical bandgap fiber. It turns out that in this case, the light guiding mechanism is based on an effect which is the optical analog of the electronic band gap in a semiconductor. The pattern of voids around the central hollow causes waves with only a fairly well-defined wavelength to be allowed to exist and propagate along the inside of the hollow space.

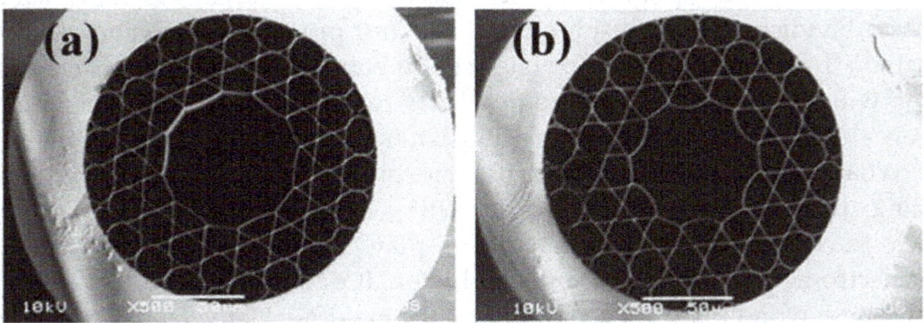

Figure 13.32: Cross-sections of two designs of an optical bandgap fiber. The scale bar is 50 μm.

An example of a real optical bandgap fiber is shown in Fig. 13.32. The main advantages of this type of fiber are that light waves propagate faster than in standard fibers and that they have very low optical losses, i.e. optical signals can propagate over very long distances.

Note the subtle difference in the appearance of the inside edge of the hollow. In Fig. 13.32a the edge is almost circular, whereas in Fig. 13.32b you can see the curved arched segments of the individual cells. It turns out that even as subtle a feature as this makes a noticeable difference in the propagation of light in these fibers. Also note how thin the walls of the holes are.

Photonic crystal fibers, and optical bandgap fibers in particular, are fabricated basically the same way as shown in Fig. 13.30, except that the preform is assembled as a set of silica capillaries with the desired geometry. Also, in order to support the hole structure, extra pressure may be applied inside during the drawing process, but you can imagine that this is easier said than done.

13.6 New Electronic Materials

There are numerous new and ongoing developments in electronic materials. I have picked just two examples.

13.6.1 Bandgap Engineering of Semiconductors

You will recall that in Chapter 12 we focused our discussion of semiconductors on elemental materials in general, and on Si in particular. For those materials, the most important electronic materials property, i.e. the bandgap, is a given.

However, elemental semiconductors, being from Group IVA in the periodic table, are a very small class of materials, even though Si has a unique place among semiconductors. When you go into the periodic table and make compounds from Group IIIA and Group VA, you can form a much larger class of materials, most of which are also semiconductors. Examples of such compounds are GaAs, GaN, GaSb, AlN, AlAs, InP, etc.

Each of these compounds has its characteristic bandgap, and the values of these bandgaps vary considerably. From our previous discussion, you can conclude that the stronger the covalent bond between the atoms involved, the larger will be the bandgap. As you will also recall, for the purpose of making purely electronic devices, let us say transistors, the bandgap is not the main issue. What is more important is whether one can dope the material, and whether doping gives rise to free charge carriers with a high mobility.

On the other hand, the bandgap itself plays a key role for the optical and optoelectronic properties of a semiconductor. It determines whether a certain wavelength of light is absorbed or not by the material. Conversely, if the material is used as a light source, the bandgap determines what the wavelength of the emitted light is. Therefore, it is of interest to have some flexibility in choosing the bandgap of a material. In addition, for subtle reasons connected with more detailed features of the bandgap, Si is not really usable as a light source.

Compound semiconductors are of great interest for yet another reason. Not only can you pick one with a suitable bandgap, but it is possible to make alloys of compound semiconductors consisting of three or even four components. This opens up a large class of additional materials with potentially interesting properties. For example, if you alloy GaAs with AlAs, you create the following compound:

$$x\,\mathrm{GaAs} + (1-x)\,\mathrm{AlAs} \rightarrow \mathrm{Al}_{1-x}\mathrm{Ga}_x\mathrm{As} \qquad (13.12)$$

Given that the bandgap is a reflection of the atomic bond energy, you would expect the bandgap of the ternary compound to be in-between the values for the individual binary compounds. This gives the material engineer an even greater flexibility in the choice of material.

The extent to which it is feasible to form an alloy $A_{1,1-x}A_{2,x}B$ from two compounds A_1B and A_2B depends on how similar the two starting materials are, just as with metal alloys. In the present case this means asking how similar the two cations A_1 and A_2 are. If the two are not similar enough, there may be a limit to the range of the parameter x over which a ternary compound can form. This amounts to saying that the solubility of A_1B in A_2B, or A_2B in A_1B, may be limited.

As regards effects at the atomic scale, first it is of course necessary that these compounds be single crystals. Second, if say A_2 replaces A_1 in A_1B, then it needs to do so by going into the same atomic site as A_1 as a substitutional impurity. Furthermore, these binary compounds all have the Zincblende crystal structure (Fig. 3.22). Thus, similarity extends to having similar lattice constants. In addition, A_1 and A_2 should be distributed evenly in the ternary compound and should not segregate.

Despite these apparent restrictions, numerous ternary semiconductors have been investigated, and still new materials are to be synthesized and investigated. Also, keep in mind that although above we considered compounds of the form $A_{1,1-x}A_{2,x}B$ as an example, it is equally possible to form compounds with two anions, $AB_{1,1-y}B_{2,y}$.

Fig. 13.33 below shows some examples of binary and ternary compound semiconductors with their bandgaps and lattice constants. A possible ternary is indicated by a dashed line between the corresponding binaries.

In order to build devices, especially integrated optoelectronics, it is not only necessary to be able to grow single crystals of compound semiconductors, but also to grow a film of one material on top of another, for the purpose of forming a semiconductor junction. Again, it is evident that this will require similar materials, in this case especially with respect to their lattice constants. If these do not match, then the interface will experience a strain which, if large enough, can lead to the formation of dislocations. Of course, these would be detrimental to the operation of an electronic device.

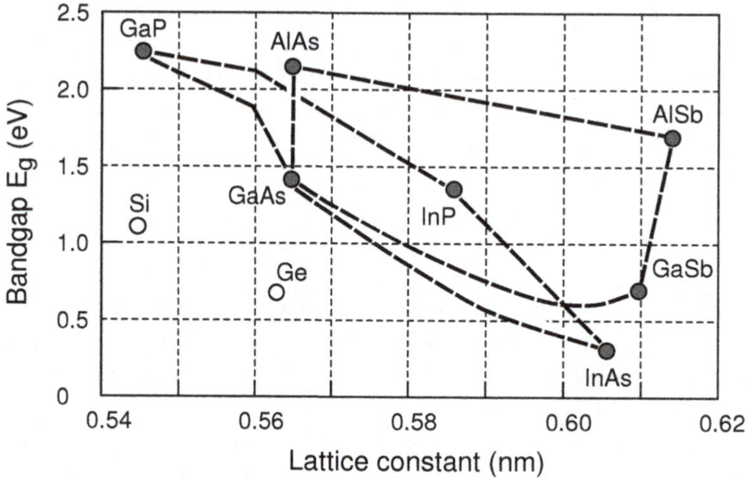

Figure 13.33: Bandgaps and lattice constants of compound semiconductors. Possible ternaries are indicated by dashed lines connecting two binaries. Si and Ge are also listed for comparison.

13.6.2 Materials and the End of Moore's Law

Moore's Law refers to the empirical observation by Gordon Moore, co-founder of Intel, that in the business of silicon integrated circuits, the size of the features of individual transistors and circuits has been decreasing steadily over time, so that the number of components on a Si chip has doubled roughly every 18-24 months.

On the basis of industry data for 1959-1965, Moore made a first prediction in 1965 regarding the evolution of integrated circuits to 1975 (Fig. 13.34). This prediction was followed closely by subsequent developments. In 1975 he made another prediction, taking into account that after 1980 further improvements would have to rely only on the two factors of increasing chip size and finer dimensions of the circuit features. Except for the exact point at which the slope of the curve changed, Moore's prediction once again has been on target for another three decades or so.

In the recent past, Moore's Law has become more a goal to be continually strived for, rather than just a prediction of a possible future. If you wanted to be a player in the semiconductor industry and did not want to fall behind, you had better stay on Moore's curve.

Progress in the Si integrated circuit industry has relied on advances in several areas, including photolithography, materials processing equipment, and circuit design. But at various stages, the introduction and development of new materials opened up crucial new opportunities. I will describe four of these below, having briefly mentioned the first two of them at the end of Chapter 12.

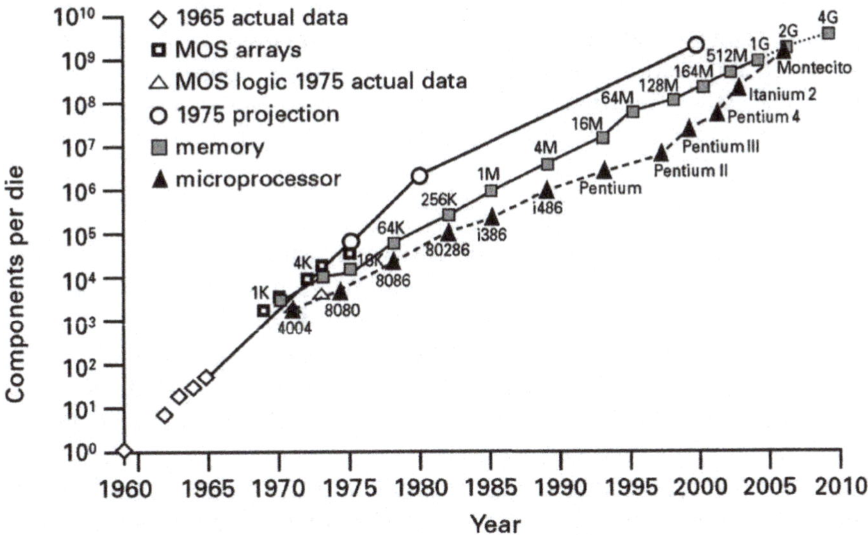

Figure 13.34: Illustration of Moore's Law. Moore's 1965 and 1975 projections, with early data (top line). Data for memory and logic circuits (two bottom lines).

Copper interconnect wiring

I indicated in Chapter 12 that Al had been the material of choice for forming the tiny metal wires that connect individual components with each other on an integrated Si circuit. In fact, for over two decades Al had been good enough, even though it was only the second best material, and it had been possible to stay on the Moore's Law curve with it.

When the change to the best material, namely Cu, became inevitable, this required a whole rethinking of materials processing for integrated circuits. Al wiring had basically been fabricated by depositing a blanket layer of Al over the entire wafer, and then etching away all the extra material in-between the desired Al wires (see Fig. 12.25). Implementing the same process with Cu, i.e. etching it away where necessary, proved to be impractical.

The solution was to etch out the wiring pattern as a set of trenches in the underlying SiO_2, and then to fill in the trenches with Cu. This latter step was unprecedented itself for the semiconductor industry, in that it involved a wet chemical deposition process, whereas before Al had been deposited essentially by evaporation in vacuum.

Integrating Cu as the metal for wiring introduced a host of additional issues related to materials compatibility. For example, whereas Al wires formed a strong bond to the underlying SiO_2, Cu did not adhere properly to SiO_2. It was necessary to develop a type of glue layer, a very thin layer of TiN or TaN put between the Cu and the SiO_2, in order to insure mechanical stability of the complete circuit. In addition, Cu could not be in contact with the Si on the wafer because it would poison the semiconducting properties of the Si. This problem was solved by use of plugs of W (tungsten) between the first layer of Cu and the Si contacts underneath (see title picture of Chapter 12).

High-k gate dielectric

From our discussion of capacitors and MOS transistors in Chapter 12, you can appreciate that what ultimately switches an MOS transistor on or off is the charge on the gate capacitor (see Fig. 12.24). For a fixed gate voltage V_g, that charge is given by Eq. 12.27 and is proportional to ϵ / l of the gate capacitor, where l is the thickness of the gate dielectric, normally SiO_2. (Note that in the semiconductor industry, it is common to use k as the symbol for the dielectric constant rather than ϵ_r. See Eq. 12.28.)

With the continued shrinking of component dimensions overall, the gate dielectric thickness l also had to shrink. The time could be foreseen where the thickness of the gate dielectric needed to be just a few nm, i.e. just a few atom layers. It became clear that under those conditions SiO_2 would not be a good enough insulator anymore. A very thin SiO_2 layer started to leak current, preventing the needed charge from staying on an MOS gate long enough.

The solution to the leakage current problem was a thicker gate dielectric. But if a larger l was to be used for the gate capacitor, then the material had to have a larger dielectric constant ϵ_r, in order to maintain the same electrical device characteristics. These requirements started a large development effort, in which many candidate materials to replace SiO_2 were evaluated.

The best solution for replacing SiO_2 as a gate dielectric was found to be HfO_2 (hafnium dioxide): ϵ_r is 25 for HfO_2 as compared to 3.9 for SiO_2. That is, a HfO_2 gate dielectric can be about six times thicker than the electrically equivalent SiO_2. The change to HfO_2 was a huge step, since it called for not only an uncommon material, but also a totally different method of depositing it onto the silicon wafers.

Low-k interconnect dielectrics

Making the on-chip wiring scheme as efficient as possible requires, as we have seen, that the metal wires be made of Cu. In addition, it is necessary to build up the wiring scheme in several layers (up to 9 layers at present), as shown in the title figures to Chapters 12 and 13. Finally, it turns out that the insulating material between the metal layers is also important. Specifically, for optimum propagation of electrical signals in the Cu wires, the dielectric constant of this so-called interlayer dielectric should be as low as possible (i.e. just the opposite of the gate dielectric).

The standard interlayer material for a long time had been SiO_2. It is being replaced by other materials, e.g. fluorine-doped SiO_2, various polymers, or carbon-doped SiO_2, all with an ϵ_r, or k, lower than the 3.9 for SiO_2. Current k values range from about 2.6 – 3.0. This development effort is ongoing. There is much room for further improvement, perhaps including porous materials, until the ultimate dielectric with $k = 1$, i.e. air, has been reached.

SiGe and strained Si

As long as the goal is to improve the speed of MOS transistors, it turns out that a subtle property of Si itself can be of help. It has been known for some time that when Si is under compression, the mobility of the holes is increased, and when Si is under tension, the mobility of the electrons is increased.

Since the hole mobility is much lower than the electron mobility (see Table 12.5), the case of Si in compression is of special interest. Bending a wafer would do it, but that is not practical. So how can one achieve Si in compression on a wafer without compromising anything else?

In order to see the solution, we take a small detour, comparing Si with Ge and SiGe alloy:

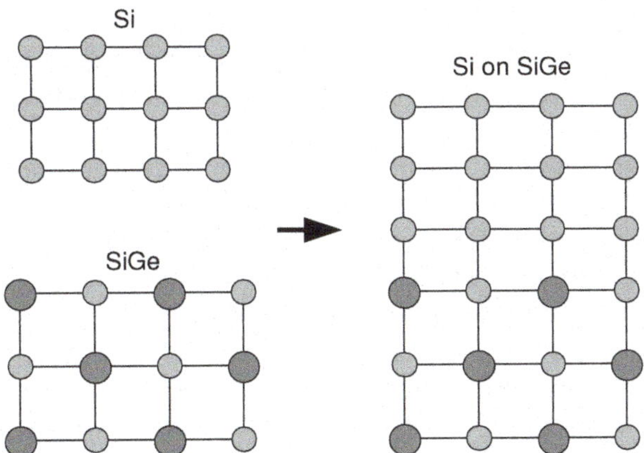

Figure 13.35: Schematic crystal structures of Si, SiGe, and Si grown on top of SiGe.

Fig. 13.35 illustrates schematically crystalline Si, the crystalline SiGe alloy, and Si grown on top of SiGe. You know that both Si and Ge are diamond-cubic, but their lattice constants are different: $a_{Si} = 0.543$ nm and $a_{Ge} = 0.566$ nm. For the alloy, a is somewhere in-between. Now, if Si is grown carefully on top of SiGe so that the two crystals are in registry, a process known as **epitaxial growth,** the Si has to adjust its lattice constant to that of the SiGe underneath. This lattice mismatch causes the Si to be in tension. That is, the Si ends up being strained. This condition can be maintained as long as the thickness of the Si is quite small. Eventually, a thicker layer of Si layer would relax by forming dislocations at the interface.

But we wanted Si in compression! Fig. 13.36 shows schematically the trick by which this can be achieved via the same lattice mismatch mechanism:

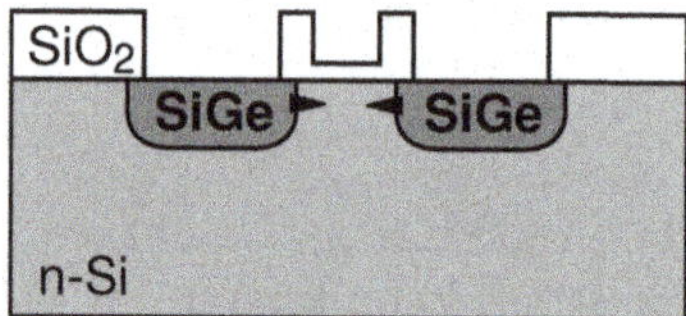

Figure 13.36: p-MOS transistor with inlaid SiGe source and drain.

The source and drain regions in the Si of this p-MOS Si transistor are first hollowed out, and then SiGe is grown in the hollows. The larger lattice constant of the SiGe squeezed into the hollows causes sideways pressure to be be exerted onto the Si channel region, and the resulting compressive strain is large enough to lead to significantly enhanced mobility of the holes.

13.7 Epilogue

Moore's Law for microelectronics invariably raises the question: Where is it going? When will it end, and how? The same can be asked, of course, in any other branch of materials science.

The interesting thing is that this question has come up repeatedly in the past. When I got into electronic materials, about 25 years ago, one could think of good reasons why within the next couple of years the then current technology would encounter serious obstacles. Yet clever scientists and engineers came up with work-around solutions for those problems. This cycle has repeated itself many times since.

In silicon microelectronics specifically, of course one can envision an "end" to the current trajectory of technical development, for example when the gate dielectric will consist of only one layer of atoms, or when an MOS transistor will be made up of only a handful of atoms. But who knows what people will figure out to do between now then: graphene electronics, large-scale integrated carbon nanotube transistors?

In 2006 Gordon Moore's answer to the "big" question above was:

> "Not very soon. I can see the complexity curve lasting for at least as long now as I ever could in the past. I always could see what we were going to do to make the next two or three technology generations happen on the curve. Today, as I speak with the Intel research and development staff members, they are looking out even further. Now I can see what we are going to do for the next four generations, which is further than we have ever been able to look out before. Many advances have to happen to make those future generations occur, but we are confident the problems are going to

> be solved in time to make it all happen. It is amazing what a group of dedicated scientists and engineers can do, as the past forty years have shown. I do not see an end in sight, with the caveat that I can only see a decade or so ahead."

My guess is that a similar conclusion will apply to other branches of materials science and engineering as well. Just have a look at a recent editorial in the Materials Research Society Bulletin (November 2013) reviewing four decades of materials development that is transforming society. The featured areas, all with far-reaching prospects, were: Oxides in electronics, organic electronics, carbon nanostructures, multiscale materials modeling, materials in tissue engineering, and materials science for nuclear-waste management.

There will always be new "stuff" to be made and investigated. So for you there is only one thing to do: Keep learning more about materials, get involved hands-on in a lab, and you will find interesting things to do for as far as you can see.

Image sources

Title picture: IBM press release (circa 2000)
An example of the multi-layer wiring scheme connecting components on a modern integrated circuit.

Fig. 13.3: reproduced with permission by the American Institute of Physics
E.W. Muller, J. Appl. Phys. 26 (1955) 732-737
`http://dx.doi.org/10.1063/1.1722081`

Fig. 13.6:
`http://www.ntmdt-tips.com/products/view/csg01`

Fig. 13.8:
`http://www.ntmdt.com/scan-gallery`

Fig. 13.11: data from
J.R. Sambles, Proc. R. Soc. London A 324 (1971) 339-351
Ph. Buffat and J. P. Borel, Phys. Rev. A 13 (1976) 2287-2296
T. Castro, R. Reifenberger, E. Choi, R. P. Andres, Phys. Rev. B42 (1990) 8548-8556

Fig. 13.16: reproduced with permission by the American Chemical Society
Synthesis, Alignment, and Magnetic Properties of Monodisperse Nickel Nanocubes, Alec P. LaGrow et al., J. Am. Chem. Soc. 134 (2012) 855-858
`http://www.victoria.ac.nz/scps/research/research-groups/nanoparticles/research-proj/catalysis`

Fig. 13.17:
W. Lu, P. Soukliassian, and J. Boeckl, MRS Bull. 37 (2012) 1119-1124

Figs. 13.18 and 13.19: reproduced with permission by Elsevier
A TEM study of soot, carbon nanotubes, and related fullerene nanopolyhedra in common fuel-gas combustion sources
L.E. Murr, K.F. Soto 50-65, Mat. Char. 55 (2005) 50-65
`http://www.sciencedirect.com/science/article/pii/S1044580305001051`

Fig. 13.20: reproduced with permission by the American Institute of Physics
Low temperature growth of ultra-high mass density carbon nanotube forests on conductive supports
H. Sugime et al. , Appl. Phys. Lett. 103 (2013) 073116
`http://scitation.aip.org/content/aip/journal/apl/103/7/10.1063/1.4818619`

Figs. 13.21, 13.22: reproduced with permission by the Am. Inst. of Physics
Fabrication and properties of ultranano, nano, and microcrystalline diamond membranes and sheets
D. K. Reinhard et al., J. Vac. Sci. Technol. B 22 (2004) 2811
`http://dx.doi.org/10.1116/1.1819928`

Fig. 13.23: adapted from
`http://www.worldautosteel.org/steel-basics/steel-types/evolving-ahss-types/`

Fig. 13.26: adapted from
Prediction of the glass forming ability in Cu-Zr binary and Cu-Zr-Ti ternary alloys
L. Ge et al., Intermetallics 16 (2008) 27-33
`http://www.sciencedirect.com/science/article/pii/S096697950700163X`

Fig. 13.27: reproduced with permission by the American Institute of Physics
Vitrification and determination of the crystallization time scales of the bulk-metallic-glass-forming liquid Zr58.5Nb2.8Cu15.6Ni12.8Al10.3
C. Hays et al., Appl. Phys. Lett. 79 (2001) 1605-1607
`http://authors.library.caltech.edu/2966/1/HAYapl01.pdf`

Fig. 13.31: reproduced with permission by the Optical Society of America
Low loss broadband transmission in hypocycloid-core Kagome hollow-core photonic crystal fiber
Y. Y. Wang et al., Opt. Lett. 36 (2011) 669-671
`http://www.opticsinfobase.org/ol/abstract.cfm?uri=ol-36-5-669`

Fig. 13.34: reproduced with permission by the Chem. Heritage Foundation, from: Moore's Law at 40, by Gordon E. Moore.
First published in: Understanding Moore's Law: Four Decades of Innovation; edited by David C. Brock (Chemical Heritage Foundation, 2006).
`http://www.chemheritage.org/Downloads/Publications/Books/Understanding-Moores-Law/Understanding-Moores-Law_Chapter-04.pdf`

References

For a set of tutorial slides on electron microscopies, see:
`http://tpm.amc.anl.gov/Lectures/Zaluzec-1-Instrumentation.ppt.pdf`

For additional information on AFM instruments, check out:
`http://www.nanoscience.com/products/afm/afm-probes/`
`http://www.tedpella.com/probes_html/`
`http://www.spmtips.com/high-resolution-afm-probes.html`

For more examples of work on shapes of small particles:
Shape Control of Colloidal Metal Nanocrystals
A. R. Tao, S. Habas, and P. Yang, Small 4 (2008) 310-325
`http://www.cchem.berkeley.edu/pdygrp/pub_files/pubpdf/159.pdf`
Shape Control of Inorganic Materials via Electrodeposition
K-S. Choi, Dalton Trans. 40 (2008) 5432-5438
`http://pubs.rsc.org/en/content/articlelanding/2008/dt/b807848c`

For an overview of new carbon-based materials, see:
Beyond Silicon: Carbon-Based Nanotechnology, MRS Bull. 35 No. 10 (2010)
Graphene: Fundamentals and Functionalities, MRS Bull. 37 No. 12 (2012)

For more information on Gum Metal, see:
Designing New Structural Materials Using Density Functional Theory: The Example of Gum Metal
H. Ikehata, N. Nagasako, S. Kuramoto, and T. Saito, MRS Bull. 31 (2006) 688-692
Gum Metal: Ultra Sensitive and Super Tough Ti Alloy
`http://www.nissey-sabae.co.jp/pdf/eng_gummetal01.pdf`

For an overview of metallic glasses, see:
J. J. Gilman, Physics Today, May 1975.
`http://dx.doi.org/10.1063/1.3068966`
J. Schroers, Physics Today, February 2013.
`http://dx.doi.org/10.1063/PT.3.1885`

For optical fiber tutorials, visit:
`http://www.rp-photonics.com/fibers.html`
`http://www.rp-photonics.com/passive_fiber_optics.html`

Exercises

1. An AFM tip in contact mode is deflected by 0.03 nm from its equilibrium position. How large is the force exerted by the sample on the tip ?

2. Discuss the resolution of an AFM for spherical particles (i.e not hemispherical). Do the corresponding geometrical calculation, as outlined in the text, or give a qualitative engineering argument.

3. For an SC cubic particle as in Fig. 13.9, what is the fraction of atoms on the cube edges, compared to the total number.

4. If the particle of Fig. 13.9 were BCC, each little cube being a unit cell, what would be the fraction of surface atoms as a function of n ? Try using the same argument as in the text. How does the BCC case compare with SC ?

5. Can one make a semi-quantitative argument to explain why the 13-atom icosahedron of Fig. 13.15 is more stable than the two 13-atom clusters of Fig. 13.14 ?
Determine and compare the average number of nearest neighbors in each of those three clusters.

6. The "carbon nanotube forest" layer in Fig. 13.20 is said by the authors to be exceptionally dense, at a mass density of 1.6 g/cm^3. Is this true ? What should this density be compared to ?

7. If the yield strength of an advanced steel for a car frame could be improved by 20%, what approximate savings in the weight of the car frame could you expect ?

8. Among the advanced steels are some containing three or more phases. How can that be, given what you know about the Fe-C phase diagram ?

9. Demonstrate that Fig. 13.24 indeed shows a unit cell, and that a single cube is not a unit cell. Verify the composition by showing how to count atoms properly. What kind of a unit cell is this ?

10. Draw a stress-strain curve for gum metal which shows plastic deformation with no strain hardening.

11. How long a fiber with a diameter of 125 μm could you draw from a silica preform weighing 250 kg if all the material in the preform could be used up in the process ?

12. The Si-on-SiGe structure in Fig. 13.35 is drawn slightly inaccurately: There are two inaccuracies. Identify and explain them. If you cannot see them on the figure, think of two effects that need to be taken into account with respect to the lattice deformation due to the stress state at the interface.

Problems

1. How sharp should the tip of an FIM sample be so that individual atoms can be resolved in a reasonable laboratory setup ?
2. What problem may arise when one tries to examine an insulating sample in an electron microscope ?
3. Based on the discussion in the text, is it likely that two neighboring atoms on a solid surface can be detected by an AFM used in contact mode ?
4. Is there a quick way of coming up with a good approximate answer to the fraction of surface atoms for any type of cubic particle, i.e. SC, BCC, and FCC ?
   Tip: Look at a unit cube and count atoms properly. Then extrapolate to many unit cubes stuck together.
5. Assuming you have answered the previous problem, to what extent is Eq. 13.8 a generally valid expression for the fraction of surface atoms in a small particle ?
   Tip: Discuss the factor 6, and the dependence on n^{-1}.
6. How would you deal with the question of the fraction of surface atoms for a spherical particle ?
7. In the text I said, on p. 405:
   One instance where we did deal with surfaces was in connection with the effects of grain boundaries.
   There is another instance elsewhere in the book where size, specifically particle size, was the critical variable. What was it? How does that instance relate to the present discussion of small particles ?
8. What is the mass density of a single-wall carbon nanotube ?
9. If a "carbon nanotube forest" layer as in Fig. 13.20 could be grown with single-wall nanotubes, all parallel and arranged in a hexagonal close-packed pattern, what would the mass density of this film be ?

Key Words

Here is a list of key words, abbreviations, and acronyms. Each term is given with the page on which it appears (in bold) and is explained for the first time. When more than one page is listed, the numbers indicate the main locations in the book where the concept is used.